Nanoelectronics

This book exhibits a unique way of explaining nanomaterials and devices and analyzing their design parameters to meet the sub-nanoregime challenges for low-power chip design. Since process variability, device sizing, and power supply scaling are ongoing challenges in very large-scale integration (VLSI) circuit designs, this book highlights the conventional and novel nanomaterials, devices and circuits, leakage current mitigation techniques, and other important trade-offs along with exhaustive analysis. More focus has been placed throughout the book on various trade-offs for high-speed and low-power VLSI devices and circuits co-design.

This book:

- Discusses advanced nano-semiconductor devices such as FinFET, nanowires, tunnel field-effect transistors, carbon nanotube field-effect transistors, and high-electron-mobility transistors.
- Presents high-performance semiconductor devices at nanoscale technology nodes for the analysis of quantum effects and their impact on circuits and systems.
- Covers power dissipation and reduction techniques for high-performance devices.
- Explains both silicon and non-silicon devices for various applications like digital logic and analog/radio frequency applications.
- Examines the difficulties and practical design approaches for extremely low-power analog-integrated circuits.

It is primarily written for senior undergraduates, graduate students, and academic researchers in the fields of electrical engineering, electronics and communications engineering, materials science, nanoscience, and nanotechnology.

Smart Engineering Systems: Design and Applications

Series Editor – Suman Lata Tripathi

Internet of Things
Robotic and Drone Technology
Edited by Nitin Goyal, Sharad Sharma, Arun Kumar Rana, Suman Lata Tripathi

Smart Electrical Grid System
Design Principle, Modernization, and Techniques
Edited by Krishan Arora, Suman Lata Tripathi and Sanjeevikumar Padmanaban

Artificial Intelligence, Internet of Things (IoT) and Smart Materials for Energy Applications
Edited by Mohan Lal Kolhe, Kailash J. Karande and Sampat G. Deshmukh

Artificial Intelligence for Internet of Things: Design Principle, Modernization, and Techniques
Edited by N. Thillaiarasu, Suman Lata Tripathi and V. Dhinakaran

Machine Learning and Internet of Things in Solar Power Generation
Edited by Prabha Umapathy, Jude Hemanth, Shelej Khera, Abinaya Inbamani and Suman Lata Tripathi

Microgrid
Design, Optimization, and Applications
Edited by Amit Kumar Pandey, Sanjeevikumar Padmanaban, Suman Lata Tripathi, Vivek Patel, Vikas Patel

For more information about this series, please visit: https://www.routledge.com/Smart-Engineering-Systems/book-series/CRCSESDA?srsltid=AfmBOop-tpQ60KY0nHiaZr0mbW2H8Py8MjxmJK1WqtKVcJ_pRW1TSqzP

Nanoelectronics

Materials to Chips Design

Edited by

Balwinder Raj, Suman Lata Tripathi,
Tarun Chaudhary and Mandeep Singh

CRC Press
Taylor & Francis Group
Boca Raton London New York

CRC Press is an imprint of the
Taylor & Francis Group, an informa business

Designed cover image: Shutterstock

First edition published 2025
by CRC Press
2385 NW Executive Center Drive, Suite 320, Boca Raton FL 33431

and by CRC Press
4 Park Square, Milton Park, Abingdon, Oxon, OX14 4RN

CRC Press is an imprint of Taylor & Francis Group, LLC

ISBN: 978-1-032-74598-5 (hbk)
ISBN: 978-1-032-98462-9 (pbk)
ISBN: 978-1-003-61279-7 (ebk)

DOI: 10.1201/9781003612797

Typeset in Sabon
by Deanta Global Publishing Services, Chennai, India

Contents

Preface

Nanoelectronic devices such as double-gate metal–oxide–semiconductor field-effect transistors (MOSFETs), fin field-effect transistors (FinFETs), nanowires, carbon nanotube field-effect transistors (CNTFETs), tunnel field-effect transistors (TFETs), ferroelectric field-effect transistors (FeFETs), gate-all-around (GAA) FETs, high-electron-mobility transistors (HEMTs), and spin-based devices are proposed and available in the literature for chip design. The study of these devices is important in sectors whose applications in chip design have a huge impact on bringing revolutionary advancements in nanoelectronic devices, circuits, and systems due to improved electronic properties of the nanoscale semiconductor devices. Currently, the knowledge and resources for the fabrication of these nanoelectronic devices are also available, as its fabrication flow is similar to conventional MOSFET. It is predicted by the International Technology Roadmap for Semiconductors (ITRS) that with the advancement of technology, the conventional fabrication flow can be used for nanoelectronic devices.

This book exhibits a unique way to explain nanomaterials, devices, analyses of their design parameters to meet the sub-nano-regime challenges for low-power chip design. Since process variability, device sizing, and power supply scaling are ongoing challenges in very large-scale integration (VLSI) circuits design, this book highlights the conventional and novel nanomaterials, devices and circuits, leakage current mitigation techniques, and other important trade-offs along with exhaustive analysis. More focus has placed throughout the book on various trade-offs for high-speed and low-power VLSI devices and circuits co-design.

Chapter 1: This chapter focuses on unconventional designs like double-gate (DG) MOSFET, tri-gate MOSFET, GAAFET, and the cylindrical surrounding double-gate (CSDG) MOSFET, and the development and prospects of MOSFET architectures are examined. These structures are suitable for upcoming communication systems, as they play a critical role in mitigating short-channel effects as devices get closer to the nanoscale domain. The field-effect transistor (FET), which was developed in 1930 in response to the demand for devices that were faster, cheaper, smaller, and less power-intensive, is the main topic of this chapter. An important breakthrough was made when Kahng and Attala created the insulated-gate FET, which used a metal–oxide–semiconductor (MOS) structure to solve the problem of surface states at the oxide–semiconductor interface. The fast expansion of wireless data rates is also covered in this chapter, along with the growing need for technology that can manage massive volumes of data while consuming little power. This chapter concludes by highlighting the significance of scaling for enhancing system performance at a lower cost, current drive, computational capacity, and integration density. It goes over the fundamentals of scaling, the significance of keeping the geometrical ratio, and the advantages of scaling, including lower latency, higher clock frequencies, and lower energy usage. With references to Moore's law and the International Technology Roadmap for Semiconductors (ITRS), the chapter ends with a debate on the future of scaling.

Chapter 2: CMOS (complementary metal–oxide–semiconductor) technology development is driven by the pursuit of higher density, improved performance, and lower power consumption. However, as advancements push toward smaller scales, challenges emerge, particularly regarding short-channel effects (SCEs) when channel lengths approach depletion-layer widths. At the nanoscale, planar MOSFETs encounter edge effects, necessitating longer channel lengths to address issues. This elongation compromises gate control and power efficiency, highlighting a critical bottleneck in traditional CMOS scaling. To overcome these challenges, innovative architectures like silicon-on-insulator (SOI) and fin field-effect transistors (FinFETs) have emerged. These architectures offer higher gate-to-channel capacitance and lower drain-to-channel capacitance, enabling better control over leakage currents, especially at the channel center. The adoption of multiple-gate structures facilitates enhanced electrostatic control, mitigating the limitations posed by short-channel effects.

Chapter 3: This study explores the field of nanoelectronics by looking into a range of state-of-the-art gadgets and cutting-edge technology. The main objective is to conduct a thorough analysis of each technology's performance by carefully contrasting important factors. The conversation covers the most recent developments in nanoscale electronic components and throws light on their special traits and capabilities. This chapter seeks to offer important insights into the current state of nanoelectronics and lay the groundwork for future advancements in this quickly developing sector by critically analyzing these technologies. We shall look into novel nanoelectric device structures in more detail later in this study, with particular emphasis on quantum dots, nanosheets, and tunneling devices. Examining these innovative architectures, this chapter provides a succinct but thorough examination of their special characteristics and uses.

Chapter 4: High-electron-mobility transistors (HEMTs) have drawn a lot of attention lately because of their improved electron transport, which guarantees high-speed and high-power applications. With superior power density, outstanding high-frequency performance, and satisfactory manufacturing efficiency, HEMT devices are challenging and eventually replacing conventional field-effect transistors (FETs). An overview of the performance of certain well-known and often-used HEMT devices is given to readers in this study. This chapter begins with an overview of the various HEMT structures and then moves on to the working concept with graphical representations. The performance of the device is examined using both analytical and numerical models from the body of current literature. In addition, some noteworthy recent studies on HEMT devices have been pointed out, and future trends have been forecasted. An in-depth understanding of current findings, potential paths, and their analytical strategy would be helpful in creating innovative HEMT devices.

Chapter 5: Metallic nanoparticles and their oxides have been the focus of extensive research due to their remarkable properties, prompting investigations for both fundamental understanding and practical applications. Various metallic nanoparticles and their oxides such as gold, silver, platinum, palladium, iron, copper, etc., have found numerous applications in a wide range of disciplines. The applications include nanotechnology, nanobiotechnology/nanobiomedicine, drug delivery, food/packaging industry, wound healing, cosmetics industry, animal husbandry, disinfection, environmental remediation, agriculture, and ceramics, to name a few. The synthesis of these nanoparticles can be achieved through physical, chemical, or biological means. While physical and chemical methods often involve toxic elements, time-consuming processes, and expensive instrumentation, biological synthesis, commonly referred to as green synthesis, offers a more environmentally friendly, cost-effective, and non-toxic alternative. Utilizing plants and microorganisms, green synthesis showcases enhanced biocompatibility in the resulting nanoparticles. Within the realm of nanotechnology, phytonanotechnology emerges as a field leveraging plant-derived materials, notably harnessing bioactive compounds and phytochemicals, to advance innovative applications. Pomegranates contain numerous minerals, vitamins, and botanical compounds in their seeds, peels, and leaves, which offer various health benefits.

Chapter 6: Memory blocks are part of many digital circuits and devices. Static random access memory (SRAM) and read-only memory (ROM) cells are frequently part of such memory. Prefabrication design and verification using EDA tools are becoming popular to avoid any failure after costly fabrication processes. This chapter provides description of basic memory design using the Tanner EDA tool. Here different technology nodes are compared for the design and simulation of SRAM and ROM blocks. Transient analysis is used for the evaluation of delay, noise margin, and power consumption. The presented work is useful in performance optimization at different transistor feature sizes and memory design topologies.

Chapter 7: By stopping long-standby-time leakage energy, power-gating embedded memory is an effective way to achieve better power efficiency in generally off-instant-on applications. Presently available non-volatile SRAM (NVSRAM) architectures featuring in-cell NVM element backup provide a practical way to perform both normal-mode computation and off-mode backup and restore (B and R). In order to achieve the optimal balance between B and R energy and area overheads, circuit enhancements are required. The field of system design is becoming more and more interested in emerging non-volatile memories. They are employed in the design of logic-in-memory circuits and in the proposition of von Neuman architecture substitutes. Logic-in-memory designs find special interest in ferroelectric memory technology based on hafnium oxide, which is entirely compatible with CMOS technologies. The ability to work with fine-grain logic in memory applications is made possible by this compatibility, as it allows for tight integration between the memory-capable element and the transistors in the system. Embedded artificial intelligence and non-volatile, energy-efficient computing for the Internet of Things are two potential applications for this technology.

Chapter 8: This comprehensive study investigates various aspects of sustainable bioplastic development and its implications. Firstly, the degradation of three bioplastics (poly(butylene succinate)-starch (PBS-starch), poly(butylene succinate) (PBS), and polylactic acid (PLA)) in soil environments was analyzed, revealing differential degradation rates correlated with the main components of bioplastics. The influence of bacterial biomass on bioplastic degradation was examined, indicating an enhanced degradation rate with increased bacterial biomass. Microbial diversity and nitrogen circulation in the soil remained unaffected by bioplastic degradation. Biodegradation tests on oil palm empty fruit bunch bioplastic composites showed higher rates than those of oxodegradable and conventional plastics. Sustainable bioplastic films made from *Prosopis juliflora* wood showed promising mechanical properties and biodegradability, making them viable alternatives to conventional plastics. Starch-based bioplastics were made from the starch of jackfruit seeds, and the mechanical characteristics of different plasticizers were investigated. Despite being weaker than polymers derived from petroleum, they are better for the environment. The study also looked into biodegradable polyvinyl alcohol-enhanced bioplastics made from maize starch and *Euphorbia antiquorum* latex.

Chapter 9: The semiconductor industry underwent a revolution with the introduction of MOSFETs, which are now the foundation of many low-power applications. Due to its high power consumption, MOSFET technology is becoming less and less common as it advances from the microscale to the nanoscale. Tunnel FET (TFET) has the benefit of eliminating the short-channel effects (SCEs) that affect MOSFETs. Because of its lower subthreshold swing (below 60 mV/dec), TFETs also require less power than

MOSFETs. Extrinsic sources of variability are mostly likely variations that show up from one chip or wafer to the next and are brought on by strain and modifications to the architecture of the device. Wafer-to-wafer, wafer-level, die-level, layout-dependent, and other variations are caused by significant process variances in the production of integrated circuits. There are several factors that cause variability in the characteristics of FETs; some are discussed in this chapter such as random dopant fluctuation (RDF), orientation, and heavy-ion irradiation. Random dopant fluctuation (RDF), which is the doping variation or fluctuation resulting from the random occupancy of impurities in the device during manufacturing, is one of the major factors influencing the device's variability among the other potential variations. The other factor is orientation; because of variations in recombination and generation rates, carrier tunneling mass, and tunneling barrier width, surface orientation affects the drain current of TFETs. Another is radiation, it is the major worry for semiconductor firms working on space devices. Radiation causes soft faults in the devices, which affects the integrated circuits' dependability and creates variations in the properties of the devices.

Chapter 10: This chapter introduces a novel 5-port MIMO antenna tailored for the 24–40 GHz frequency range, with a focus on applications such as n257, n258 satellite-to-ground communication, and n259, n260 NR 5G with material-based analysis. Fr4 and Roger substrate are used to analyze the approaches for the proposed antenna, and the Roger substrate is suitable for desirable outcomes. The antenna design is inspired by congruent circular rings incorporating fractal elements, resulting in a compact size and peak gain enhancement of 7 dBi, utilizing a Roger substrate with a height of 1.6 mm and dimensions of 25×25 mm^2. The proposed MIMO configuration exhibits remarkable performance metrics, featuring an envelope correlation coefficient (ECC) below 0.012, a diversity gain surpassing 9.666 dB, and improved isolation among its ports. In terms of frequency coverage, the MIMO antenna spans a wide bandwidth, encompassing 5.5 GHz for port 1, 3.4, 4.2 GHz for port 2, 3.4, 4.2 GHz for port 3, 3.4, 4.2 GHz for port 4, and 3.4, 5 GHz for port 5 within the 24–40 GHz spectrum. This broad coverage ensures robust communication capabilities, aligning with the requirements of advanced 5G applications and satellite communication.

Chapter 11: Two-dimensional (2D) nanomaterials are gaining a lot of attention these days in the field of electronics. 2D nanomaterials can be utilized to improve the performance of electromagnetic radiators (antennas). They offer advantages in antennas due to their lightweight nature, high electrical conductivity, tunable properties, broad bandwidth, and flexibility for conformal integration. This chapter aims to show how 2D nanomaterials like graphene can be utilized in improving the performance of antennas operating in the terahertz (THz) frequency range. Four antenna configurations, namely Antenna-A, Antenna-B, Antenna-C, and Antenna-D are presented in this work. It has been shown in this work that the use of graphene film significantly improves the performance of Antenna-B compared to Antenna-A. As the antenna is to be mounted on a silicon chip, a graphene layer is used in the ground plane for better heat transfer in Antenna-C. To further improve airflow and thermal convection, a hole/airway has been created in Antenna-D. Thus, it can help prevent overheating and performance degradation of other electronic components placed on a silicon chip. The electromagnetic and thermal behavior of Antenna-D mounted on a silicon chip is also shown in this chapter.

Chapter 12: This review study presents a thorough examination of the properties and applications of nanomaterials in semiconductor devices. Specifically, we delve into the characteristics of nanowires, carbon nanotube field-effect transistors (CNTFETs), quantum dots, and other nanoscale materials. Our discussion encompasses a diverse range of applications across semiconductor devices, including GAAFETs (gate all-around field-effect transistors), MOSFETs (metal–oxide–semiconductor field-effect transistors), photovoltaic cells, nanowire array solar cells, quantum dot lasers, solar cells, and LEDs (light-emitting diodes). The exploration of nanowires elucidates their unique structural and electrical properties, emphasizing their pivotal role in enhancing transistor performance. Carbon nanotube field-effect transistors are scrutinized for their exceptional electrical conductivity and potential to overcome traditional transistor limitations. Quantum dots, with their quantum confinement effects, are discussed in detail for their applications in novel devices such as quantum dot lasers. This chapter provides a comprehensive overview of how these nanomaterials are strategically incorporated into semiconductor devices. GAAFETs and MOSFETs benefit from the superior electronic properties of nanowires and CNTFETs, thereby enhancing device efficiency and scalability. Photovoltaic cells, nanowire array solar cells, and solar cells, in general, demonstrate improved performance through the integration of nanomaterials, capitalizing on their unique optoelectronic properties.

Chapter 13: This chapter introduces an innovative MUX-Based CMOS 8-Bit Encoder device designed to boost the binary conversion speed of Flash analog-to-digital converters (ADCs) specifically tailored for high-speed 5G communication systems and ultra-wideband (UWB) applications. The proposed encoder device employs a novel combination of a Wallace tree structure and a binary search algorithm to overcome limitations associated with conventional encoding methods. Through comprehensive simulations conducted in a 180 nm CMOS technology, the presented design demonstrates exceptional performance metrics. The achieved results include a compact track with an area of 3161 μm^2, a minimal delay of 0.4090 ns, a low power consumption of 23.32 μW, and an impressive figure of merit at 13.28 fJ. These outcomes signify a remarkable improvement in conversion speed and a substantial reduction in power consumption, illustrating the potential for optimizing ADC performance in critical communication systems and UWB applications. While traditional encoding methods, such as binary ROM-based encoders, offer simplicity in design, they often compromise on speed and power efficiency. The proposed MUX-Based CMOS 8-Bit Encoder stands out as a compelling solution, striking a balance between simplicity and efficiency. By incorporating a Wallace tree structure and a binary search algorithm, this design not only enhances the conversion speed but also mitigates the drawbacks associated with traditional encoding methods. The presented results underscore the potential of this encoder device to revolutionize ADC performance, making it a promising candidate for applications in high-speed 5G communication systems and UWB scenarios.

About the editors

Balwinder Raj (MIEEE'2006) is currently working as Associate Professor at the National Institute of Technology Jalandhar, India. He has more than 15 years of teaching and research experience. He did BTech in Electronics Engineering (PTU Jalandhar), MTech in Microelectronics (PU Chandigarh), and PhD in VLSI Design (IIT Roorkee), India, in 2004, 2006, and 2010, respectively. For further research work, the European Commission awarded him the "Erasmus Mundus" Mobility of Life research fellowship for postdoc research work at the University of Rome, Tor Vergata, Italy, in 2010–2011. Dr Raj received India4EU (India for European Commission) Fellowship and worked as visiting researcher at KTH University, Stockholm, Sweden, October–November 2013. He also visited Aalto University in Finland as visiting researcher during June 2017.

He had received Best Teacher Award from the Indian Society for Technical Education (ISTE), New Delhi, in 26 July 2013. Dr Raj received the Young Scientist Award from the Punjab Academy of Sciences during 18th Punjab Science Congress held on 9 February 2015. He has also received research paper award in the International Conference on Electrical and Electronics Engineering held at Pattaya, Thailand, from 11 to 12 July 2015. Dr Raj has authored/co-authored 8 books, 15 book chapters, and more than 150 research papers in peer-reviewed international/national journals and conferences. His areas of interest in research are classical/non-classical nanoscale semiconductor device modeling; nanoelectronics and their applications in hardware security; sensors and circuit design; FinFET-based memory design; low-power VLSI design; digital/analog VLSI design; and FPGA implementation.

Suman Lata Tripathi is working as Professor in Lovely Professional University with more than 21 years of experience in academics and research. She has completed her PhD in the area of microelectronics and VLSI design from MNNIT, Allahabad, India. She did her MTech in Electronics Engineering from UP Technical University, Lucknow, and BTech in Electrical Engineering from Purvanchal University, Jaunpur, India. She is also a remote post-doc researcher at Nottingham Trent University, London,

UK, in the year 2022. She has published more than 102 research papers in refereed Springer, Elsevier, IEEE, and IOP science journals; conference proceeding; and e-books. She has also published 13 Indian patents and 4 copyrights. She has guided four PhD scholars and four are under submission stage. She has organized several workshops, summer internships, and expert lectures for students. She has worked as a session chair, conference steering committee member, editorial board member, and peer reviewer in international/national IEEE, Springer, Wiley, etc., journal and conferences. She has received the "Research Excellence Award" in 2019 and "Research Appreciation Award" in 2020 and 2021 at Lovely Professional University, India. She is recipient of IGEN Women's for Green Technology "Women's Achievers Award" on International Women's Day, 8 March 2023. She had received the best paper at IEEE ICICS-2018. She has also received funded project from SERB DST under the scheme TARE in the area of microelectronic devices. She has edited and authored more than 19 books in different areas of Electronics and Electrical Engineering. She is associated for editing work with top publishers like Elsevier, CRC Taylor and Francis, Wiley-IEEE, SP Wiley, Nova Science, and Apple Academic Press. She is also working as book series editor for title, "Smart Engineering Systems" CRC Press, "Engineering System Design for Sustainable Developments", and "Decentralized Systems & Next Generation Internet" Wiley-Scrivener, and conference series editor for "Conference Proceedings Series on Intelligent Systems for Engineering Designs" CRC Press Taylor & Francis. She is serving as academic editor of *Journal of Electrical and Computer Engineering* (Scopus/WoS, Q2), *International Journal of Reconfigurable Computing* (Scopus, Q3), and *Active and Passive Electronic Component* (Scopus, Q4), Hindawi. She is associated as senior member IEEE, Fellow IETE, and Life member ISC and is continuously involved in different professional activities along with academic work. Her area of expertise includes microelectronics device modeling and characterization, low-power VLSI circuit design, VLSI design of testing, and advanced FET design for IoT, embedded system design, reconfigurable architecture with FPGAs, and biomedical applications.

Tarun Chaudhary did her BTech in Electronics and Communication Engineering from UIET, Panjab University, Chandigarh, in 2010; MTech. in VDAT (ECE) from NIT Hamirpur, Himachal Pradesh, in 2012; and PhD from NIT Hamirpur, Himanchal Pradesh, India, in the year 2017. During her PhD she has worked on the designing and mathematical modeling of vertical FET. She is currently working as Assistant Professor in ECE Department, Dr B R Ambedkar National Institute of Technology Jalandhar, India. She has worked as a faculty member under SMDP C2SD Project from October 2017 till November 2019. She is guiding PhD and PG scholars in the areas of nanoscale devices and circuits, and FPGA-based design. She has 5 book chapters and more than 25 research papers in peer-reviewed international/national journals and conferences. She has organized several STCs and workshops in the domain of VLSI design. Of the three students who are pursuing their PhD under her supervision, all are working on TFETs, NCFETs, and nanowire-based sensors. Her present research is largely focused on the design and modeling of nanoscale devices, TFETs, junctionless devices, and low-power VLSI design circuits.

Mandeep Singh has done his BTech in Electronics and Communication Engineering, MTech ECE (VLSI DESIGN) from Punjabi University, Patiala, and PhD from NIT Jalandhar, Punjab, India. He has three years of teaching experience to teach undergraduate and master students in National Institute of Technology Uttarakhand, India; Indian Institute of Information Technology (IIIT) Surat, India; Punjab Technical University, India. Currently he is working as Postdoc Fellow in the Indian Institute of Technology Kanpur, UP, India. He has published 3 books, 10 book chapters and various research papers in the domain of VLSI Design and Circuit. He is a versatile individual and has participated in multiple workshops and conferences worldwide (such as USA, Germany), underscoring his commitment to cutting-edge research. His areas of research are semiconductor device modelling, memory designs, CNT and Nanowire, FeFET, Ferroelectric Memory, and low power VLSI design.

Contributors

Neeraj Adhya
NIT Jalandhar
Punjab, India

Debanjali Barman Roy
Ajeenkya DY Patil University
Pune, Maharashtra, India

Dipjyoti Borah
NIT Jalandhar
Punjab, India

Tarun Chaudhary
Dr B R Ambedkar National Institute of Technology Jalandhar
Jalandhar, Punjab, India

Debanjan Das
NIT Jalandhar
Punjab, India

Susanta Das
Ajeenkya DY Patil University
Pune, Maharashtra, India

Prashant Deshmukh
Dr B R Ambedkar National Institute of Technology Jalandhar
Punjab, India

Sandeep Singh Gill
NITTTR
Chandigarh, India

Shubham Kumar Gupta
Dr B R Ambedkar National Institute of Technology Jalandhar
Punjab, India

Sagar Gurung
NIT Jalandhar
Punjab, India

Yadnyesh Khapekar
Ajeenkya DY Patil University
Pune, Maharashtra, India

Sourabh Khemot
NIT Jalandhar
Punjab, India

Banoth Krishna
NITTTR
Chandigarh, India

Amod Kumar
NITTTR
Chandigarh, India

Ankit Kumar
NIT Jalandhar
Punjab, India

Deepak Kumar
NIT Jalandhar
Punjab, India

Mukesh Kumar
NIT Jalandhar
Punjab, India

Durbadal Mandal
National Institute of Technology
Durgapur, West Bengal, India

Ashish Maurya
Kanpur Institute of Technology
Kanpur, Uttar Pradesh, India

Bellamkonda Prudhvi
NIT Jalandhar
Punjab, India

Arun Raj
National Institute of Technology
Durgapur, West Bengal, India

Balwinder Raj
Dr B R Ambedkar National Institute of Technology Jalandhar
Jalandhar, Punjab, India

Sunil Sankathala
Ajeenkya DY Patil University
Pune, Maharashtra, India

Shobhit Saxena
Chaudhary Charan Singh University (CCSU), Meerut
Uttar Pradesh, India

Aditya Shrivastav
Ajeenkya DY Patil University
Pune, Maharashtra, India

Mandeep Singh
Dr B R Ambedkar National Institute of Technology Jalandhar
Jalandhar, Punjab, India

Ramendra Singh
NIT Jalandhar,
Punjab, India

Shilpa Sonali
NIT Jalandhar
Punjab, India

Devanshi Soni
Ajeenkya DY Patil University
Pune, Maharashtra, India

Brajbhushan Tripathi
NIT Jalandhar
Punjab, India

Suman Lata Tripathi
Lovely Professional University
Phagwara, Punjab

Aryan Vasa
Ajeenkya DY Patil University
Pune, Maharashtra, India

Pindi Vishal
NIT Jalandhar
Punjab, India

Yogesh Kumar Yadav
Dr B R Ambedkar National Institute of Technology Jalandhar
Punjab, India

Acknowledgment

We would like to extend our sincerest thanks to Prof Binod Kumar Kanaujia, Director of the National Institute of Technology Jalandhar, for providing us with this opportunity and helping to create the environment for the start of this book project. We are also thankful to Dr Ramesh Kumar Sunkaria, Head of the Department of Electronics and Communication Engineering, NIT Jalandhar, for guidance and support. We have been fortunate in receiving support, assistance, and encouragement from our colleagues and scholars in the Department of Electronics and Communication Engineering at NIT Jalandhar.

We would like to thank and acknowledge the efforts and support of all contributors for completing this book. This book is based on research work carried out by a number of authors. Many people have contributed greatly to our book *Nanoelectronics: Materials to Chips Design.* We are grateful to a number of friends and colleagues for encouraging us to start work on this book. We, as the editors, would like to acknowledge all of them for their valuable help and generous ideas in improving the quality of this book. With our feelings of gratitude, we would like to introduce them in turn. The first mention is the authors and reviewers of each chapter of this book. Without their outstanding expertise, constructive reviews, and devoted effort, this comprehensive book would become something without content. The second mention is the CRC Press, Taylor and Francis staff for their constant encouragement, continuous assistance, and untiring support. Without their technical support, this book would not be completed. The third mention is the editors' families for being the source of continuous love, unconditional support, and prayers not only for this work but throughout our lives. Last but not least, we express our heartfelt thanks to the Almighty for bestowing upon us the courage to face the complexities of life and complete this work.

July 2024
Balwinder Raj
Suman Lata Tripathi
Tarun Chaudhary
Mandeep Singh

Chapter 1

Design and analysis of advanced MOSFET

Neeraj Adhya, Mandeep Singh, Prashant Deshmukh, Ramendra Singh, Balwinder Raj, and Tarun Chaudhary

1.1 INTRODUCTION

In this chapter, we have looked back at new non-conventional metal–oxide–semiconductor field-effect transistor (MOSFET) structures and state-of-the-art dimension scaling, in particular the double-gate (DG) MOSFET, tri-gate MOSFET, gate-all-around (GAA) FET, and the cylindrical surrounding double-gate (CSDG) MOSFET. As the device's dimension becomes closer to the nanoscale regime, in order to reduce short-channel effects in upcoming devices, these methods are suggested for forthcoming generations of communication systems [1–4].

When the giant vacuum tube was first introduced in 1930, there was a demand for smaller, less expensive, faster, and less power-hungry devices. This contributed to the advancement of the field-effect transistor (FET). The existence of surface states at the oxide–semiconductor interface, which stops the electric field from permeating the semiconductor material, prevented it from demonstrating its structure, though. This problem was solved by Kahng and Attala, who combined the use of three layers – metal (M), oxide (O), and semiconductor (S) – to create the insulated-gate FET. Furthermore, the quantity of surface states is decreased by the developed silicon dioxide layer on the semiconductor surface, which is why MOSFET is named.

As illustrated in Figure 1.1 [1], observations made over the past 25 years indicate that the wireless data rate doubles every 18 months and is getting close to wired communication systems' capability. Thus, new technologies that can handle large amounts of data while consuming less power overall are required to satisfy the demands of the wireless network of the future. For this reason, several GHz channel bandwidths can be obtained in the THz range of the frequency spectrum, which spans from 300 GHz to 3 THz. High data rates, up to multiple gigabits per second, are possible with this frequency. THz technology research has recently expanded into considerably broader applications, including wireless sensors and communications, medical imaging, and more [5–9]. In addition, two further critical needs for the next generation of communication systems are fast processing speed and low power consumption, both of which must be efficiently satisfied.

DOI: 10.1201/9781003612797-1

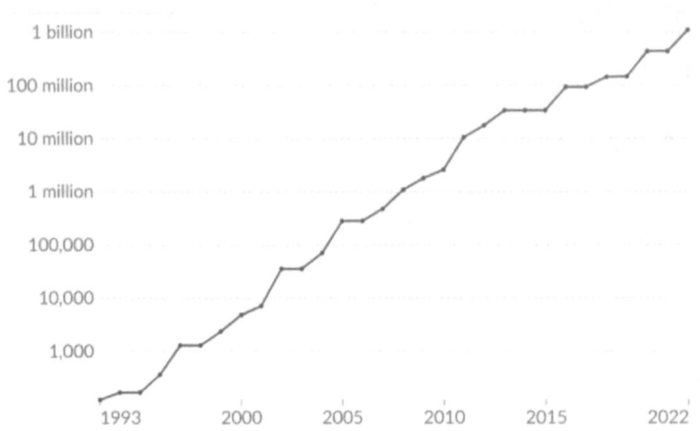

Figure 1.1 Moore's law [47].

Utilizing scaling is imperative for enhancing system dependability, current drive, computational capacity, and integration density at a significantly reduced cost [2]. The fundamental principle behind scaling is to simultaneously reduce all dimensions and produce a smaller transistor with performance comparable to the larger one [3, 4]. Nonetheless, it is important to precisely maintain the geometrical ratio, which is the ratio of the vertical dimension (channel length) to the horizontal dimension (channel width) [10–15]. Furthermore, it has been shown that scaling can reduce latency and advance clock frequency. Reference [5] reports a reduction in energy consumption with active power per transition in addition to a doubling of the device density [16–21].

Through the International Technology Roadmap for Semiconductors (ITRS) [8], the semiconductor industries have thus promoted the projection of the scaling theory in addition to the observations made by Moore's law [6, 7]. Figure 1.1 shows that Moore's law will continue to be followed until 2022 when transistor doubling occurs. But after that, the rate of progress (i.e., doubling) will decrease to once every three years. The general idea for scaling is shown in Figure 1.1, where the scaling factor, oxide thickness, depletion depth, gate length, and wiring width are represented, respectively, by a, W, Wd, Lg, and tox [22–27].

1.2 SHORT-CHANNEL EFFECTS (SCES)

The device is considered short-scaled when the MOSFET channel length, or the distance between the source and drain within the MOSFET, is comparable to the source and drain depletion depth beneath a gate with zero drain–source voltage. The numerous short-channel effects are collected in

Table 1.1. Short-channel effects take precedence over device performance as the device's dimensions considerably decrease [11]. These effects have been attributed to the following two physical phenomena:

1. Constraints imposed on the electron drift characteristics within the channel.
2. Variation in the threshold voltage due to the shortened channel length.

While preserving the improved device performance in the nanoscale domain, the decrease of SCE is crucial. As a result, a number of techniques

Table 1.1 Various short-channel effects

Reference	*Effect*	*Reason*
[17, 18]	Velocity saturation	Current in short-channel devices becomes independent of gate-to-source voltage and saturates due to carrier velocity saturation.
[12–14]	Drain-induced barrier lowering (DIBL)	The channel's potential barrier lowers as drain voltage increases, enabling electrons to flow between the source and the drain.
[15, 16]	Surface scattering	Extension of the depletion layer in the channel leads to an increase in the electric field component, resulting in greater surface mobility dependence with decreasing channel length.
[18, 19]	Hot electron effect	High-energy electrons are trapped as they penetrate the oxide layer that causes oxide charging, affecting device performance by raising the threshold voltage and impairing gate control.
[9]	Threshold voltage roll-off	Close proximity of source and drain in short-channel devices alters threshold voltage, requiring less voltage for MOSFET operation.
[20]	Punch-through	Depletion regions of the drain and source extend into the channel, rendering applied gate bias ineffective in regulating current flow.
[21, 22]	Oxide-tunneling current	Reduced oxide thickness increases the electric field, forcing more current through the gate terminal.
[23]	Reverse short-channel effect	Point defects accumulate at the surface edge, raising channel doping and threshold voltage with scaling until short-channel effects predominate.
[24, 25]	Gate-induced drain leakage (GIDL)	Overlap zone depletion region due to drain's electric field may invert the surface, affecting carrier flow during channel formation.
[26, 27]	Mobility degradation	Increased scattering and rough surface contribute to reduced mobility, particularly due to velocity saturation perpendicular to the gate.

have produced unfavorable outcomes like increased leakage and performance degradation. The electric field lines that emerge from the drain and travel to the source are terminated as a result of an increase in channel doping [28–35]. When the impurity dispersion increases due to this high amount of doping, the carrier's low field mobility deteriorates and the drive current is further reduced [28]. Additionally, it causes more band-to-band tunneling across the reverse-biased drain junction and gate-induced drain leakage (GIDL). Furthermore, statistical fluctuations in the channel dopants have been found to cause a notable variance in threshold voltage, primarily in the nanoscale regime [29]. Reducing the thickness of the oxide is crucial for enhancing the gate control over the channel; however, when the oxide approaches 2 nm, this leads to tunneling, leakage current, and an additional rise in standby power [10, 30, 31]. To decrease drain coupling to the source barrier, the depth of the source/drain junction must be lowered; yet, this requires an increase in doping density, which has a limit in terms of solid solubility. As a result, as the series resistance increases, the device's performance decreases [32, 33].

1.3 SILICON-ON-INSULATOR (SOI) MOSFETS

As seen in Figure 1.2(a), silicon-on-insulator (SOI) MOSFETs are created on wafers made up of a crystalline silicon film separated from the bulk substrate by a layer of SiO_2 [36–47]. When compared to bulk MOSFETs, SOI MOSFETs provide a number of benefits, including better latch onto free operation, less leakage current, less short-channel effects, and increased switching speeds because of lower drain–body capacitance [48, 49].

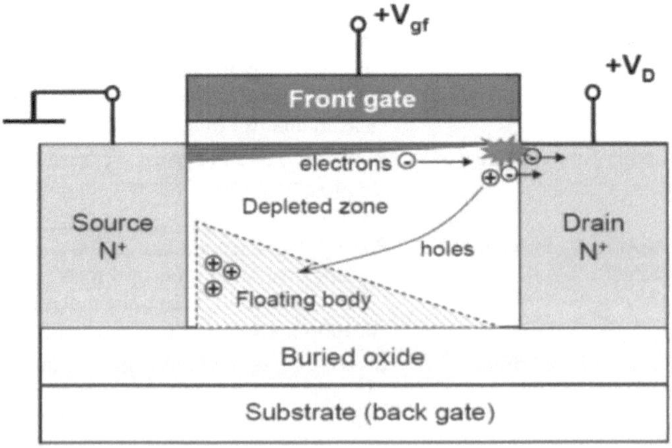

Figure 1.2 Schematic illustration of the kink effect in the [49] n-channel PD-SOI MOSFET: (a) Physical processes resulting in the kink effect. (b) Schematic diagram of SOI [49].

There are two types of SOI MOSFETs:

1. Partially depleted devices.
2. Fully depleted devices.

1.3.1 Partially depleted devices

The silicon channel/body section of the partially depleted (PD) SOI devices is not completely penetrated by the depletion region. The silicon thickness of PD-SOI devices is more than 0.15 µm. The term "partially depleted SOI" refers to the fact that the depletion zone, which reaches into the device's body beneath the gate at the source–body and drain–body connections, does not completely exhaust all of the charges in the body. Devices with partial depletion are frequently referred to as thick films. Despite these advantages, PD-SOI MOSFET devices have some unfavorable side effects, such as kink [50–56] in the output current–voltage characteristics, as seen in Figure 1.2(a). Devices with partially depleted SOI MOSFETs exhibit the kink effect.

1.3.2 Fully depleted devices

The maximum depletion layer width is greater than the thickness of the silicon film in FD-SOI MOSFETs, also known as thin-film SOI MOSFETs. In this scenario, the Si film will be completely depleted both below and above the threshold (with the possible exception of a thin accumulation or inversion layer at the back silicon film surface in the event that a significant positive or negative bias is applied to the substrate) [57–63] (Figure 1.3).

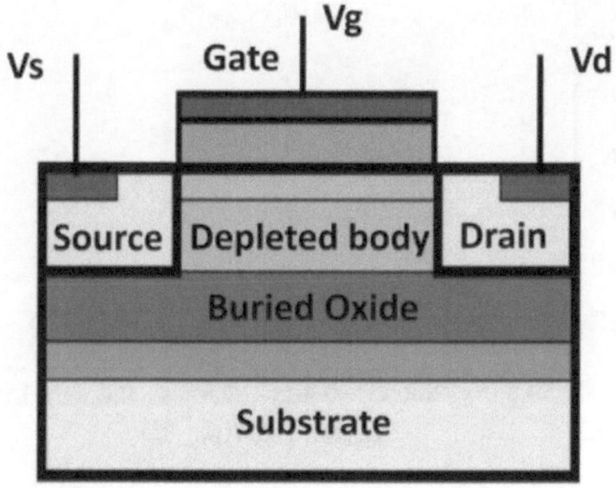

Figure 1.3 Proposed FD-SOI MOSFET.

As a result, in an FD-SOI MOSFET, the depletion layer's charge remains constant and does not grow when the gate voltage increases. When compared to partially depleted devices, FD-SOI MOSFETs have better electrical properties [49].

The top Si layer in PD-SOI devices has a thickness of about 300 nm. Such devices can be made in a manner that is completely compatible with bulk silicon technology. Consequently, it is simple to translate the design for bulk silicon devices to PD-SOI technology [50]. Because fabricating FD-SOI devices requires creating an ultrathin body, more advanced technology is required. In PD-SOI devices, the threshold voltage is comparatively less susceptible to variations in the thickness of the silicon film. The Si film thickness affects the threshold voltage in an FD-SOI device [51].

As shown in Figure 1.4, FD-SOI MOSFETs have a smaller sub-threshold swing than bulk Si or PD-SOI devices. Because of this, FD-SOI devices can employ a lower threshold voltage without raising the off-state leakage current, which is beneficial for ultralow-power applications [64–71] (Figures 1.5, 1.6 and Table 1.2).

1.4 PROPOSED DOUBLE-GATE MOSFET

The ultrathin silicon layer is sandwiched between two gates, known as the front and back gates, in the planar DG MOSFET, an expansion of the single-gate MOSFET [34]. The extra gate greatly improves the electrostatic gate control over the channel, and these gates work effectively to lower the SCE and lower the potential barrier by preventing the drain electric field

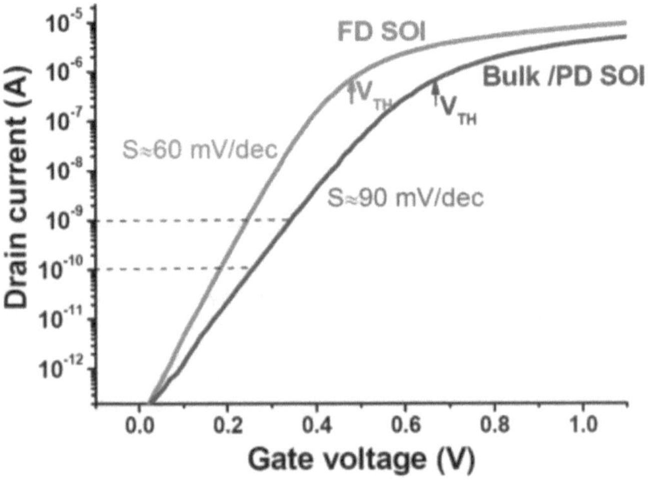

Figure 1.4 Sub-threshold characteristics of fully depleted SOI MOSFET and bulk/partially depleted SOI MOSFET [49].

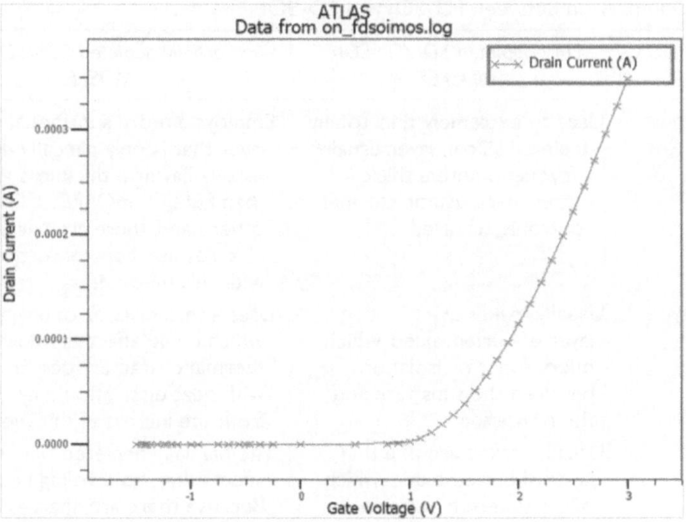

Figure 1.5 I_d vs V_{gs}.

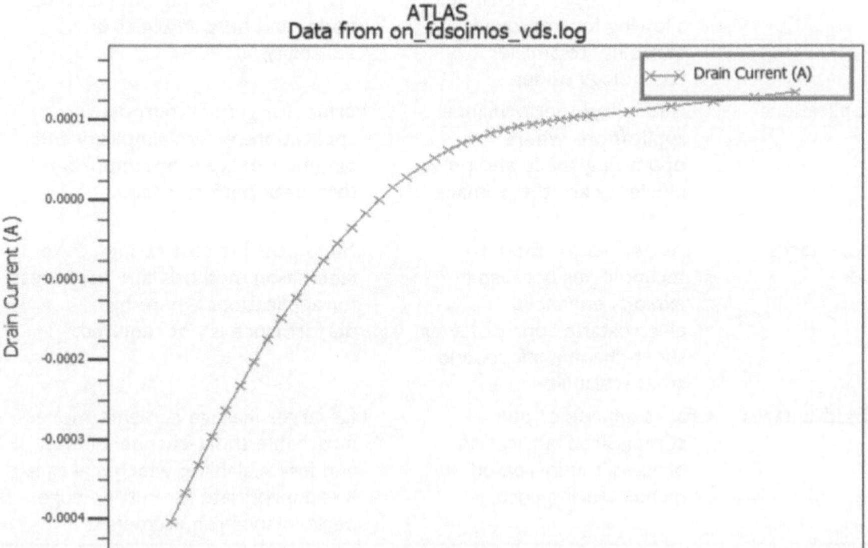

Figure 1.6 I_d vs V_{ds}.

Table 1.2 Comparison between FD-SOI and PD-SOI

Aspect	*Fully depleted SOI (FD-SOI) MOSFET*	*Partially depleted SOI (PD-SOI) MOSFET*
Body thickness	Uses an extremely thin, totally depleted silicon layer, usually a few nanometers thick. Better electrostatic channel control is achieved.	Employs a more substantial silicon layer that is only partially depleted, usually having a thickness greater than FD-SOI MOSFETs. On the other hand, there are certain electrostatic control restrictions with this method.
Substrate type	Usually makes use of a thin layer of buried oxide, which offers superior isolation between the substrate and the transistor.	Uses a thicker layer of buried oxide, which could affect the device's thermal characteristics. Problems with heat dissipation may also arise from the increased thickness.
Threshold voltage control	Usually makes use of a thin layer of buried oxide, which offers superior isolation between the substrate and the transistor.	The partially depleted silicon layer affects threshold voltage control. Because there are charge carriers that aren't completely depleted, adjusting the threshold voltage becomes more difficult.
Short-channel effects	Reduces short-channel effects significantly due to the entirely depleted channel, allowing for improved scalability to smaller technology nodes.	Increased short-channel effects as a result of partial depletion, which may limit the scaling of technology nodes and have an effect on scalability.
Applications	Used in high-performance applications where optimizing speed and power efficiency are the primary goals.	Perfect for general-purpose applications where simplicity and pragmatism are more important than peak performance.
Advantages	It is perfect for modern technologies because it exhibits enhanced electrostatic control, fewer short-channel effects, and great scalability.	Offers possible cost savings, easier fabrication methods, and usefulness for applications where high performance is not required.
Disadvantages	Faces difficulties due to complicated fabrication process and increased manufacturing costs.	Has larger leakage currents, more noticeable short-channel effects, and less scalability, which makes it less appropriate for cutting-edge performance requirements.

lines from getting too close to the source [72–77]. Additionally, the issue of random dopant fluctuation is eliminated because of the undoped or lightly doped body. Additionally, when these two gates control the charge concurrently, two pathways for the current flow are generated and both gates contribute to inversion carriers, which have a high drive capability. Both gates are switched simultaneously in the common mode of operation [78–85]. In other modes of operation, on the other hand, a conducting plane is created by applying back-gate bias and switching the front-gate, which adds a parameter to the circuit design. As a result, two different kinds of structures may exist (Figure 1.7).

1.4.1 Symmetric DG MOSFET

The work function of both gates is the same. However, the gate's work function allows for the adjustment of the threshold voltage [86–91]. Unlike bulk devices, where carrier conduction occurs across the whole volume of the material, the thin silicon substrate is subject to volume inversion conditions [35].

1.4.2 Asymmetric DG MOSFET

The work function of each gate varies. Without the use of specific gate material, the threshold voltage can be changed by adjusting the thickness of the body and the gate oxide. Two oxide thicknesses, two distinct gate biases, and two different flat-band voltages are present in the asymmetric double-gate device. The silicon bands in an asymmetric DG MOSFET with a mid-gap gate electrode are not flat in the sub-threshold domain [35].

The two operating modes and various architectures of the DG MOSFET are displayed in Figure 1.8. The sub-threshold slope of the four-terminal driven DG MOSFETs is not optimal, though. This is because if the potential of the non-switching gate is fixed, the potential across the silicon film does not move together with the switching gate as a whole [92–97]. In comparison to the widely utilized three-terminal driven DG MOSFET, the four-terminal driven DG MOSFET likewise exhibits a worse short-channel effect (Figures 1.9–1.11).

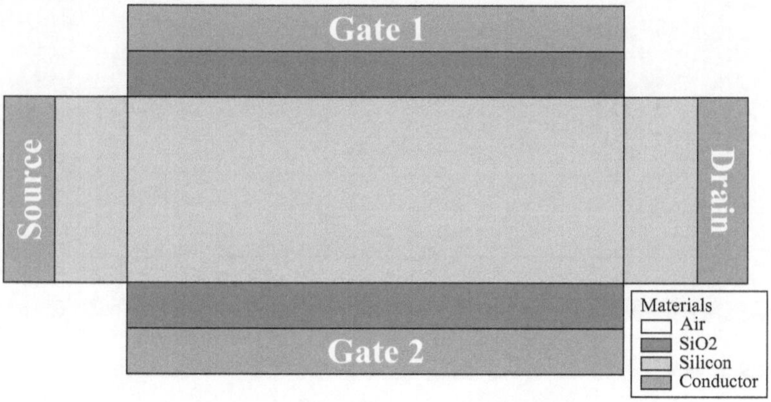

Figure 1.7 Proposed double-gate MOSFET.

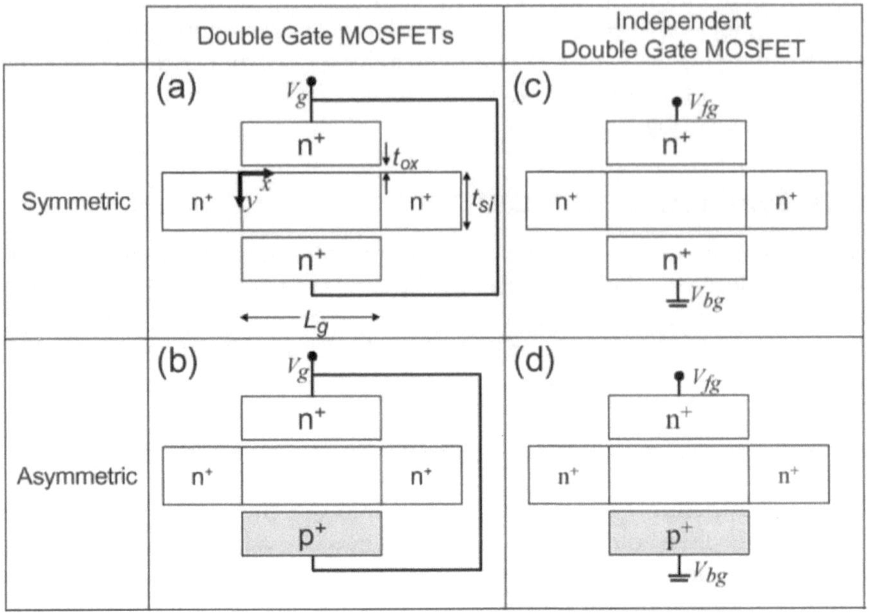

Figure 1.8 Schematics of DG MOSFET structures for tied gates and separated gates [36].

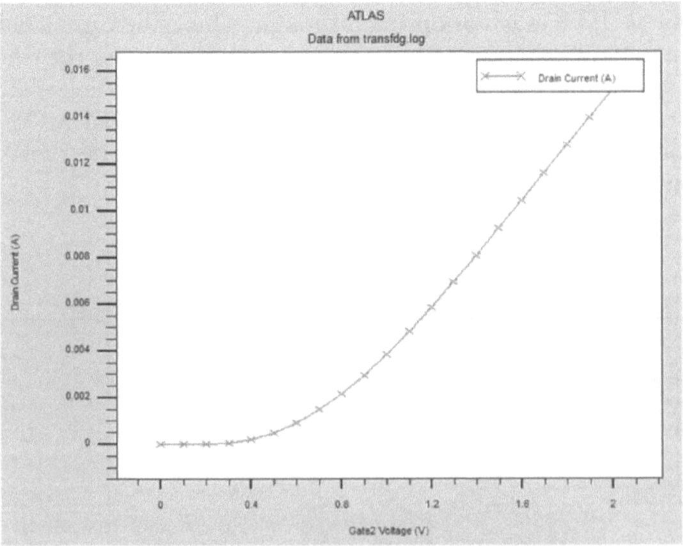

Figure 1.9 I_d vs V_{gs}.

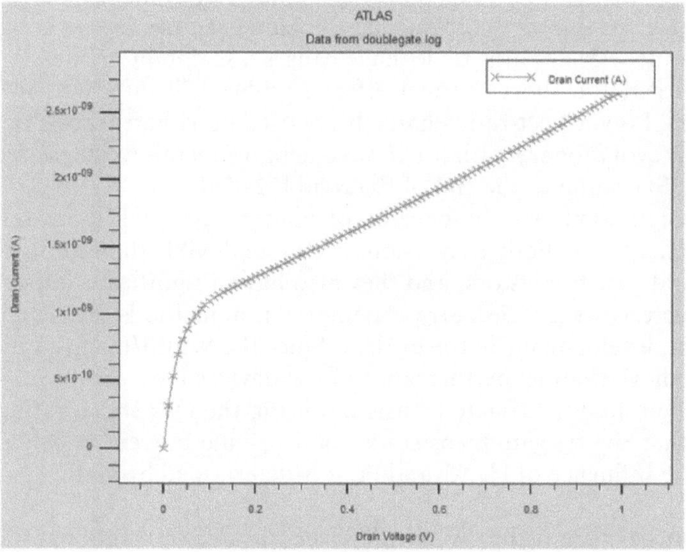

Figure 1.10 I_d vs V_{ds}.

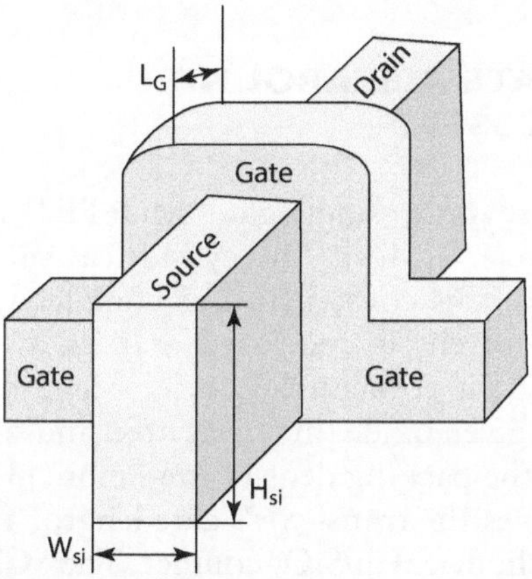

Figure 1.11 Tri-gate transistor [58].

1.5 TRI-GATE MOSFET

An integrated transistor with three gates positioned around the three substrate phases to cover the top, side 1, side 2, and bottom sides is known as a tri-gate transistor. Gate 1, gate 3, and gate 2, respectively, acquire side 1, side 2, and top. As seen in Figure 1.11, gates 1 and 2 have physical lengths of L and W, while gate 3 has a length of L and a width of W. Because of the rapid changes in technology in order to keep up with Moore's law, cell sizes are getting smaller [98–102]. Transistors have been scaled up more and more in order to improve chip packaging. However, because of the growing short-channel effect, parasitic leakage causes a scalability issue.

In contrast, the short-channel effect significantly improved in tri-gate transistors. Low-aspect-ratio channels are used in tri-gate transistor designs to offer an evolutionary path for complementary metal–oxide–semiconductor (CMOS) scaling to the end of the road [52–54].

The design combines the benefits of planar MOSFETs with multi-gate transistors. Tri-gate body dimensions are more flexible than those of single, double, and gate transistors, and they also have a significant impact on the V–I characteristics [55, 56]. Stress memorization methods can help us make significant developments in the present. Since the Width/Length ratio determine the short-channel performance of the devices [103–108].

It can be found in tri-gate transistors using the (W+2H)/L ratio. Due to the fact that the tri-gate transistor's total driving current depends on the cumulative influence of H, W, and H, it is determined by adding gates 1, 2, and 3 [57].

Thus, rather than being given by W over L as in a conventional transistor, it is given by (W+2H) over L. This demonstrates unequivocally that increasing the transistor's height will result in a higher drive current [57].

1.6 PROPOSED GATE-ALL-AROUND NANOWIRE MOSFET

The circular structure of a single-gate MOSFET is the surrounding-gate MOSFET, also referred to as the cylindrical surrounding MOSFET. Because the channel in this construction is completely encircled by gates, the device's leakage current is decreased and its overall performance is increased. Additionally, it reduces the device's length when compared to DG MOSFETs with a given oxide thickness (tox) and silicon body thickness (tsi), which enhances the packing density even more [37]. The height of the gate material determines the transistor's gate length, and the current flows vertically along the cylindrical Si/SiO_2 contact. The SCE is decreased by the tight capacitive coupling in every direction. Figure 1.12 depicts the basic surrounding-gate MOSFET construction.

For low-power and high-speed applications, the surrounding-gate MOSFET is very successfully scaled down to the nanoscale domain [38]. The surrounding-gate MOSFET, which is highly useful for high-speed and low-power circuit applications, has been examined by Rustagi et al. [39]. A DC model for surrounding-gate MOSFETs has been created by Iniguez et al. [40]. It is based on the charge-controlled model and the channel current's continuity, which is maintained throughout all operating zones [109–113] (Figures 1.13 and 1.14 and Table 1.3).

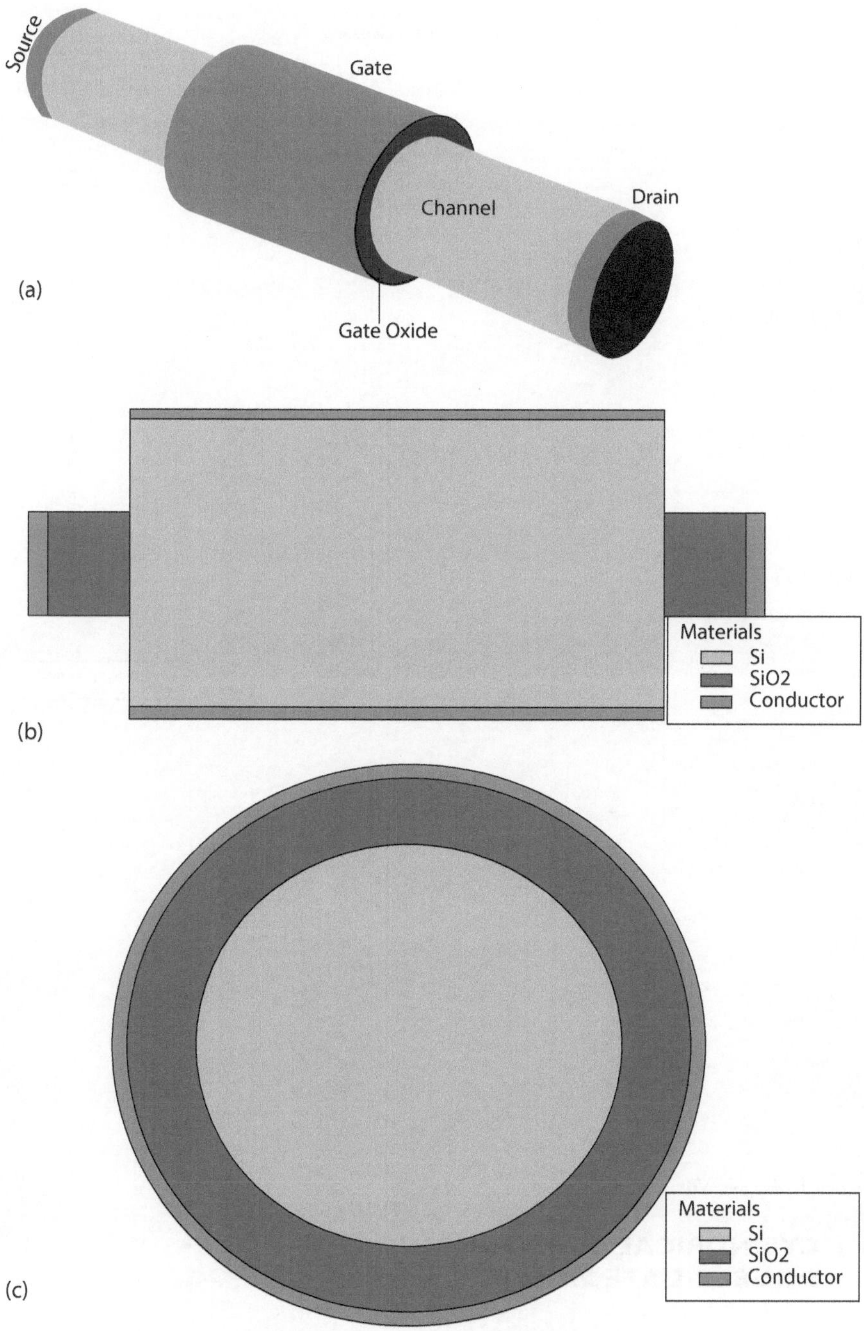

Figure 1.12 Proposed schematic of (a) DG MOSFET, (b) three-dimensional CSDG MOSFET, and (c) side view of CSDG MOSFET.

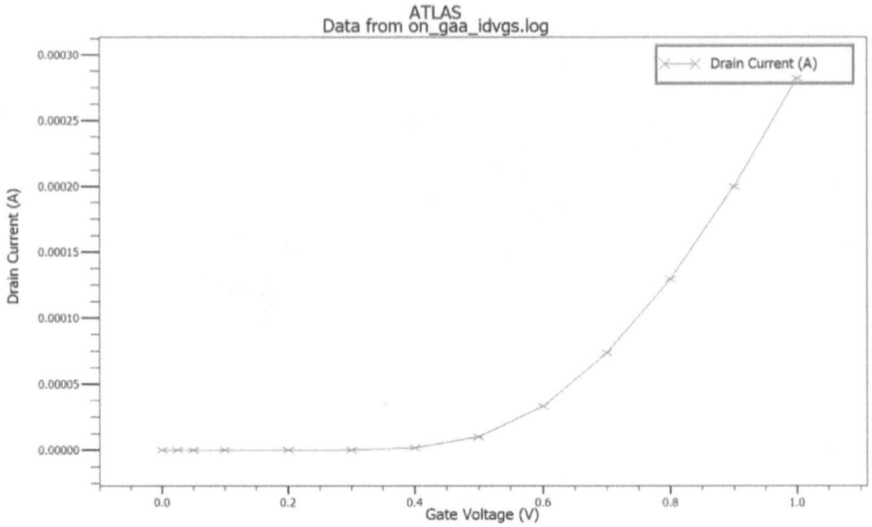

Figure 1.13 I_d vs V_{gs}.

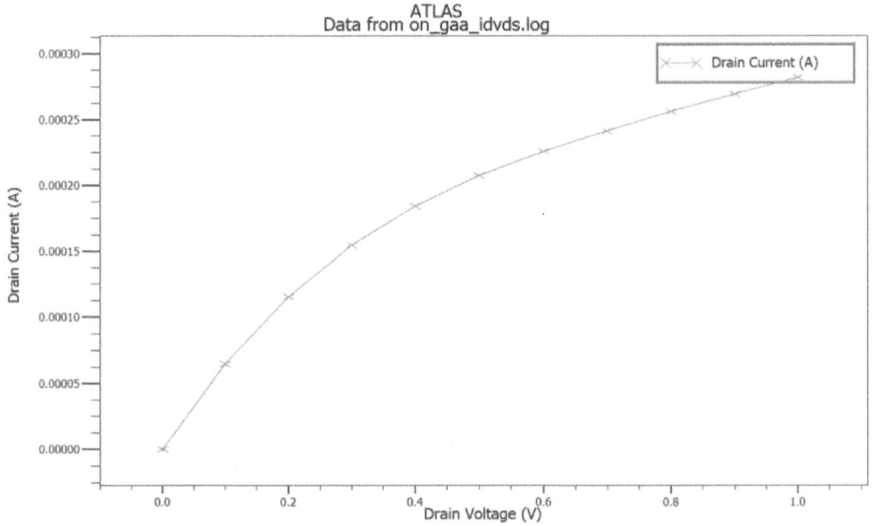

Figure 1.14 I_d vs Vds.

1.7 CYLINDRICAL SURROUNDING DOUBLE-GATE MOSFET

The high-frequency capacity of the transistors has reached the GHz and THz regimes with the development of surrounding-gate MOSFETs in the nanometer scale regime, making them suitable for radio frequency (RF)

Table 1.3 Challenges at different technology nodes [59]

Node	*Issue*	*Solution*	*Best Device*
<0.1 μm	Short-channel effects (SCE), low drive current	• Strained SiGe • Metal gate • High K dielectric	Bulk MOSFET
0.1 μm–32 nm	Power, leakage current	• Ultrathin body SOI MOSFET	SOI MOSFET
32 nm–10 nm	SCEs are prominent	• Use of multi-gate material • Stacked oxide	FinFET
<10nm	Power, area, and cost	• Vertically stacked GAA • Work function engineering • High K dielectric	GAA

circuit applications [41, 42]. The CSDG MOSFET's structure is the DG MOSFET's rotational structure. Because the gate surrounds the channel on both the internal and external sides of the device, the device's performance is further enhanced and leakage current is effectively reduced [114–119]. Figure 1.15 illustrates the full-gate control over the channel that Srivastava et al. [43, 44] discussed and suggested in 2011. The model is called the CSDG MOSFET. This construction is built around a hollow silicon cylinder, and it has internal and exterior gates on either side of the channel. The CSDG MOSFET is a promising device for Very Large-Scale Integration (VLSI) and Ultra Large-Scale Integration (ULSI) technology because of its low OFF-state current.

Following the design of this CSDG MOSFET, the authors [43, 44] simulated the available parameters and created the layout. It covers the fundamentals of the circuit element parameters needed for the radio frequency sub-systems of integrated circuits, including power or voltage gain, oxide thickness, polysilicon resistance, number of bulk capacitors, drain current,

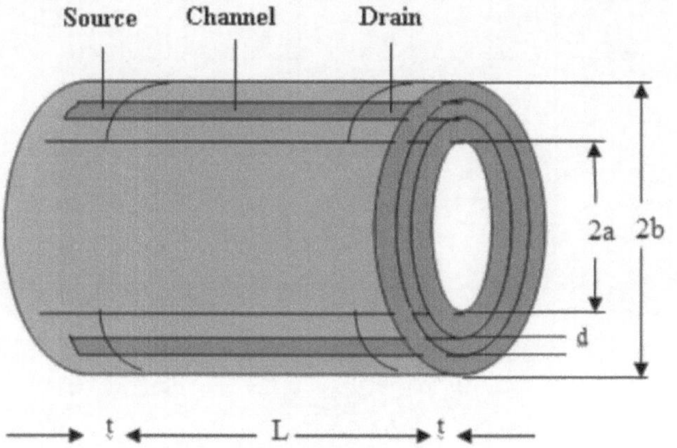

Figure 1.15 Three-dimensional CSDG MOSFET [43].

output voltage, threshold voltage, capacitances, resistances at switch ON-state, and energy stored [120–125]. Figure 1.16 is used to illustrate these concepts. By employing suitable formulas for the channel charge densities in terms of applied voltages, the model appears clear. The charge control model sufficiently takes into consideration the channel charge distribution in the Si film. Reference [45] presents the physics-based model of the CSDG nanowire MOSFET. The authors of this chapter described a threshold voltage model in the strong inversion zone utilizing an extrapolation method [126–132]. They discovered that the cylindrical surrounding-gate MOSFET is not as effective as the CSDG MOSFET. Reference [46] presents the technology computer-aided design (TCAD) analysis of a CSDG MOSFET that includes an inner charge control gate. With greater RF performance and enhanced electrostatic integrity, the CSDG MOSFET was introduced. It has been possible to reach an astounding cutoff frequency of more than 2.3 THz (Tables 1.4 and 1.5).

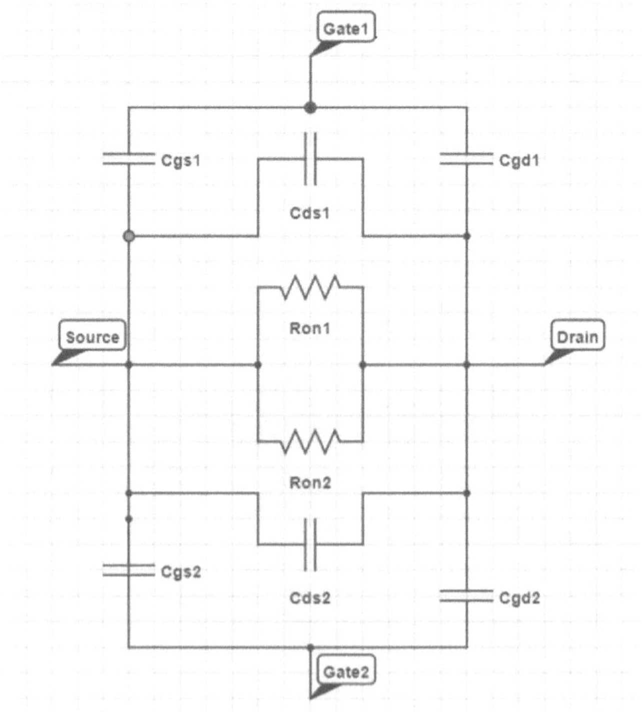

Figure 1.16 Proposed circuit of CSDG MOSFET at switch ON-state.

Table 1.4 Simulation parameter used for proposed device

Parameter	*Bulk MOSFET*	*Double-gate MOSFET*	*SOI MOSFET*	*Gate-all-around MOSFET*
Oxide thickness	0.03	0.005	0.017	0.004
Channel length	6	0.1	1	0.05
n-type doping	1e21	1e21	1e20	1e20
p-type doping	1e17	1e16	1e17	1e16
Gate work function	4.1 eV	4.6 eV	4.6 eV	4.8 eV
Gate oxide material	SiO_2	SiO_2	SiO_2	SiO_2

1.8 CONCLUSION AND FUTURE RESEARCH

The communication sector is becoming more and more dependent on the creation of equipment that can operate at high frequencies and data rates. Furthermore, the gadgets' reduced dimensions play a crucial role in the communication market's advancement. Conventional scaling, a technique for advancing MOS technology into the nanoscale regime, has been hindered by a number of factors, including process variability, channel carrier mobility deterioration, and the SCE. These are of utmost importance to circuit designers, and as a result, the channel potential's drain is significantly controlled [133–138].

The DG MOSFET, surrounding-gate MOSFET and CSDG MOSFET are highlighted extensively in this analysis because they eliminate short-channel effects by scaling down to the smallest channel length achievable for a selected oxide thickness. As a result, the main focus of the analysis is on the device characterization and modeling for high-frequency applications of DG MOSFET, surrounding-gate MOSFET, and CSDG MOSFET [139–143]. The practicality and upcoming difficulties of the device for usage in integrated circuits/systems on chips are also covered in this chapter. This would demonstrate high packaging density and improved performance in the THz high-frequency region of the spectrum, where the data rate reaches 0.01 TB/s, in terms of unity-gain frequency and maximum oscillation frequency.

Consequently, reducing the drain's impact on the channel potential may be necessary, since this has led many scientists and researchers to consider alternative, non-conventional MOSFET topologies such as SOI, pie-gate, and omega-gate MOSFETs [144–149].

Beyond the parameters that have been examined, there are other possible issues with the CSDG MOSFET, such as the difficult procedure involved in fabricating these kinds of devices, which will be covered in later communications. Moreover, the maximum cutoff frequency is seen in dual material gate, gradual channel cylindrical MOSFETs. The advantages of the dual

Table 1.5 Comparative analysis of MOSFET technologies

Feature	*Double-gate MOSFET*	*SOI MOSFET*	*Gate-all-around (GAA) MOSFET*	*Tri-gate MOSFET*
Advantage	• Improved gate control due to two gates, enhancing on/off current ratios. • Better immunity to short-channel effects. • Enhanced device scalability. • Improved electrostatic control. • Reduced leakage current. • Potential for low standby power.	• Reduced short-channel effects due to the thin buried oxide layer • Improved isolation. • Reduced power consumption. • Enhanced thermal performance. • Reduced variability. • Improved latch-up immunity. • Suitable for high-temperature environments.	• Enhanced gate control due to multiple gates, allowing for superior electrostatic control • Excellent scalability, supporting Moore's law. • Potential for better power efficiency. • Improved device reliability. • Low variability and improved statistical performance. • Better control of leakage current.	• Superior gate control with three gates, leading to significantly reduced short-channel effects. • Improved performance and power efficiency. • Enhanced electrostatic control and reduced leakage current. • Superior control in the sub-threshold region for low-power operation.
Disadvantage	• Manufacturing complexity due to double-gate structure. • Moderate scalability challenges at smaller technology nodes. • Increased sensitivity to process variations. • Potential challenges in 3D integration.	• Increased fabrication complexity, particularly in manufacturing SOI wafers. • Limited scalability at smaller nodes. • Higher manufacturing cost. • Potential for floating body effect. • Impact of radiation on floating body.	• Higher manufacturing complexity due to multiple gate structures. • Complex integration processes. • Challenges in gate patterning for small nodes. • High cost of multiple patterning techniques.	• Complex manufacturing processes, potentially leading to higher production costs. • Requires advanced lithography techniques. • Potential for increased gate leakage current.

(*Continued*)

Table 1.5 (Continued) Comparative analysis of MOSFET technologies

Feature	*Double-gate MOSFET*	*SOI MOSFET*	*Gate-all-around (GAA) MOSFET*	*Tri-gate MOSFET*
Applications	• General-purpose digital and analog circuits. • High-speed processors. • Low-power devices. • Static random access memory (SRAM) and dynamic random access memory (DRAM) cells. • High-frequency RF applications.	• High-performance applications (processors, memory). • RF and analog circuits. • Low-power applications. • High-speed communication. • Image sensors. • Power amplifiers.	• High-performance processors. • High-speed communication systems. • Low-power and IoT (Internet of Things) devices. • Memory devices (SRAM, DRAM). • Analog and RF applications.	• High-performance processors. • High-speed communication systems. • Low-power and IoT devices. • Memory devices (SRAM, DRAM). • Analog and RF applications.

material gate junctionless transistor include its high maximum oscillation frequency, high ON-state drain current, and high unity-gain frequency. These criteria are transferable to further studies. There is a lot of room for analysis regarding the different uses of CSDG MOSFETs in the near future.

REFERENCES

1. S. Cherry, "Edholm's law of bandwidth," *IEEE Spectrum*, vol. 41, no. 7, pp. 58–60, July 2004.
2. H. Iwai, "Silicon MOSFET scaling beyond 0.1 micron," *Proceeding of 21st Int. Conf. on Microelectronics*, Nis, Yugoslavia, 14–17 Sep. 1997, pp. 11–18.
3. R. H. Dennard, F. H. Gaensslen, H. N. Yu, V. Leo Rideout, E. Bassous, and A. R. Leblanc, "Design of ion-implanted MOSFET's with very small physical dimensions," *IEEE Journal of Solid-State Circuits*, vol. 9, no. 5, pp. 256–268, Oct. 1974 and *Proceeding of the IEEE*, vol. 87, no. 4, pp. 668–678, Apr. 1999.
4. B. Davari, R. H. Dennard, and G. G. Shahidi, "CMOS scaling for high performance and low power: the next ten years," *Proceeding of the IEEE*, vol. 83, no. 4, pp. 595–606, Apr. 1995.
5. G. G. Shahidi, "Challenges of CMOS scaling at below 0.1 µm," *Proceeding of 12th Int. Conf. on Microelectronics*, Tehran, 31 Oct.–2 Nov. 2000, pp. 5–8.
6. P. K. Bondyopadhyay, "Moore's law governs the silicon revolution," *Proceeding of the IEEE*, vol. 86, no. 1, pp. 78–81, Jan. 1998.
7. G. E. Moore, "Cramming more components into integrated circuit," *IEEE Solid-State Circuits Society Newsletter*, vol. 20, no. 3, Sep. 2006, *Proceeding of the IEEE*, vol. 86, no. 1, pp. 82–85, Jan. 1998, and *Electronics*, vol. 38, no. 8, pp. 11–13, Apr. 1965.
8. International Technology Roadmap for Semiconductors (ITRS), http://public.itrs.net.
9. G. Baccarani, M. R. Woedeman, and R. H. Dennard, "Generalized scaling theory and its application to a ¼ µm MOSFET design," *IEEE Transactions on Electron Devices*, vol. 31, no. 4, pp. 452–462, May 1984.
10. Y. Taur, D. A. Buchanan, W. Chen, D. J. Frank, K. E. Ismail, S. Hsien Lo, G. A. Sai Halasz, R. G. Viswanathan, H. J. C. Wann, S. J. Wind, and H. S. Wong, "CMOS scaling into the nanometer regime," *Proceeding of the IEEE*, vol. 85, no. 4, pp. 486–504, Apr. 1997.
11. M. Shahnawaz, "Drain current (I-V) model of asymmetric double gate MOSFETs," Master of Electronics and Telecommunication Engineering Thesis, Jadavpur University Kolkata, India, 2010.
12. R. R. Troutman, "VLSI limitation from drain-induced barrier lowering," *IEEE Transactions on Electron Devices*, vol. 26, no. 4, pp. 461–469, Apr. 1979 and *IEEE Journal of Solid-State Circuits*, vol. 14, no. 2, pp. 383–391, Apr. 1979.
13. S. G. Chamberlain and S. Ramanan, "Drain-induced barrier-lowering analysis in VSLI MOSFET devices using two-dimensional numerical simulations," *IEEE Transactions on Electron Devices*, vol. 33, no. 11, pp. 1745–1753, Nov. 1986.

14. M. A. Karim, S. Venugopalan, Y. S. Chauhan, D. Lu, A. Niknejad, and C. Hu, "Drain induced barrier lowering (DIBL) effect on the intrinsic capacitance of nano-scale MOSFET," *Nanotechnology* 2011 and *Electronics, Devices, Fabrication, MEMS, Fluidics and Computational*, vol. 2, pp. 814–817, 2011.
15. S. Jin, Massimo V. Fischetti, and T. W. Tang, "Modelling of surface-roughness scattering in ultrathin-body SOI MOSFETs," *IEEE Transactions on Electron Devices*, vol. 54, no. 9, pp. 2191–2203, Sep. 2007.
16. C. L. Alexander and A. Asenov, "Statistical MOSFET current variation due to variation in surface roughness scattering," *Proc. of IEEE Int. Conf. on Simulation of Semiconductor Processes and Devices (SISPAD)*, Osaka, Japan, 8–10 Sep. 2011, pp. 275–278.
17. H. Iwai, M. R. Pinto, C. S. Rafferty, J. E. Oristian, and R.W Dutton, "Velocity saturation effect on short-channel MOS transistor capacitance," *IEEE Electron Device Letters*, vol. 6, no. 3, pp. 120–122, Mar. 1985.
18. M. Rodder, "Leakage-current-induced hot-carrier degradation of p-channel MOSFETs," *IEEE Electron Device Letter*, vol. 9, no. 11, pp. 573–575, Nov. 1988.
19. T. Tsuchiya and J. Frey, "Relationship between hot-electrons/holes and degradation of p- and n-channel MOSFET's," *IEEE Electron Device Letters*, vol. 6, no. 1, pp. 8–11, Jan. 1985.
20. X. Zhao, D. E. Ioannou, W. C. Jenkins, H. L. Hughes, and S. T. Liu, "Hot electron induced punch-through (HEIP) in p-channel SOI MOSFET's," *Proc. of IEEE Int. SOI Conf.*, Stuart, FL, USA, 5–8 Oct. 1998, pp. 83–84.
21. M. U. Erdogan, M. C. Chang, C. Bowen, A. Chatterjee, J. Seitchik, and H. Shichijo, "Analysis of gate tunneling current in ultra-thin oxide MOSFET's," *Proc. of 56th IEEE Device Research Conf. Digest*, Charlottesville, VA, USA, 22–24 June 1998, pp. 14–15.
22. M. F. Karim, S. Shaari, and B. Y. Majlis, "Gate tunneling current in thin oxide MOSFET," *Proc. of IEEE Int. Conf. on Semiconductor Electronics*, Penang, Malasia, 26–28 Nov. 1996, pp. 287–292.
23. N. Collaert, P. Verheyen, and K. De Meyer, "On the reverse short channel effect in deep submicron heterojunction MOSFET's and its impact on the current-voltage behaviour," *IEEE Transactions on Electron Devices*, vol. 47, no. 6, pp. 1214–1220, June 2000.
24. N. Maouhoub and K. Rais, "Characterization of defect density created by stress in the gate to drain overlap region using GIDL current model in n-channel MOSFET," *Journal of Computational Electronics*, vol. 10, no. 1, pp. 141–143, June 2011.
25. V. Nathan and N. C. Das, "Gate-induced drain leakage current in MOS devices," *IEEE Transactions on Electron Devices*, vol. 40, no. 10, pp. 1888–1890, Oct. 1993.
26. A. S. Roy, C. C. Enz, and J. M. Sallese, "Charge, current, and noise partitioning in MOSFET in the presence of mobility degradation," *IEEE Electron Device Letters*, vol. 27, no. 8, pp. 674–677, Aug. 2006.
27. M. L. P. Tana, M. A. Riyadi, and V. K. Arora, "Mobility degradation in a nano-MOSFET due to ballistic and high-field effects," *Proc. of IEEE Int. Conf. on Enabling Science and Nanotechnology*, Johor Bahru, Malaysia, 5–7 Jan. 2012, pp. 1–2.

28. J. Moers, "Turning the world vertical: MOSFETs with current flow perpendicular to the wafer surface," *Applied Physics A*, vol. 87, no. 3, pp. 531–537, 2007.
29. T. Tanaka, T. Usuki, T. Futatsugi, Y. Momiyama, and T. Sugii, "Vth fluctuation induced by statistical variation of pocket dopant profile," *IEEE Int. Electron Devices Meeting*, San Francisco, CA, USA, 10–13 Dec. 2000, pp. 271–274.
30. B. Yu, H. Wang, C. Riccobene, Q. Xiang, and M. R. Lin, "Limits of gate-oxide scaling in nanotransistors," *Proc. of the Symposium on VLSI Technology Digest of Technical Papers*, Honolulu, HI, USA, 13–15 June 2000, pp. 90–91.
31. W. Haensch, E. J. Nowak, R. H. Dennard, P. M. Solomon, A. Bryant, O. H. Dokumaci, A. Kumar, X. Wang, J. B. Johnson, and M. V. Fischetti, "Silicon CMOS devices beyond scaling," *IBM Journal of Research and Development*, vol. 50, no. 4–5, pp. 339–361, July–Sep. 2006.
32. H. S. P. Wong, "Beyond the conventional transistor," *IBM Journal of Research and Development*, vol. 46, no. 2–3, pp. 133–168, Mar.–May 2002.
33. K. Shin, "Technologies for enhancing multi-gate Si MOSFET performance," Ph.D. Thesis, University of California, Berkeley, 2006.
34. S. Oda and D. Ferry, *Silicon Nanoelectronics*, CRC Press, Boca Raton, FL, USA, 2006.
35. S. Subramaniam, R. N. Awale, and S. M. Joshi, "Drain current models for single-gate MOSFETs & undoped symmetric & asymmetric double-gate SOI MOSFETs and quantum mechanical effects: a review," *International Journal of Engineering Science and Technology*, vol. 5, no. 1, pp. 96–105, Jan. 2013.
36. J. W. Han, C. J. Kim, and Y. K. Choi, "Universal potential model in tied and separated double-gate MOSFETs with consideration of symmetric and asymmetric structure," *IEEE Transactions on Electron Devices*, vol. 55, no. 6, pp. 1472–1479, June 2008.
37. A. O. Conde and F. J. G. Sanchez, "Unification of asymmetric DG, symmetric DG and bulk undoped-body MOSFET drain current," *Solid-State Electronics*, vol. 50, no. 11–12, pp. 1796–1800, Nov.–Dec. 2006.
38. Y. Yamamoto, T. Hidaka, H. Nakamura, H. Sakuraba, and F. Masuoka, "Decananometer surrounding gate transistor (SGT) scalability by using an intrinsically-doped body and gate work function engineering," *IEICE Transactions on Electronics*, vol. E-89-C, no. 4, pp. 560–567, 2006.
39. S. C. Rustagi, N. Singh, W. W. Fang, K. D. Buddharaju, S. R. Omampuliyur, S. H. G. Teo, C. H. Tung, G. Q. Lo, N. Balasubramanian, and D. L. Kwong, "CMOS inverter based on gate-all-around silicon nanowire MOSFETs fabricated using top–down approach," *IEEE Electron Device Letters*, vol. 28, no. 11, pp. 1021–1024, Nov. 2007.
40. B. Iniguez, D. Jimenez, J. Roig, H. A. Hamid, L. F. Marsal, and J. Pallares, "Explicit continuous model for long-channel undoped surrounding gate MOSFETs," *IEEE Transactions on Electron Devices*, vol. 52, no. 8, pp. 1868–1873, Aug. 2005.
41. V. M. Srivastava, "Circuit perspective of terahertz double-gate MOSFET circuits for switch," *Proc. of 12th IEEE India Int. Conference (INDICON)*, Delhi, India, 17–20 Dec. 2015, pp. 1–4.

42. A. Lazaro and B. Iniguez, "RF and noise model of gate-all around MOSFETs," *Semiconductor Science and Technology*, vol. 23, no. 7, pp. 1–8, July 2008.
43. V. M. Srivastava, K. S. Yadav, and G. Singh, "Design and performance analysis of cylindrical surrounding double gate MOSFET for RF switch," *Microelectronics Journal*, vol. 42, no. 10, pp. 1124–1135, Oct. 2011.
44. V. M. Srivastava, K. S. Yadav, and G. Singh, "An approach for the design of cylindrical surrounding double-gate MOSFET," *Proc. of IEEE Int. Symposium on Microwave, Antenna, Propagation and EMC Technologies for Wireless Communications (MAPE)*, Beijing, China, 1–3 Nov. 2011, pp. 313–316.
45. J. H. K. Verma, S. Halder, R. S. Gupta, and M. Gupta, "Modeling and simulation of subthreshold behaviour of cylindrical surrounding double-gate MOSFET for enhanced electrostatic integrity," *Superlattices and Microstructures*, vol. 88, pp. 354–364, Dec. 2015.
46. J. H. K. Verma, S. Halder, R. S. Gupta, and M. Gupta, "Impact of inner charge control gate on cylindrical surrounding gate MOSFET for improved electrostatic integrity and RF performance," *Advanced Science, Engineering and Medicine*, vol. 8, no. 2, pp. 95–101, Feb. 2016.
47. "What is Moore's law?" Our world in data, 28 Mar. 2023, ourworldindata .org/moores-law.
48. J. B. Kuo, and S. C. Lin, *Low-Voltage SOI CMOS VLSI Devices and Circuits*, 1st edition, John Wiley & Sons, South America, 2001.
49. T. E. Rudenko, A. Nazarov, and V. S. Lysenko, "The advancement of silicon-on-insulator (SOI) devices and their basic properties," *Semiconductor Physics, Quantum Electronics and Optoelectronics*, vol. 23, pp. 227–252, 2020.
50. K. Bernstein and J. Norman, *SOI Circuit Design Concepts*, Kluwer Academic Publishers, United States, 2000.
51. J. Chen, R. Solomon, T. Y. Chan, P. K. Ko, and C. Hu, "Threshold voltage and C-V characteristics of SOI MOSFET's related to Si film thickness variation on SIMOX wafers," *IEEE Transactions on Electron Devices*, vol. 39, no. 10, pp. 2346–2353, 1992.
52. X. Sun, Q. Lu, V. Moroz, H. Takeuchi, G. Gebara, and J. Wetzel, "Tri-gate bulk MOSFET design for CMOS scaling to the end of the roadmap," *IEEE Electron Device Letters*, vol. 29, no. 5, pp. 491–493, May 2008.
53. F. Lime and B. Guillaumot, "Investigation of electron and hole mobilities in MOSFETs with TiN/HfO_2/SiO_2 gate stack," *Proc. of 33rd Int. Conf. on European Solid State Device Research*, 16–18 Sep. 2003, pp. 247–250.
54. S. Kubicek, J. Chen, A. Ragnarsson, R. J. Carter, V. Kaushik, and K. De Meyer, "Investigation of poly Si/HfO/sub 2/gate stacks in a self-aligned 70 nm MOS process flow," *Proc. of 33rd Int. Conf. on European Solid-State Device Research*, 16–18 Sep. 2003, pp. 251–254.
55. B. Doyle, B. Boyanov, S. Datta, M. Doczy, S. Hareland, B. Jin, J. Kavalieros, T. Linton, R. Rios, and R. Chau, "Tri-gate fully-depleted CMOS transistors: fabrication, design and layout," *Proc. of Symp. on VLSI Technology, Digest of Technical Papers*, 10–12 June 2003, pp. 133–134.

56. M. Saitoh, Y. Nakabayashi, K. Ota, K. Uchida, and T. Numata, "Performance improvement by stress memorization technique in trigate silicon nanowire MOSFETs," *IEEE Electron Device Letters*, vol. 33, no. 3, pp. 8–10, Jan. 2012.
57. V. Srivastava, and S. Singh, "Analysis and design of tri-gate MOSFET with high dielectrics gate," *International Journal of Intelligent Systems and Applications*, vol. 4, pp. 16–22, 2012.
58. P. McLellan, "Intel's 22nm Process. Atom, ARM, Apple," Semiwiki, 15 June 2019, semiwiki.com/ip/arm/527-intels-22nm-process-atom-arm-apple.
59. M. S. Narula and A. Pandey, "A comprehensive review on FinFET, gate all around, tunnel FET: concept, performance and challenges," *2022 8th International Conference on Signal Processing and Communication (ICSC)*, Noida, India, 2022, pp. 554–559.
60. A. Bhardwaj, P. Kumar, B. Raj, and S. Anand, "Design and performance optimization of doping-less vertical nanowire TFET using gate stack technique," *Journal of Electronic Materials (JEMS)*, vol. 41, no. 7, pp. 4005–4013, 2022.
61. J. Singh and B. Raj, "Tunnel current model of asymmetric MIM structure levying various image forces to analyze the characteristics of filamentary memristor," *Applied Physics A*, vol. 125, no.3, pp. 203.1–203, 11 Feb. 2019.
62. C. Goyal, J. S. Ubhi, and B. Raj, "Low leakage zero ground noise nanoscale full adder using source biasing technique," *Journal of Nanoelectronics and Optoelectronics*, vol. 14, pp. 360–370, Mar. 2019.
63. G. Singh, R. K. Sarin, and B Raj, "A novel robust exclusive-OR function implementation in QCA nanotechnology with energy dissipation analysis," *Journal of Computational Electronics*, vol. 15, no. 2, pp. 455–465, June 2016.
64. G. Wadhwa, P. Kamboj, and B. Raj, "Design optimisation of junctionless TFET biosensor for high sensitivity," *Advances in Natural Sciences: Nanoscience and Nanotechnology*, vol. 10, pp. 045001, 2019.
65. P. Bansal and B. Raj, "Memristor modeling and analysis for linear dopant drift kinetics," *Journal of Nanoengineering and Nanomanufacturing*, vol. 6, pp. 1–7, 2016.
66. D. V. Ashima and B. Raj," Performance analysis of charge plasma induced graded channel Si nanotube," *Journal of Engineering Research (JER)*, vol. 9, pp. 146–154, Aug 2021.
67. A. S. Tomar, V. K. Magraiya, and B. Raj, "Scaling of access and data transistor for high performance DRAM cell design," *Quantum Matter*, vol. 2, pp. 412–416, Oct. 2013.
68. N. Jain and B. Raj, "Parasitic capacitance and resistance model development and optimization of raised source/drain SOI FinFET structure for analog circuit applications," *Journal of Nanoelectronics and Optoelectronics*, vol. 13, pp. 531–539, Apr. 2018.
69. S. Singh, S. K. Vishvakarma, and B Raj, "Analytical modeling of split-gate junction-less transistor for a biosensor application," *Sensing and Bio-Sensing*, vol. 18, pp. 31–36, Apr. 2018.
70. M. Gopal and B. Raj, "Low power 8T SRAM cell design for high stability video applications," *ITSI Transaction on Electrical and Electronics Engineering*, vol. 1, no. 5, pp. 91–97, 2013.

71. B. Raj, J. Mitra, Deepak K. Bihani, V. Rangharajan, A. K. Saxena, and S. Dasgupta, "Analysis of noise margin, power and process variation for 32 nm FinFET based 6T SRAM cell," *Journal of Computer (JCP)*, vol. 5, no. 6, pp. 1–8, 2010.
72. D. Sharma, R. Mehra, and B. Raj, "Comparative analysis of photovoltaic technologies for high efficiency solar cell design," *Superlattices and Microstructures*, vol. 153, pp. 106861, May 2021.
73. P. Kaur, A. S. Buttar, and B. Raj, "A comprehensive analysis of nanoscale transistor based Biosensor: a review," *Indian Journal of Pure and Applied Physics*, vol. 59, pp. 304–318, Apr. 2021.
74. D. Yadav, B. Raj, and B. Raj, "Design and simulation of low power microcontroller for IoT applications," *Journal of Sensor Letters*, vol. 18, pp. 401–409, May 2020.
75. S. Singh and B. Raj, "A 2-D analytical surface potential and drain current modeling of double-gate vertical t-shaped tunnel FET," *Journal of Computational Electronics*, vol. 19, pp. 1154–1163, Apr. 2020.
76. J. Singh and B. Raj, "An accurate and generic window function for non-linear memristor model," *Journal of Computational Electronics*, vol. 18, no. 2, pp. 640–647, June 2019.
77. M. Kaur, N. Gupta, S. Kumar, B. Raj, and A. K. Singh, "Comparative RF and crosstalk analysis of carbon based nano interconnects," *IET Circuits, Devices & Systems*, vol.15, no. 6, pp. 493–503, Feb. 2021.
78. N. Kandasamy, F. Ahmad, D. Ajitha, B. Raj, and N. Telagam, "Quantum dot cellular automata based scan flip flop and boundary scan register," *IETE Journal of Research*, vol. 66, pp. 535–548, 2020.
79. J. Singh and B. Raj, "Design and investigation of 7T2M NVSARM with enhanced stability and temperature impact on store/restore energy," *IEEE Transactions on Very Large Scale Integration Systems*, vol. 27, no. 6, pp. 1322–1328 June 2019.
80. A. K. Bhardwaj, S. Gupta, B. Raj, and A. Singh, "Impact of double gate geometry on the performance of carbon nanotube field effect transistor structures for low power digital design," *Computational and Theoretical Nanoscience*, vol. 16, pp. 1813–1820, 2019.
81. N. Jain, and B. Raj, "Thermal stability analysis and performance exploration of asymmetrical dual-k underlap spacer (ADKUS) SOI FinFET for security and privacy applications," *Indian Journal of Pure & Applied Physics (IJPAP)*, vol. 57, pp. 352–360, May 2019.
82. N. Jain and B. Raj, "Dual-k spacer region variation at the drain side of asymmetric SOI FinFET structure: performance analysis towards the analog/RF design applications," *Journal of Nanoelectronics and Optoelectronics*, vol. 14, pp. 349–359, Mar. 2019.
83. J. Singh, S. Sharma, and B. Raj, "Mamta Khosla analysis of barrier layer thickness on performance of In1-xGaxAs based gate stack cylindrical gate nanowire MOSFET," *JNO, ASP*, vol. 13, pp 1473–1477, Oct. 2018.
84. N. Jain, and B. Raj, "Analysis and performance exploration of High-k SOI FinFETs over the conventional low-k SOI FinFET toward analog/RF design," *Journal of Semiconductors (JoS)*, vol.39, no. 12, pp. 124002-1–7, Dec. 2018.

85. C. Goyal, J. S. Ubhi, and B. Raj, "A reliable leakage reduction technique for approximate full adder with reduced ground bounce noise," *Journal of Mathematical Problems in Engineering*, vol. 2018, Article ID 3501041, p. 16, 15 Oct. 2018.
86. J. S. Anuradha, B. Raj, and Mamta Khosla, "Design and performance analysis of nano-scale memristor-based nonvolatile SRAM," *Journal of Sensor Letter*, vol. 16, pp. 798–805, Oct. 2018.
87. G. Wadhwa, and B. Raj, "Parametric variation analysis of charge-plasma-based dielectric modulated JLTFET for biosensor application" *IEEE Sensor Journal*, vol. 18, no. 15, pp. 6070–6077, 1 Aug. 2018.
88. J. Singh, and B. Raj, "Comparative analysis of memristor models for memories design", *JoS, IoP*, vol. 39, no. 7, pp. 074006, 1–12 July 2018.
89. D. Yadav, S. S. Chouhan, S. K. Vishvakarma, and B. Raj, "Application specific microcontroller design for IoT based WSN," *Sensor Letter*, vol. 16, pp. 374–385, May 2018..
90. G. Singh, R. K. Sarin, and B. Raj, "Fault-tolerant design and analysis of quantum-dot cellular automata based circuits," *IEEE/IET Circuits, Devices & Systems*, vol. 12, pp. 638–64, 2018.
91. J. Singh and B. Raj, "Modeling of mean barrier height levying various image forces of metal insulator metal structure to enhance the performance of conductive filament based memristor model," *IEEE Nanotechnology*, vol. 17, no. 2, pp. 268–267, Mar. 2018.
92. A. Jain, S. Sharma, and B. Raj, "Analysis of triple metal surrounding gate (TM-SG) III–V nanowire MOSFET for photosensing application," *Opto-Electronics Journal*, vol. 26, no. 2, pp. 141–148, May 2018.
93. A. Jain, S. Sharma, and B. Raj, "Design and analysis of high sensitivity photosensor using cylindrical surrounding gate MOSFET for low power sensor applications," *Engineering Science and Technology, an International Journal*, vol. 19, no. 4, pp. 1864–1870, Dec. 2016.
94. G. Saiphani Kumar, A. Singh, and B. Raj, "Design and analysis of gate all around CNTFET based SRAM cell design," *Journal of Computational Electronics*, vol. 17, no. 1, pp. 138–145, Mar. 2018.
95. G. P. Singh, B. S. Sohi, and B. Raj, "Material properties analysis of graphene base transistor (GBT) for VLSI analog circuits," *Indian Journal of Pure & Applied Physics (IJPAP)*, vol. 55, pp. 896–902, Dec. 2017.
96. S. Kumar, and B. Raj, "Estimation of stability and performance metric for inward access transistor based 6T SRAM cell design using n-type/p-type DMDG-GDOV TFET," *IEEE VLSI Circuits and Systems Letter*, vol. 3, no. 2, pp. 25–39, June 2017.
97. S. Sharma, A. Kumar, M. Pattanaik, and B Raj, "Forward body biased multimode multi-threshold CMOS technique for ground bounce noise reduction in static CMOS adders," *International Journal of Information and Electronics Engineering*, vol. 3, no. 3, pp. 567–572, 2013.
98. H. Singh, P. Kumar, and B. W. Raj, "Performance analysis of majority gate SET based 1-bit full adder," *International Journal of Computer and Communication Engineering (IJCCE)*, vol. 2, no. 4, pp. 2010–3743, 2013.

99. A. K. Bhardwaj, S. Gupta, and B. Raj, "Investigation of parameters for Schottky Barrier (SB) height for Schottky barrier based carbon nanotube field effect transistor device," *Journal of Nanoelectronics and Optoelectronics*, vol. 15, pp. 783–791, July 2020.
100. P. Bansal, and B. Raj, "Memristor: a versatile nonlinear model for dopant drift and boundary issues," *JCTN*, vol. 14, no. 5, pp. 2319–2325, May 2017.
101. N. Jain and B. Raj, "An analog and digital design perspective comprehensive approach on fin-FET (Fin-Field Effect transistor) technology: a review," *Reviews in Advanced Sciences and Engineering (RASE)*, vol. 5, pp. 1–14, 2016.
102. B. Raj, A. K. Saxena, and S. Dasgupta, "Analytical modeling for the estimation of leakage current and subthreshold swing factor of nanoscale double gate FinFET device," *Microelectronics International*, vol. 26, pp. 53–63, 2009.
103. Soniya, S. Singh, G. Wadhwa, and B. Raj, "An analytical modeling for dual source vertical tunnel field effect transistor", *International Journal of Recent Technology and Engineering (IJRTE)* vol. 8, no. 2, pp. 603–608, July 2019.
104. S. Singh and B. Raj, "Design and analysis of hetrojunction vertical T-shaped tunnel field effect transistor," *Journal of Electronics Material*, vol. 48, no. 10, pp. 6253–6260, Oct. 2019.
105. C. Goyal, J. S. Ubhi, and B. Raj, "A low leakage CNTFET based inexact full adder for low power image processing applications," *International Journal of Circuit Theory and Applications*, vol. 47, no. 9, pp. 1446–1458, Sep. 2019.
106. B. Raj, A. K. Saxena, and S. Dasgupta, "A compact drain current and threshold voltage quantum mechanical analytical modeling for FinFETs," *Journal of Nanoelectronics and Optoelectronics (JNO)*, vol. 3, no. 2, pp. 163–170, 2008.
107. G. Wadhwa and B. Raj, "An analytical modeling of charge plasma based tunnel field effect transistor with impacts of gate underlap region," *Superlattices and Microstructures*, vol. 142, pp. 106512, June 2020.
108. S. Singh and B. Raj,"Modeling and simulation analysis of SiGe hetrojunction double GateVertical t-shaped tunnel FET," *Superlattices and Microstructures*, vol. 142, p. 106496, June 2020.
109. N. Jain, and B. Raj, "Device and circuit co-design perspective comprehensive approach on FinFET technology: a review," *Journal of Electron Devices*, vol. 23, no. 1, pp. 1890–1901, 2016.
110. S. Kumar and B. Raj, "Analysis of I_{ON} and Ambipolar current for dual-material gate-drain overlapped DG-TFET," *Journal of Nanoelectronics and Optoelectronics*, vol. 11, pp. 323–333, June 2016.
111. N. Anjum, T. Bali, and B. Raj, "Design and simulation of handwritten multiscript character recognition," *International Journal of Advanced Research in Computer and Communication Engineering*, vol. 2, no. 7, pp. 2544–2549, July 2013.
112. K. Singh and B. Raj, "Performance and analysis of temperature dependent multi-walled carbon nanotubes as global interconnects at different technology nodes," *Journal of Computational Electronics*, vol. 14, no. 2, pp. 469–476, June 2015.

113. S. Kumar and B. Raj, "Compact channel potential analytical modeling of DG-TFET based on evanescent–mode approach," *Journal of Computational Electronics*, vol. 14, no. 2, pp. 820–827, July 2015.
114. K. Singh and B. Raj, "Temperature dependent modeling and performance evaluation of multi-walled CNT and single-walled CNT as global interconnects," *Journal of Electronic Materials*, vol. 44, no. 12, pp 4825–4835, Dec. 2015.
115. V. K. Sharma, M. Pattanaik, and B. Raj, "INDEP approach for leakage reduction in nanoscale CMOS circuits," *International Journal of Electronics*, vol. 102, no.2, pp. 200–215, 2014.
116. K. Singh and B. Raj, "Influence of temperature on MWCNT bundle, SWCNT bundle and Copper interconnects for nanoscaled technology nodes," *Journal of Materials Science: Materials in Electronics.*, vol. 26, no. 8, pp. 6134–6142, 2015.
117. N. Anjum, T. Bali, and B. Raj, "Design and simulation of handwritten Gurumukhi and Devanagri numerical recognition," *International Journal of Computer Applications*, vol. 73, no. 12, pp. 16–21, 2013.
118. S. Khandelwal, V. Gupta, B. Raj, and R. D. Gupta, "Process variability aware low leakage reliable nano scale DG-FinFET SRAM cell design technique," *Journal of Nanoelectronics and Optoelectronics*, vol. 10, no. 6, pp. 810–817, Dec. 2015.
119. V. K. Sharma, M. Pattanaik, and B. Raj, "ONOFIC approach: low power high speed nanoscale VLSI circuits design," *International Journal of Electronics*, vol. 101, no. 1, pp. 61–73, 2014.
120. S. Khandelwal, Balwinder Raj, and R. D. Gupta, "FinFET based 6T SRAM cell design: analysis of performance metric, process variation and temperature effect," *Journal of Computational and Theoretical Nanoscience*, vol. 12, pp. 2500–2506, 2015.
121. S. Singh, S. Yadav, J. Rahul, A. Srivastava, and B. Raj, "Impact of HfO_2 in graded channel dual insulator double gate MOSFET," *Journal of Computational and Theoretical Nanoscience*, vol. 12, no. 6, pp. 950–953, Apr. 2015.
122. V. K. Sharma, M. Pattanaik, and B. Raj, "PVT variations aware low leakage INDEP approach for nanoscale CMOS circuits," *Microelectronics Reliability*, vol. 54, pp. 90–99, 2014.
123. B. Raj, A. K. Saxena, and S. Dasgupta, "Quantum mechanical analytical modeling of nanoscale DG FinFET: evaluation of potential, threshold voltage and source/drain resistance," *Journal of Material Science in Semiconductor Processing*, vol. 16, no. 4, pp. 1131–1137, 2013.
124. M. Gopal, S. S. D. Prasad, and B. Raj, "8T SRAM cell design for dynamic and leakage power reduction," *International Journal of Computer Applications*, vol. 71, no. 9, pp. 43–48, June 2013.
125. M. Pattanaik, B. Raj, S. Sharma, and A. Kumar, "Diode based trimode multi-threshold CMOS technique for ground bounce noise reduction in static CMOS adders," *Advanced Materials Research*, vol. 548, pp 885–889, 2012.

126. B. Raj, A. K. Saxena, and S. Dasgupta, "Nanoscale FinFET based SRAM cell design: analysis of performance metric, process variation, underlapped FinFET and temperature effect," *IEEE Circuits and System Magazine*, vol. 11, no. 2, pp. 38–50, 2011.
127. V. K. Sharma, M. Pattanaik, and B. Raj, "Leakage current ONOFIC approach for deep submicron VLSI circuit design," *International Journal of Electrical, Computer, Electronics and Communication Engineering*, vol. 7, no. 4, pp. 239–244, 2013.
128. P. Kaur, S. S. Gill, and B. Raj, "Comparative analysis of OFETs materials and devices for sensor applications," *Journal of Silicon*, vol 14, pp. 4463–4471, 2022.
129. S. K. Sharma, P. Kumar, B. Raj, and B. Raj, "$In_{1-x}Ga_xAs$ double metal gate-stacking cylindrical nanowire MOSFET for highly sensitive photo detector," *Journal of Silicon*, vol. 14, pp. 3535–3541, 2022.
130. B. Raj, A. K. Saxena, and S. Dasgupta, "Analytical modeling of quasi planar nanoscale double gate FinFET with source/drain resistance and field dependent carrier mobility: a quantum mechanical study," *Journal of Computer (JCP)*, vol. 4, no. 9, pp. 1–8, 2009.
131. S. Bhushan, S. Khandelwal, and B. Raj, "Analyzing different mode FinFET based memory cell at different power supply for leakage reduction," *Seventh International Conference on Bio-Inspired Computing: Theories and Application, (BIC-TA 2012) Advances in Intelligent Systems and Computing*, vol. 202, pp. 89–100, 2013.
132. J. Singh, and B. Raj, "Temperature dependent analytical modeling and simulations of nanoscale memristor," *Engineering Science and Technology, an International Journal*, vol. 21, pp. 862–868, Oct. 2018.
133. S. Singh, S. Bala, B. Raj, and B. Raj, "Improved sensitivity of dielectric modulated junctionless transistor for nanoscale biosensor design," *Sensor Letter*, vol. 18, pp. 328–333, Apr. 2020.
134. V. Kumar, S. K. Vishvakarma, and B. Raj, "Design and performance analysis of ASIC for IoT applications," *Sensor Letter ASP*, vol. 18, pp. 31–38, Jan. 2020.
135. A. Jaiswal, R. K. Sarin, B. Raj, and S. Sukhija, "A novel circular slotted microstrip-fed patch antenna with three triangle shape defected ground structure for multiband applications," *Advanced Electromagnetic (AEM)*, vol. 7, no. 3, pp. 56–63, Aug. 2018.
136. G. Wadhwa, and B. Raj, "Label free detection of biomolecules using charge-plasma-based gate underlap dielectric modulated junctionless TFET," *Journal of Electronic Materials (JEMS)*, vol. 47, no. 8, pp 4683–4693, Aug. 2018.
137. G. Singh, R. K. Sarin, and B. Raj, "Design and performance analysis of a new efficient coplanar quantum-dot cellular automata adder," *Indian Journal of Pure & Applied Physics (IJPAP)*, vol. 55, pp. 97–103, Feb. 2017.
138. P. Kaur, V. Pandey, and B Raj, "Comparative study of efficient design, control and monitoring of solar power using IoT," *Sensor Letter*, vol. 18, pp. 419–426, May 2020.
139. A. K. Bhardwaj, S. Gupta, and B. Raj, "Development & analysis of compact model for double gate Schottky barrier CNTFET," *Journal of Nanoelectronics and Optoelectronics*, vol. 15, pp. 1199–1208, Aug. 2020.

140. G. Wadhwa, P. Kamboj, J. Singh, and B. Raj, "Design and investigation of junctionless DGTFET for biological molecule recognition," *Transactions on Electrical and Electronic Materials*, vol. 22, pp. 282–289, 2021.
141. S. K. Verma, S. Singh, G. Wadhwa, and B. Raj, "Detection of biomolecules using charge-plasma based gate underlap dielectric modulated dopingless TFET," *Transactions on Electrical and Electronic Materials (TEEM)*, vol. 21, pp. 528–535, June 2020.
142. N. Jain and B. Raj, "Impact of underlap spacer region variation on electrostatic and analog/RF performance of symmetrical high-k SOI FinFET at 20 nm channel length," *Journal of Semiconductors (JoS)*, vol. 38, no. 12, pp. 122002, Dec. 2017.
143. S. Singh and B. Raj, "Analytical modeling and simulation analysis of T-shaped III–V heterojunction vertical T-FET," *Superlattices and Microstructures*, vol. 147, p. 106717, Nov. 2020.
144. G. Singh, R. K. Sarin, and B. Raj, "Design and analysis of area efficient QCA based reversible logic gates," *Journal of Microprocessors and Microsystems*, vol. 52, pp. 59–68, May 2017.
145. S. Singh and B. Raj, "Parametric variation analysis on hetero-junction vertical t-shape TFET for supressing ambipolar conduction," *Indian Journal of Pure and Applied Physics*, vol. 58, pp. 478–485, June 2020.
146. Soniya, S. Singh, G. Wadhwa, and B. Raj, "Design and analysis of dual source vertical tunnel field effect transistor for high performance," *Transactions on Electrical and Electronics Materials*, vol. 21, pp. 74–82, Oct. 2019.
147. M. Kaur, N. Gupta, S. Kumar, B. Raj, and A. K. Singh, "RF performance analysis of intercalated graphene nanoribbon based global level interconnects," *Journal of Computational Electronics*, vol. 19, pp. 1002–1013, June 2020.
148. G. Wadhwa and B. Raj, "Design and performance analysis of junctionless TFET biosensor for high se*nsitivity," *IEEE Nanotechnology*, vol. 18, pp. 567–574, 2019.
149. J. Singh and B. Raj, "Enhanced nonlinear memristor model encapsulating stochastic dopant drift," *JNO*, vol. 14, pp. 958–963, 2019.

Chapter 2

FinFET and its application

Debanjan Das, Sagar Gurung, Mandeep Singh, Tarun Chaudhary, and Balwinder Raj

2.1 INTRODUCTION

The evolution of standard metal–oxide–semiconductor field-effect transistors (MOSFETs) encounters heightened complexities with the progression of technology into smaller nanoscale dimensions, especially under the 32 nm technology threshold. Managing the amplified challenges of handling large channel outputs becomes increasingly intricate, and efforts to minimize the length as well as the size of MOS devices often result in compromised performance and escalated consumption of power. A potential remedy presents itself in the form of Fin-On MOSFET devices, which offer improvements in device speed, packing density, and reductions in power demands, noise, and the requisite transistors [1]. However, notable concerns arise, including issues such as oxide breakdown, particularly as the insulator size diminishes. This necessitates the deployment of costly heat sinks and cooling devices due to the heightened heat production in emerging integrated circuits (ICs) [1, 2]. The intricacies of modeling challenges add an additional layer of complexity to the analysis and prediction of the final stages of devices undergoing size reduction [3].

The discourse delves into alternative transistor structures such as fin field-effect transistors (FinFETs) and gate-all-around (GAA) transistors, showcasing their potential to overcome the limitations of traditional MOSFETs. Additionally, the discussion underscores the significance of nanowires, with particular focus on silicon nanowire FinFETs (SNWFT). These structures are posited as potential solutions due to their promise of high performance, effective gate control, and enhanced travel characteristics [4]. In the comparative analysis of various transistor types, including MOSFETs, FinFETs, and nanowires, the review emphasizes the pivotal role of material-based structures in achieving the desired performance benchmarks [4, 5]. This chapter is meticulously organized into sections covering the historical context of FinFETs, a comprehensive literature review, an in-depth comparison of different transistor types, and a concluding section [6]. This structured approach provides a thorough exploration of the challenges and

DOI: 10.1201/9781003612797-2

advancements in transistor technology, offering valuable insights into the evolving landscape of semiconductor devices [7].

2.2 HISTORY

The word "FinFET" was first used to describe a dual-gated non-planar transistor built on a silicon-on-insulator (SOI) substrate. It came from the University of California. The FinFET was dubbed "FinFET" because, unlike the single-gate planar MOSFET as shown in Figure 2.1a, it had two gates controlling the channel and wings created on the silicon surface [8]. 1989 saw the introduction of the first iteration in Japan, dubbed the "Depleted Lean Channel Transistor" or "DELTA" transistor [8, 9].

FinFET structure shown in Figure 2.1b integrates a vertical channel called "Fin", whereas planar MOSFET structure has a horizontal channel. In the triple-gate multigate MOSFET, the fin length multiplied by the fin's end height and width determines the width of the channel. The main differences are seen in the number of gates and the channel geometry. A FinFET's vertical channel offers advantages over planar CMOS, including faster switching times and a higher current size [10].

Advanced architectures like the independent-gate FinFET and short-gate FinFET were designed to improve electrostatic control over the channel. These architectures provide better electrical features, protection against transient channel effects, and better charging control [10]. There are two types of FinFETs: bulk and SOI. Each has advantages and disadvantages with regard to heating, junction capacitance power, and integration procedures. The advantages of FinFET technology over conventional CMOS include high driving speed, reduced leakage, lower energy consumption, the absence of random dopant fluctuations, better mobility, and suitability for transistors measuring below 28 nm [9–11]. However, as technology advances, there is a paradigm shift toward the gate-all-around FET with nanowires or nanosheets, signifying a transition from

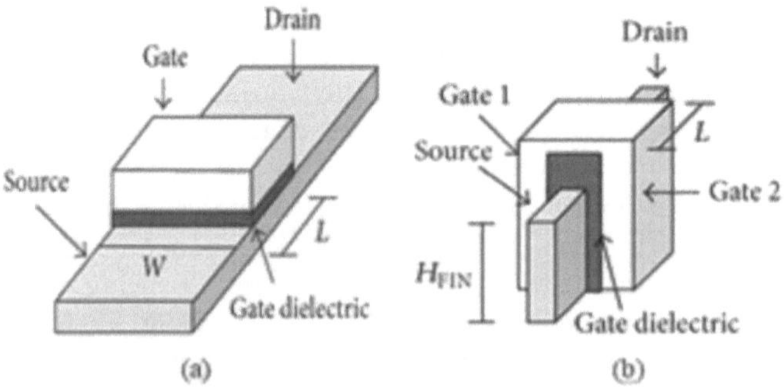

Figure 2.1 (a) Planar MOSFET and (b) FinFET [35].

three-dimensional gates to fully closed gates [12]. This evolutionary progression is seen as a response to the limitations of FinFETs and planar transistors [13].

In comparing FinFETs and SOI-FinFETs, both exhibit advantages in mitigating short-channel effects (SCEs) faced by MOSFETs. Strengths of SOI-FinFETs include a streamlined manufacturing process and compatibility with high-cost layout libraries. Conversely, FinFETs offer higher drive current and the availability of strain engineering. Nonetheless, they also entail manufacturing complexities and expenses. This continuous evolution marks the transition from FinFETs to gate-all-around FETs with nanowires or nanosheets [14].

2.3 BASICS OF FINFET

FinFET, an acronym for fin field-effect transistor, epitomizes a revolutionary leap in transistor architecture, boasting a multi-gate design with a distinctive three-dimensional arrangement. This groundbreaking structure involves the encapsulation of the device's body in a thin silicon film, sculpting a channel reminiscent of a "fin". The consequential conducting channel not only showcases the unique aesthetics but, more importantly, serves to curtail leakage current, thereby elevating the overall operational efficiency of FinFETs [15–17].

FinFETs, akin to MOSFETs, encompass both n-channel and p-channel configurations [18–24]. The symbol for the n-channel and p-channel FinFET is illustrated in Figure 2.2.

2.4 CLASSIFICATION OF FINFET

2.4.1 Bulk and SOI-FinFET

Distinguishing itself into two primary variants, SOI within the realm of FinFET technology, two primary variants stand out: SOI-FinFET and bulk FinFET, each shaping the device's core composition differently [24–32]. Bulk FinFETs, housed on wafers with shared substrates among all fins,

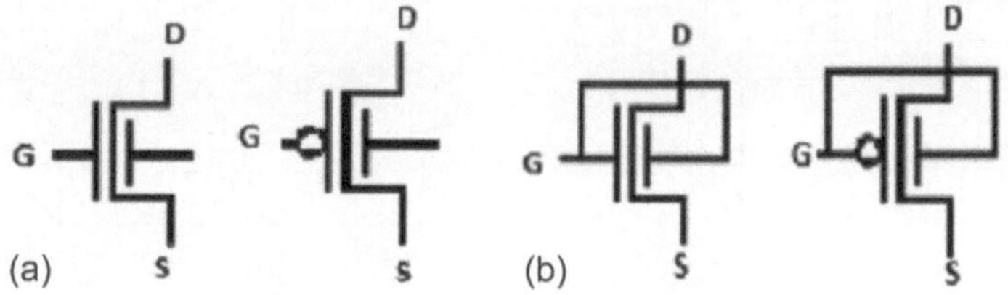

Figure 2.2 (a) Independent-gate FinFET and (b) short-gated FinFET [36].

offer a seamless transition from bulk MOSFETs. Conversely, SOI FinFETs feature isolated fins on a planar substrate, distinguishing them structurally. Figure 2.3 elegantly illustrates the comparison between these two, showcasing their distinct design approaches.

In the industry, bulk FinFET is being positioned as a less expensive option to SOI-FinFET; however, current debates center on comparing the two technologies' costs and levels of complexity. Although using bulk silicon wafers initially saves money, manufacturing bulk fins involves six to eight extra processes, which begs the question of whether it is feasible to retain exact control over fin variations and dimensions in a production environment [33–41]. The difficulties encountered during the production of junctions in bulk FinFET are similar to those in its SOI counterpart, with the additional complexity of junctions being deeper than the fin height since a buried oxide is not present. This deeper location raises the possibility of bulk punch-through and makes it more difficult to regulate the short-channel effect [18]. Halo implantation and channel-doping are required to overcome these problems, but their 3D optimization introduces additional complexity. Moreover, the advantages of the FinFET architecture are diminished by the higher doping concentration in the channel, which increases random dopant fluctuation (RDF) [42–48]. But the real distinguishing feature of the FinFET is its dual-gate architecture, which offers separate control over each gate. This crucial component results in a significant decrease in leakage current and an improvement in power efficiency (Table 2.1).

2.4.2 Terminal-based FinFETs

FinFET is divided into two primary groups:

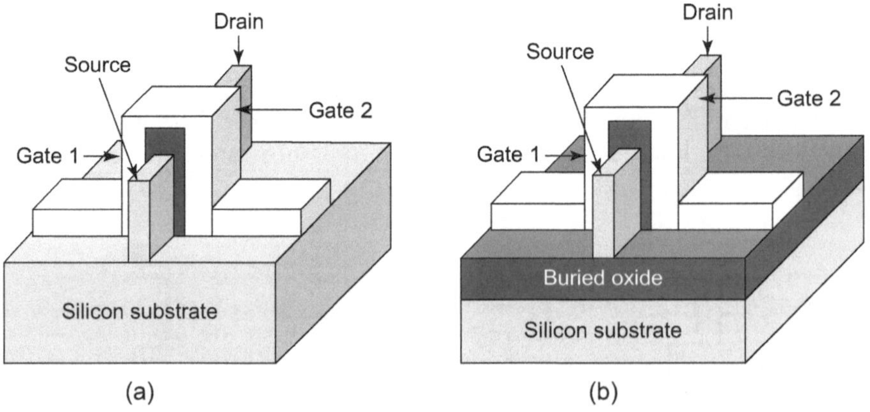

Figure 2.3 (a) Bulk and (b) SOI-FinFET [35].

Table 2.1 Comparisons between different FinFET and SOI-FinFET MOSFET performance matrices [24]

	FinFET	*MOSFET*	*SOI-FinFET*
DIBL (mV)	00.0000530	47.00	0.000050
Sub-threshold swing (V/dec)	0.982	0.70	0.062
Leakage current	8.62×10^{-8} nA (low)	High	7.77×10^{-8} nA (very low)
SCEs	Low	High	Very less
Output current	0.150 mA	10.0 uA	0.10 mA

2.4.2.1 Short-gated (SG) or three terminal FinFETs

FinFET functions in a single-gate mode when the gate is shortened, showing less sensitivity to short-channel phenomena than the independent-gate mode, such as threshold voltage roll-off, drain-induced barrier lowering, and sub-threshold swing deterioration [49–56] given in Figure 2.4. The ability to control each gate independently or combine them via short-circuiting adds even more design flexibility to FinFET gates. This arrangement offers better control and successfully reduces the influence of narrow channels. It is distinguished by two electrically connected gates and wrapped thin silicon bodies. For overall high system performance, dynamic metal–oxide–semiconductor (MOS) logic circuits are frequently used in the field of Very Large-Scale Integration (VLSI) for high-quality applications. But noise is a problem for dynamic logic circuits, especially in deep sub-jurisdictions, which makes them more prone to logical errors [19]. These circuits exhibit various noise sources, including exchange, exhaust, cross-sectional, power, ground, and substrate noise. FinFET, with its diverse design options and heightened channel control, not only ensures lower power consumption but also enhances noise immunity in circuit configurations [57–65]. These circuits show a variety of noise sources, such as substrate, ground, exchange, exhaust, and cross-sectional noise. FinFET guarantees lower power consumption and improves noise immunity in circuit topologies with its wide range of design possibilities and enhanced channel management.

2.4.2.2 Independent-gate FinFET (IG FinFET)

When two gates function independently of one another in the independent-gate (IG) configuration of FinFET technology, the structure is referred to as an independent-gate (IG) mode FinFET shown in Figure 2.5. Novel transistor topologies are possible because each gate can be independently controlled. Compared to current devices, this unique structure offers alternatives in the VLSI industry [66–74]. Comparing it to planar transistors or traditional FinFETs, however, reveals an interesting trade-off between power consumption and delay [19].

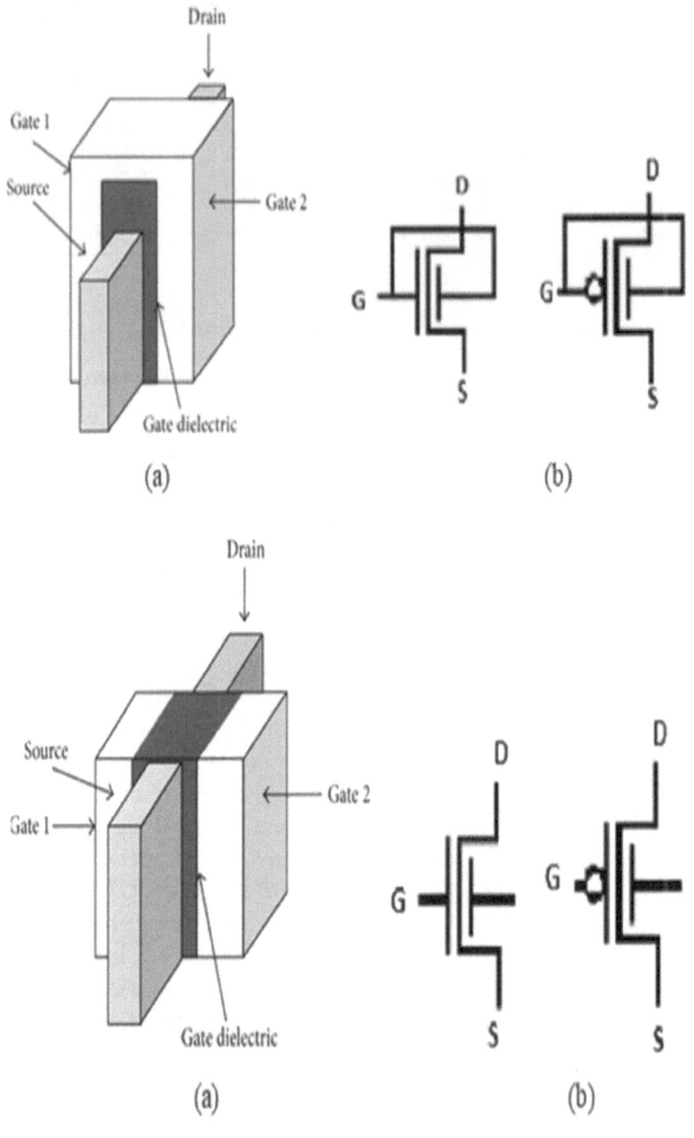

Figure 2.4 Short-gated FinFET with symbol [36].

The vertical-gate configuration roughly correlates with the FinFET's width. This design's main goals are to reduce power consumption, improve on-state current amplification, minimize leakage currents both on and off, and handle short-channel effects [75–82]. Together, these qualities solidify FinFETs' status as pivotal components in the advancement of semiconductor technology, offering vital efficiency gains for contemporary integrated circuits.

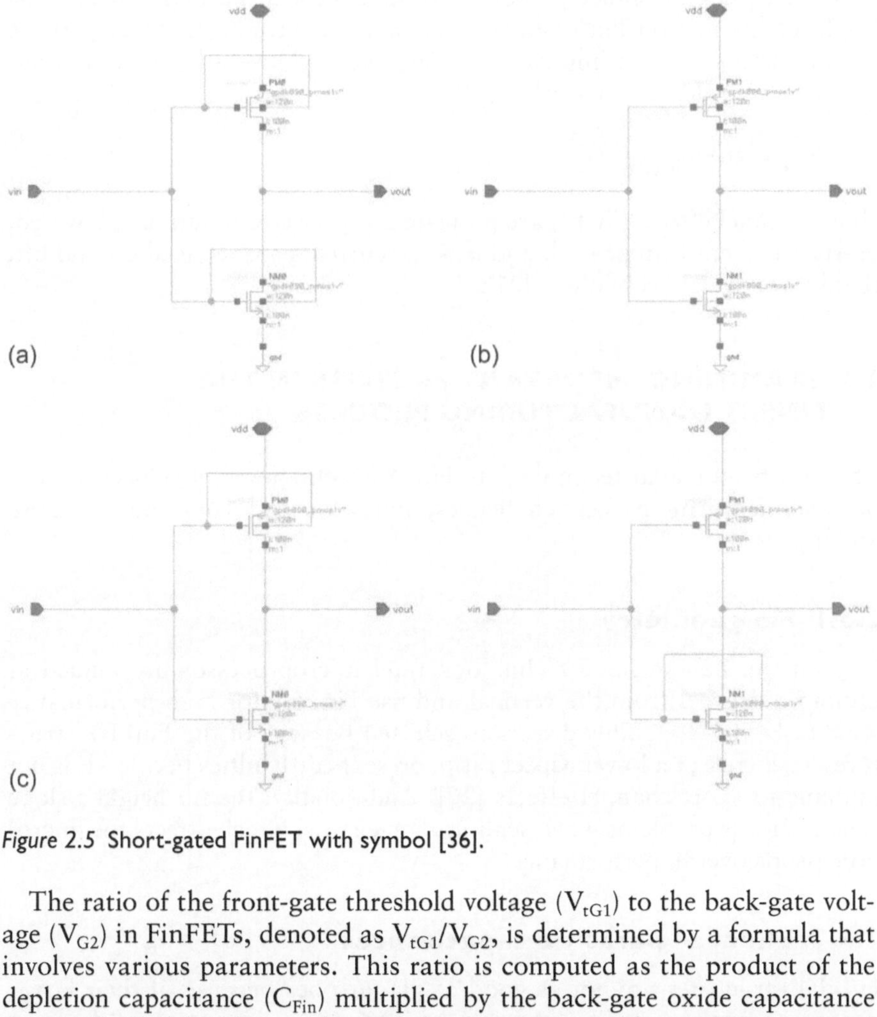

Figure 2.5 **Short-gated FinFET with symbol [36].**

The ratio of the front-gate threshold voltage (V_{tG1}) to the back-gate voltage (V_{G2}) in FinFETs, denoted as V_{tG1}/V_{G2}, is determined by a formula that involves various parameters. This ratio is computed as the product of the depletion capacitance (C_{Fin}) multiplied by the back-gate oxide capacitance (C_{ox2}), divided by the product of the front-gate oxide capacitance (C_{ox1}) and the sum of the depletion capacitance and back-gate oxide capacitance [83–88]. Further, this value is multiplied by the ratio of the front gate-oxide thickness (t_{ox1}) to the sum of the back-gate oxide thickness (t_{ox2}) and the width of the fin structure (Fin_{WIDTH}). Essentially, it describes the nuanced interrelation between gate voltages, different capacitances (depletion and oxide), and oxide layer characteristics, integral in understanding the behavior and properties of FinFET devices in semiconductor technology [20].

$$\frac{V_{tG1}}{V_{G2}} = -\frac{C_{Fin}\,C_{ox2}}{C_{ox1}\left(C_{Fin} + C_{ox2}\right)} \approx -\frac{3t_{ox1}}{3t_{ox2} + Fin_{WIDTH}}$$

When it comes to reducing short-channel effects in FinFET devices, the height of the fins (H(Fin)) and the thickness of the silicone body (tsi) are critical factors [7]. It is imperative to keep the fins' height higher than their thickness in order to reduce these effects.

$$W_{eff} = 2nFin_{height}$$

Short-channel effects (SCEs) are mitigated, power consumption is lowered, on-state current is improved, and leakage currents are reduced on and off, thanks to FinFET technology [21].

2.5 EXAMINING IMPORTANT FACTORS IN THE FINFET MANUFACTURING PROCESS

The shift from planar technology to FinFET technology introduces numerous manufacturing process challenges, and in the following some of these are discussed.

2.5.1 Fin geometry

Fins on the 22 nm node technology Intel microprocessor are angled at around 8 degrees from the vertical and use FinFET for high-performance logic tasks [89–96]. This direction, selected because of the FinFET structure's resilience at a lower aspect ratio, presents difficulties because it is not immune to short-channel effects [22]. Additionally, the fin height's slope causes serious problems when scaling gate lengths, which affects the microprocessor's overall performance.

2.5.2 Effect of parasitic capacitances

FinFET structures are not as good as planar ones because of their larger parasitic capacitances. Elevated device capacitances are a result of the front and back gates' greater overlap area [97–103]. Variations in fin height are used to lessen this; a drop in fin pitch and an increase in fin height reduce the parasitic capacitance of the FinFET [23, 24]. The conformal mapping technique can be used to model these parasitic capacitances analytically [25–29].

2.5.3 Device doping

Although light doping is advised for improved control over leakage current, the FinFET technology ideally uses an undoped channel [25]. The fin's geometry is negatively impacted by the high dopant concentration needed for the source/drain area, which also raises the device's series resistance. In the source and drain regions, in-situ epitaxial growth is carried out, either with or without the fin structure removed, to solve this problem.

2.5.4 Orientation of fin

Because of the fin position in the <100> direction, the current normally travels along the <110> sidewall surface. Electron flow in the <110> direction is slower than in the <100> direction in many instances. On the other hand, FinFETs display the opposite behavior, with increased electron mobility in the <110> direction due to the quantum confinement effect [25].

2.5.5 Reliability of FinFET

A lower transverse field in the device can be achieved by totally depleted FinFET operation, which also improves N-type metal oxide semiconductor (NMOS) reliability for dielectric breakdown. The observed instability in the threshold voltage is linked to a transition from 32 nm planar technologies to 22 nm FinFET technologies. Remarkably, PMOS reliability in FinFET devices doesn't change [30].

2.5.6 Incorporating FinFETs into the system

Compatibility with conventional CMOS techniques is advantageous with FinFETs, where strict control over fabrication stages during gate construction and Fin patterning is essential, going beyond the level of accuracy needed in planar technology. For Fin width < 10 nm, a uniformity of about 1 nm is required to maintain low threshold voltage variations for LG < 20 nm, which means that strict process variation standards must be followed [104–109]. The complicated gate stack and metal gate structures of FinFETs are formed by the stack of HF-based dielectric and the deposition of TiN/TaN/W/La_2O_3/Dy_2O_3/Al_2O_3, which present challenges in the gate patterning process step [31]. Fin channels continue to be connected underneath the oxide, resulting in sub-Fin leakage even when shallow trench isolation (STI) oxide is used for fundamental body isolation between neighboring fins. Self-aligned punch-through stop layers (PTSL) at the bottom of the fin are used to regulate leakage in order to remedy this [110–116].

2.6 DIVERSE APPLICATIONS UTILIZING FINFET TECHNOLOGY

FinFET technology is widely used in a broad range of electronic products, such as high-end networks, wearable technologies, tablets, smartphones, home computers, laptops, and automobile systems. It serves as a versatile switch and amplifier, in addition to being a transistor [117–122]. In micromechatronics, the use of FinFET-based switches instead of conventional MOSFET switches presents a number of design and scaling alternatives for actuator control circuits [32]. A notable advantage of FinFETs lies in their remarkably low leakage current (Ioff), making them highly suitable for efficient load switching applications. Furthermore, their structural flexibility and precise control over the channel make FinFETs ideal for low-power designs, showcasing resilience to high noise. Emerging as promising alternatives to conventional MOSFETs, both nFinFET and pFinFET configurations are utilized in various digital applications [32–34]. FinFETs have an impact not only on conventional electronics but also on biological applications. They enable the development of double gate metal semiconductor field effect transistor (DGMSFET)-based biosensors with improved survival capacities by integrating nano-gap cavity regions beneath the side gate.

2.7 FINFET LOGIC

2.7.1 Inverter

Four unique configurations are produced by designing inverters (INV) with various combinations of SG (standard-gate) and IG (independent-gate) FinFETs: SG, low-power (LP), IGn, and IGp INV, as shown in Figure 2.6a and 2.6b. SG n/p FinFETs are used in an SG INV, which guarantees a very small footprint and excels in area and propagation delay. When compared to LP INV, it frequently experiences noticeably larger leakage current [123–128]. With its focus on low-power consumption, LP INV reduces leakage problems while sacrificing space and propagation delay performance. At least one IG FinFET is included in the IGn INV design, which offers a viable compromise given in Figure 2.6 c and 2.6d. The arrangement is more complex, resulting in a 36% area overhead when compared to twice the size of a minimum-sized SG INV; however, the characteristics of area, latency, and leakage are intermediate. The normalized area, delay, and leakage of these setups are compared in Table 2.2, which demonstrates that SG INV dominates in area and propagation delay while having greater leakage, whereas LP INV struggles in these areas [38]. Overall, IGn INV shows promise for applications where trade-offs between power, area, and performance are important factors because of its balanced performance over a range of measures.

2.7.2 NAND

NAND2 gates display SG (standard-gate) and LP (low-power) configurations, much like inverters, with all transistors being IG (independent-gate) or SG FinFETs as shown in Figure 2.7 a and 2.7b. However, due to the inherently larger structure of a NAND gate, accommodating more

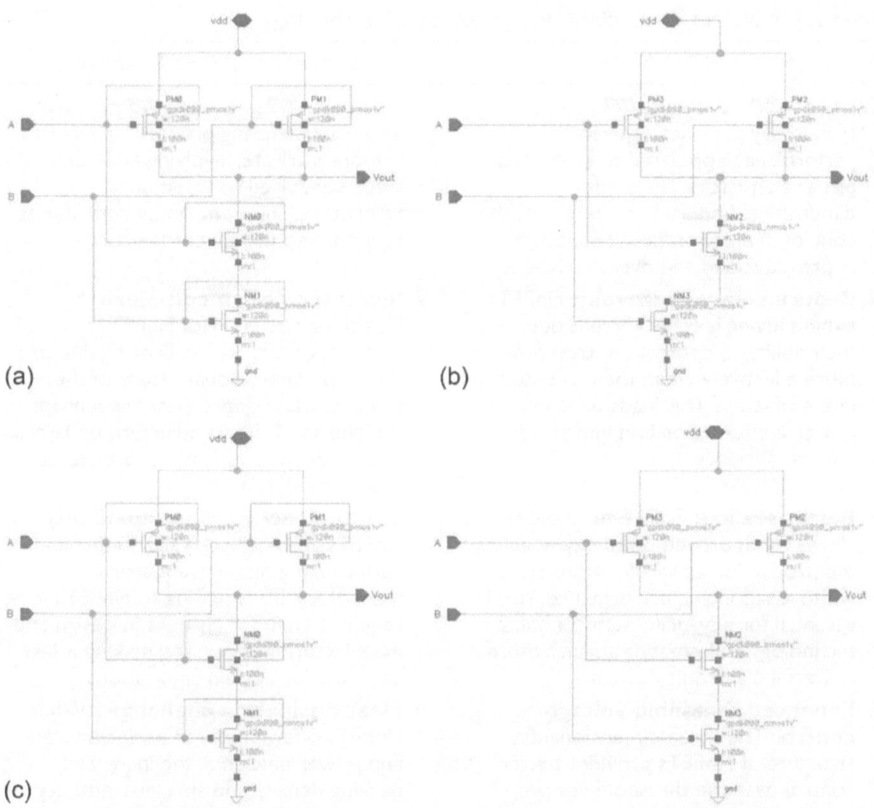

Figure 2.6 **(a) Short-gated; (b) low-power-gate FinFET inverter; (c) n-Type independent gate; (d) p-type independent-gate FinFET inverter.**

transistors compared to an inverter, there are increased possibilities and options for integrating both SG and IG FinFETs within its design given in Figure 2.7c and 2.7d. This expanded transistor count within the NAND gate presents additional opportunities for exploring diverse combinations and arrangements of SG and IG FinFETs, potentially leading to a wider array of configurations that can optimize various aspects such as power efficiency, area utilization, propagation delay, and leakage current [38].

2.8 CONCLUSION

In summary, the potential of FinFET technology to propel semiconductor advancements toward smaller nodes and higher performance levels is evident, yet it is not without its formidable challenges. Problems with strain booster integration, access resistance reduction, and the meticulous control

Table 2.2 Advantages and disadvantages of FinFET technology

Advantages of FinFET technology	*Disadvantages of FinFET technology*
1. **Improved performance:** FinFET technology provides superior performance compared to traditional planar transistors. The three-dimensional fin structure enhances the control of electron flow, resulting in improved speed and overall efficiency.	1. **Complex manufacturing process:** The manufacturing process for FinFETs is more intricate, involving additional steps compared to traditional transistors. This complexity contributes to increased manufacturing costs.
2. **Reduced leakage current:** FinFETs exhibit lower leakage current due to their ability to control electron flow more effectively when the transistor is in the off-state. This leads to lower power consumption and improved energy efficiency.	2. **Increased design complexity:** Designing circuits with FinFET technology is more challenging due to the three-dimensional nature of the transistor. Designers must account for the effects of the fin structure on layout and interconnects, leading to increased design complexity.
3. **Better scaling:** FinFET technology allows for improved scalability, enabling the production of smaller transistors without sacrificing performance. This is essential for advancing semiconductor technology and creating smaller, more powerful electronic devices.	3. **Limited backward compatibility:** FinFETs are not backward compatible with existing planar transistor technology. Transitioning to FinFETs may require significant changes in design and manufacturing processes, making it less seamless for certain applications.
4. **Enhanced threshold voltage control:** The three-dimensional fin structure of FinFETs provides better control over the threshold voltage, allowing for more precise management of power consumption without compromising performance.	4. **Heat dissipation challenges:** While FinFETs offer improved performance and power efficiency, the increased packing density and smaller transistor sizes can lead to challenges in heat dissipation. Managing heat is crucial to prevent thermal issues and maintain circuit reliability.
5. **Improved variability and reliability:** FinFET technology exhibits better variability control, ensuring more consistent and reliable performance across a batch of transistors. This is crucial for manufacturing yields and overall circuit reliability.	5. **Cost:** The complexity of the manufacturing process, additional processing steps, and potential yield issues can contribute to higher production costs for devices using FinFET technology.

required for fin parameters present significant hurdles that demand innovative solutions and collaborative efforts from the industry.

Despite these obstacles, the semiconductor sector has a commendable track record of overcoming challenges with current planar technology, showcasing its resilience and adaptability. This history of successful navigation through technological complexities underscores the industry's capacity to innovate and adapt in the face of adversity.

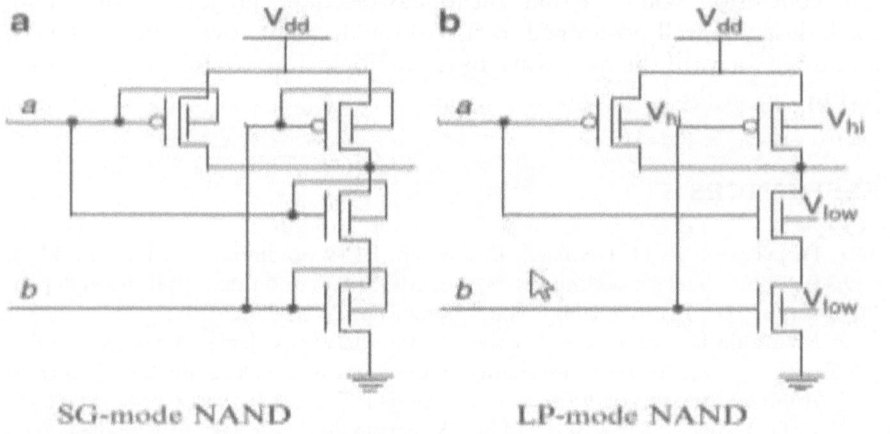

Figure 2.7 Short gate (a) and low power (b) in a FinFET NAND circuit.

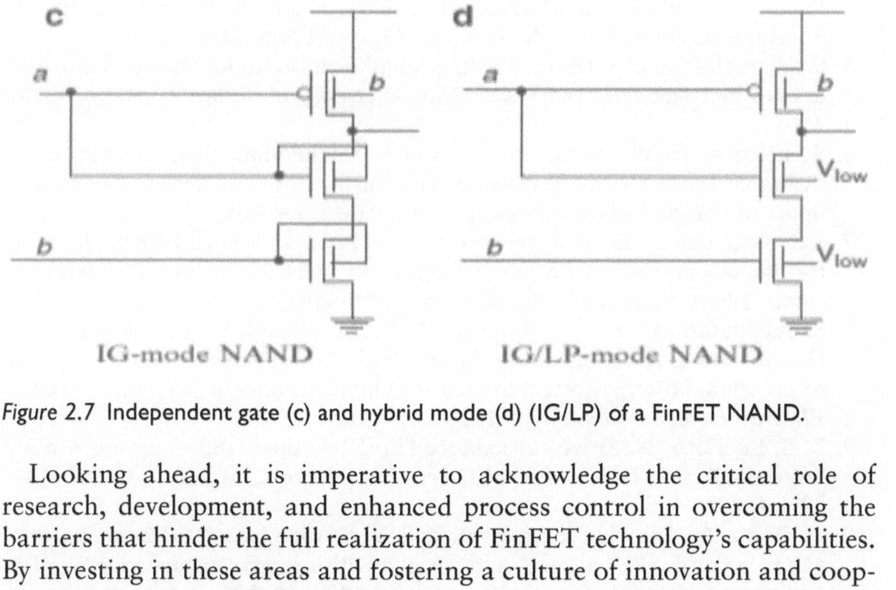

Figure 2.7 Independent gate (c) and hybrid mode (d) (IG/LP) of a FinFET NAND.

Looking ahead, it is imperative to acknowledge the critical role of research, development, and enhanced process control in overcoming the barriers that hinder the full realization of FinFET technology's capabilities. By investing in these areas and fostering a culture of innovation and cooperation, the semiconductor industry can effectively address the challenges posed by FinFET technology.

Moreover, by leveraging collective expertise and collaborative efforts, the industry can accelerate progress toward resolving these challenges and unlocking the transformative potential of FinFET technology. This not only ensures continued advancements in semiconductor technology but also opens up new possibilities for addressing evolving needs in various sectors, from consumer electronics to automotive and beyond.

In conclusion, while the road ahead may be challenging, the semiconductor industry is well-positioned to rise to the occasion, overcome obstacles, and drive forward the next wave of technological innovation with FinFET technology at the forefront.

REFERENCES

1. D. J. Frank, R. H. Dennard, E. Nowak, P. M. Solomon, Y. Taur, and H. S. P. Wong, "Device scaling limits of Si MOSFETs and their application dependence," *IEEE Proceedings*, vol. 89, no. 3, pp. 259–287, 2001.
2. M. Rahimian, and A. A. Orouji, "Nanoscale fully-depleted SOI MOSFETs with a changed current mechanism: a reduction in leakage current," *Current Applied Physics*, vol. 12, no. 5, pp. 1366–1371, 2012.
3. Y.-C. Yeo, T.-J. King, and C. Hu, "Selection guide for various gate dielectrics based on leakage considerations and modeling of MOSFET gate leakage," *IEEE Transactions on Electron Devices*, vol. 50, no, 4, pp. 1027–1035, 2003.
4. H. C. Liu, L. Li, M. Buchanan, Z. R. Wasilewski, A. K. Jonscher, and M. Ershov, "Negative capacitance effect in semiconductor devices," *IEEE Transactions on Electron Devices*, vol. 45, no. 10, pp. 2196–2206.
5. S. Salahuddin, and S. Datta, "Voltage amplification for low power nanoscale devices by negative capacitance," *Nano Letters,* vol. 8, no. 2, pp. 405–210, 2008.
6. A. I. Khan, C. W. Yeung, C. Hu, and S. Salahuddin, "Ferroelectric negative capacitance MOSFET: capacitance adjustment & antiferroelectric operation," in *IEDM Technical Digest*, pp. 11.3.1–11.3.4, 2011.
7. S. George et al. "Implemented device circuit codesign of FEFET-based logic for processors operating at low voltages," in *IEEE Computer Society Proc. Annu. Symp. VLSI (ISVLSI)*, 2016, pp. 649–654.
8. S. Salahuddin, C. Hu, R. Ramesh, Z. Lu, A. Sachid, S. Khandelwal, J. P. Duarte, and A. I. Khan, "Short-channel FinFETs with an external connection to an epitaxial ferroelectric capacitor exhibit negative capacitance," *IEEE Electron Device Letters*, 37(1):111–114, 2016.
9. K.-S. Li et al., "Negative-capacitance FinFET with sub-60mV swing and no hysteresis," in *IEDM Technical Digest*, Washington, DC, USA, pp. 22.6.1–22.6.4, 2015.
10. T. Dutta, G. Pahwa, A. Agarwal, and Y. S. Chauhan, "Effect of process variances on FinFET devices and circuits with negative capacitance," *IEEE Electron Device Letters*, vol. 39, no. 1, pp. 147–150, 2018.
11. P. Su, and H.-P. Lee, "Reduced fin-LER-induced fluctuations in FinFETs with negative capacitance," *IEEE Electron Device Letters*, vol. 38, no. 10, pp. 1492–1495, 2017.
12. J.-P. Raskin, D. Lederer, D. Flandre, T. Ming Chung, and V. Kilchytska, "Wideband simulations and characterization of the analog/RF performance of multiple gate SOI devices," *IEEE Electron Device Letters*, vol. 53, no. 5, pp. 1088–1095, 2006.

13. A. Kumar, "Performance of a multigate FinFET at the nanoscale in terms of analog and RF," *Superlattices and Microstructures*, vol. 100, pp. 1073–1080, December 2016.
14. E. Nowak et al., "Going beyond the 65nm node FinFET-DGCMOS," in *IEEE Custom Integrated Circuit Conference (CICC)*, 2003, p. 339.
15. Y. Taur, and T.H. Ning, *Fundamentals of Modern VLSI Devices*, 2nd ed., New York: Cambridge University Press, 1998.
16. C. Hu, "Scaling limits and projection for gate oxides," in *IEEE Electron Devices Meeting*, December 1996, pp. 319–322.
17. J. Chen, T. Y. Chan, I. C. Chen, P. K. Ko, and C. Hu, "Subbreakdown drain leakage current in MOSFET," *IEEE Electron Device Letters*, vol. 8, no. 11, pp. 515–517, 1987.
18. R. Cutrone, K. Chumakov, J. Lednicky, G. Dunn, P. Rizzo, M. Bocchetta, ... M. Carbone, "Following 1961, several oral poliovirus vaccinations were tainted with pathogenic SV40," *Cancer Research*, vol. 65, no. 22, pp. 10088–5472, 2005.
19. B. Bhowmick, and R. K. Maurya, "Examining FinFET devices and considering the difficulties in circuit design," *Silicon*, vol. 13, 2021. doi:10.1007/s12633-021-01366-z.
20. J. G. Fossum, and H. Lim, "Threshold voltage of MOSFETs made on thin-film silicon on insulator," *IEEE Transactions on Electron Devices,* vol. 30–10, p. 1244, 1983.
21. A. Bryant, P. Oldiges, J. Chang, M. Guillorn, C. H. Lin, and W. Haensch, "FinFET or trigate are two non-planar device architectures for 15 nm nodes," in *IEEE International Silicon on Insulator Conference Proceedings (SOI '10)*, 2010, pp. 1–2.
22. Y. Li, and W-H. Chen, "Effect of fin angle on electrical characteristics of nanoscale bulk FinFET," in *Proc. of NSTI-Nanotechnology Conference and Trade Exhibition*, vol. 3, pp. 20–23, 2006.
23. M. Guillorn et al., "An AC perspective on the FinFET performance advantage at 22 nm," in *Symp. VLSI Technol. Technical Digest*, pp. 12–13, 2008.
24. P. Teja, S. K. Sinha, and A. Durgesh, "MOSFET, MESFET, FinFET, and SOI-FinFET Comparison," *International Journal of Emerging Technologies and Innovative Research*, vol. 7, no. 11, pp. 1155–1116, 2020.
25. C.-Y. Kang et al., "Effect of layout and process parameter on device/circuit performance and variability for 10nm node FinFET technology," in *Symp. VLSI Technol. Technical Dig.*, pp. T90–T91, 2013.
26. J. P. Colinge, *FinFETs and Other Multigate Transistors*, Springer, New York, 2008.
27. W. Wu, S. S. Member, and M. Chan, "Analysis of geometry-dependent parasitics in multifin double-gate FinFETs," *IEEE Transactions on Electron Devices*, vol. 54, April 2007.
28. A. Bansal, B. C. Paul, and K. Roy, "Modeling and optimization of fringe capacitance of nanoscale DGMOS devices," *IEEE Transactions on Electron Devices*, vol. 52, no. 2, pp. 256–262, February 2005.

29. S. Agrawal, and J. G. Fossum, "A physical model for fringe capacitance in double-gate MOSFETs with non-absurd source/drain junctions and gate underlap was developed," *IEEE Transactions on Electron Devices*, vol. 57, no. 5, pp. 1069–1075, May 2010.
30. S. M. Sharma, S. Dasgupta, and M. V. Kartikeyan, "Successive conformal mapping technique to extract inner frequency capacitance of underlap DGFinFET and its variation with geometrical parameters," *IEEE Transactions on Electron Devices,* vol. 64, no. 2, 2017.
31. S. Ramey, et al., "Intrinsic transistor reliability improvement from 22 nm tri-gate technology," in *IEEE International Reliability Physics Symposium*, 2013.
32. K. Okano, T. Izumida, H. Kawasaki, et al., "Process integration technology and device characteristics of CMOS FinFET on bulk silicon substrate with sub-10 nm fin width and 20 nm gate length," in *IEEE Electron Device Meeting*, 2005, p. 820.
33. D. K. Panda, and T. R. Lenka, "Compact thermal noise model with two subbands included for N-polar GaN MOS-HEMT enhancement mode, including 2DEG density solution," *IET Circuits, Devices & Systems*, vol. 12, no. 6, pp. 810–816, 2018.
34. P. Ghosh, S. Haldar, R. S. Gupta, and M. Gupta, "An analysis of the GME CGT MOSFET's intermodulation distortion and linearity performance for RFIC design," *Electrical Engineering Transactions*, vol. 59, no. 12, pp. 3263–3268, December 2012.
35. A. Lazzaz, K. Bousbahi, and M. Ghamnia, "Modeling and simulation of DG SOI N FinFET 10 nm utilizing Hafnium Oxide," in *21st International Conference on Nanotechnology (NANO), IEEE,* 2021, pp. 177–180.
36. *Proc. IEEE Conference on Emerging Devices and Smart Systems (ICEDSS 2017)*, Mahendra Engineering College, Tamil Nadu, India, March 3–4, 2017. 31.00; 978-1-5090-5555-5/17.
37. *International Conference on Innovative and Emerging Trends in Engineering and Technology*, Chennai, Tamil Nadu, India, vol. 1, May 2016.
38. Z. Lu, and J. G. Fossum, "Short-channel effects in independent-gate FinFETs," *IEEE Electron Device Letters*, vol. 28, no. 2, February 2007.
39. Shradhya Singh, S. K. Vishvakarma, and B Raj, "Analytical modeling of split-gate junction-less transistor for a biosensor application," *Sensing and Bio-Sensing*, vol. 18, pp. 31–36, April 2018.
40. Maisagalla Gopal, and Balwinder Raj, "Low power 8T SRAM cell design for high stability video applications," *ITSI Transaction on Electrical and Electronics Engineering*, vol. 1, no. 5, pp. 91–97, 2013.
41. Balwinder Raj, Jatin Mitra, Deepak Kumar Bihani, Rangharajan V, A.K. Saxena, and S. Dasgupta, "Analysis of noise margin, power and process variation for 32 nm FinFET based 6T SRAM cell," *Journal of Computer*, vol. 5, no. 6, pp.1–8, 2010.
42. Divya Sharma, Rajesh Mehra, and B Raj, "Comparative analysis of photovoltaic technologies for high efficiency solar cell design," *Superlattices and Microstructures*, vol. 153, pp. 106861, May 2021.

43. Pawandeep Kaur, Avtar Singh Buttar, and B. Raj, "A comprehensive analysis of nanoscale transistor based biosensor: a review," *Indian Journal of Pure and Applied Physics,* vol. 59, pp. 304–318, April 2021.
44. Divya Yadav, Balwant Raj, and B. Raj, "Design and simulation of low power microcontroller for IoT applications," *Journal of Sensor Letters*, vol. 18, pp. 401–409, May 2020.
45. Shailendra Singh, and B. Raj, "A 2-D analytical surface potential and drain current modeling of double-gate vertical t-shaped tunnel FET," *Journal of Computational Electronics*, vol. 19, pp.1154–1163, April 2020.
46. Jeetendra Singh, and B. Raj, "An accurate and generic window function for non-linear memristor model," *Journal of Computational Electronics*, vol. 18, no. 2, pp. 640–647, June 2019.
47. Manjit Kaur, Neena Gupta, Sanjeev Kumar, B. Raj, and Arun Kumar Singh, "Comparative RF and crosstalk analysis of carbon based nano interconnects," *IET Circuits, Devices & Systems*, vol. 15, no. 6, pp. 493–503, February 2021.
48. Nehru Kandasamy, Firdous Ahmad, D. Ajitha, B. Raj, and Nagarjuna Telagam "Quantum dot cellular automata based scan flip flop and boundary scan register," *IETE Journal of Research*, vol. 66, pp. 535–548, 2020.
49. Jeetendra Singh, and B. Raj, "Design and investigation of 7T2M NVSARM with enhanced stability and temperature impact on store/restore energy," *IEEE Transactions on Very Large Scale Integration Systems*, vol. 27, no. 6, pp. 1322–1328, June 2019.
50. Anil Kumar Bhardwaj, Sumeet Gupta, B. Raj, and Amandeep Singh, "Impact of double gate geometry on the performance of carbon nanotube field effect transistor structures for low power digital design," *Computational and Theoretical Nanoscience*, vol. 16, pp. 1813–1820, 2019.
51. Neeraj Jain, and B. Raj, "Thermal stability analysis and performance exploration of asymmetrical Dual-k underlap Spacer (ADKUS) SOI FinFET for security and privacy applications," *Indian Journal of Pure & Applied Physics*, vol. 57, pp. 352–360, May 2019.
52. Neeraj Jain, and Balwinder Raj, "Dual-k spacer region variation at the drain side of asymmetric SOI FinFET structure: performance analysis towards the Analog/RF design applications," *Journal of Nanoelectronics and Optoelectronics*, vol. 14, pp. 349–359, March 2019.
53. Jeetendra Singh, Sanjeev Sharma, B. Raj, and Mamta Khosla, "Analysis of barrier layer thickness on performance of $In_{1-x}Ga_xAs$ based gate stack cylindrical gate nanowire MOSFET," *Journal of Nanoelectronics and Optoelectronics*, vol. 13, pp. 1473–1477, October 2018.
54. Neeraj Jain, and B. Raj, "Analysis and performance exploration of high-k SOI FinFETs over the conventional low-k SOI FinFET toward analog/RF design," *Journal of Semiconductors*, vol. 39, no. 12, p. 124002-1–7, December 2018.
55. Candy Goyal, Jagpal Singh Ubhi, and B. Raj, "A reliable leakage reduction technique for approximate full adder with reduced ground bounce noise," *Journal of Mathematical Problems in Engineering*, vol. 2018, Article ID 3501041, pp. 16, 15 October 2018.
56. Anuradha, Jeetendra Singh, B. Raj, and Mamta Khosla, "Design and performance analysis of nano-scale memristor-based nonvolatile SRAM," *Journal of Sensor Letter*, vol. 16, pp. 798–805, October 2018.

57. Girish Wadhwa, and B. Raj, "Parametric variation analysis of charge-plasma-based dielectric modulated JLTFET for biosensor application," *IEEE Sensor Journal*, vol. 18, no. 15, pp. 6070–6077, August 2018.
58. Jeetendra Singh, and B. Raj, "Comparative analysis of memristor models for memories design," *Journal of Semiconductors*, vol. 39, no. 7, pp. 074006-1–12, July 2018.
59. Divya Yadav, Shailesh Singh Chouhan, Santosh Kumar Vishvakarma, and B. Raj, "Application specific microcontroller design for IoT based WSN," *Sensor Letter*, vol. 16, pp. 374–385, May 2018.
60. Gurmohan Singh, R. K. Sarin, and B. Raj, "Fault-tolerant design and analysis of quantum-dot cellular automata based circuits," *IEEE/IET Circuits, Devices & Systems*, vol. 12, pp. 638–664, 2018.
61. Jeetendra Singh, and B. Raj, "Modeling of mean barrier height levying various image forces of metal insulator metal structure to enhance the performance of conductive filament based memristor model," *IEEE Nanotechnology*, vol. 17, no. 2, pp. 268–267, March 2018.
62. Aakash Jain, Sanjeev Sharma, and B. Raj, "Analysis of triple metal surrounding gate (TM-SG) III-V nanowire MOSFET for photosensing application," *Opto-Electronics Journal*, vol. 26, no. 2, pp. 141–148, May 2018.
63. Aakash Jain, Sanjeev Sharma, and B. Raj, "Design and analysis of high sensitivity photosensor using cylindrical surrounding gate MOSFET for low power sensor applications," *Engineering Science and Technology, an International Journal*, vol. 19, no. 4, pp. 1864–1870, December 2016.
64. G. Saiphani Kumar, Amandeep Singh, and B. Raj, "Design and analysis of gate all around CNTFET based SRAM cell design," *Journal of Computational Electronics*, vol. 17, no. 1, pp 138–145, March 2018.
65. Gurinder pal Singh, B. S. Sohi, and Balwinder Raj, "Material properties analysis of graphene base transistor (GBT) for VLSI analog circuits," *Indian Journal of Pure & Applied Physics (IJPAP)*, vol. 55, pp. 896–902, December 2017.
66. Sunil Kumar, and B. Raj, "Estimation of stability and performance metric for inward access transistor based 6T SRAM cell design using n-type/p-type DMDG-GDOV TFET," *IEEE VLSI Circuits and Systems Letter*, vol. 3, no. 2, pp. 25–39, June 2017.
67. Shashikant Sharma, Anjan Kumar, Manisha Pattanaik, and B. Raj, "Forward body biased multimode multi-threshold CMOS technique for ground bounce noise reduction in static CMOS adders," *International Journal of Information and Electronics Engineering*, pp. 567–572, vol. 3, no. 3, 2013.
68. Hamendra Singh, Pankaj Kumar, and Balwinder Raj, "Performance analysis of majority gate SET based 1-bit full adder," *International Journal of Computer and Communication Engineering (IJCCE)*, vol. 2, no. 4, pp. 2010–3743, 2013.
69. Anil Kumar Bhardwaj, Sumeet Gupta, and B. Raj, "Investigation of parameters for Schottky barrier (SB) height for Schottky barrier based carbon nanotube field effect transistor device," *Journal of Nanoelectronics and Optoelectronics*, vol. 15, pp. 783–791, July 2020.

70. Priya Bansal, and B. Raj, "Memristor: a versatile nonlinear model for dopant drift and boundary issues," *Journal of Computational and Theoretical Nanoscience*, vol. 14, no. 5, pp. 2319–2325, May 2017.
71. Neeraj Jain, and B. Raj, "An analog and digital design perspective comprehensive approach on Fin-FET (Fin-Field effect transistor) technology - a review," *Reviews in Advanced Sciences and Engineering (RASE)*, vol. 5, pp. 1–14, 2016.
72. Balwinder Raj, A. K. Saxena, and S. Dasgupta, "Analytical modeling for the estimation of leakage current and subthreshold swing factor of nanoscale double gate FinFET device," *Microelectronics International*, vol. 26, pp. 53–63, 2009.
73. Shailendra Singh Soniya, Girish Wadhwa, and Balwinder Raj, "An analytical modeling for dual source vertical tunnel field effect transistor," *International Journal of Recent Technology and Engineering*, vol. 8, no. 2, pp. 603–608, July 2019.
74. Shailendra Singh, and B. Raj, "Design and analysis of hetrojunction vertical T-shaped tunnel field effect transistor," *Journal of Electronics Material*, vol. 48, no. 10, pp 6253–6260, October 2019.
75. Candy Goyal, Jagpal Singh Ubhi, and B Raj, "A low leakage CNTFET based inexact full adder for low power image processing applications," *International Journal of Circuit Theory and Applications*, vol. 47, no. 9, pp. 1446–1458, September 2019.
76. B. Raj, A. K. Saxena, and S. Dasgupta, "A compact drain current and threshold voltage quantum mechanical analytical modeling for FinFETs," *Journal of Nanoelectronics and Optoelectronics*, vol. 3, no. 2, pp. 163–170, 2008.
77. Girish Wadhwa, and B. Raj, "An analytical modeling of charge plasma based tunnel field effect transistor with impacts of gate underlap region," *Superlattices and Microstructures*, vol. 142, pp.106512, June 2020.
78. Shailendra Singh, and B. Raj, "Modeling and simulation analysis of SiGe hetrojunction double gate vertical t-shaped tunnel FET," *Superlattices and Microstructures*, vol. 142, pp. 106496, June 2020.
79. Neeraj Jain, and Balwinder Raj, "Device and circuit co-design perspective comprehensive approach on FinFET technology: a review," *Journal of Electron Devices*, vol. 23, no. 1, pp 1890–1901, 2016.
80. Sunil Kumar, and B. Raj, "Analysis of I_{ON} and Ambipolar current for dual-material gate-drain overlapped DG-TFET," *Journal of Nanoelectronics and Optoelectronics*, vol. 11, pp. 323–333, June 2016.
81. Naveed Anjum, Tarun Bali, and B. Raj, "Design and simulation of handwritten multiscript character recognition," *International Journal of Advanced Research in Computer and Communication Engineering*, vol. 2, no. 7, pp. 2544–2549, July 2013.
82. Karmjit Singh, and B. Raj, "Performance and analysis of temperature dependent multi-walled carbon nanotubes as global interconnects at different technology nodes," *Journal of Computational Electronics*, vol. 14 no. 2, pp. 469–476, June 2015.
83. Sunil Kumar, and B. Raj, "Compact channel potential analytical modeling of DG-TFET based on evanescent–mode approach," *Journal of Computational Electronics*, vol. 14, no. 2, pp.820–827, July 2015.

84. Karmjit Singh, and B. Raj, “Temperature dependent modeling and performance evaluation of multi-walled CNT and single-walled CNT as global interconnects,” *Journal of Electronic Materials*, vol. 44, no. 12, pp. 4825–4835, December 2015.
85. V. K. Sharma, M. Pattanaik, and B. Raj, “INDEP approach for leakage reduction in nanoscale CMOS circuits,” *International Journal of Electronics*, vol. 102, no. 2, pp. 200–215, 2014.
86. Karmjit Singh, and B. Raj, “Influence of temperature on MWCNT bundle, SWCNT bundle and copper interconnects for nanoscaled technology nodes,” *Journal of Materials Science: Materials in Electronics*, vol. 26, no. 8, pp. 6134–6142, 2015.
87. Naveed Anjum, Tarun Bali, and B. Raj, “Design and simulation of handwritten Gurumukhi and Devanagri numerical recognition,” *International Journal of Computer Applications,* vol. 73, no. 12, pp. 16–21, 2013.
88. S. Khandelwal, V. Gupta, B. Raj, and R. D. Gupta, “Process variability aware low leakage reliable nano scale DG-FinFET SRAM cell design technique,” *Journal of Nanoelectronics and Optoelectronics*, vol. 10, no. 6, pp. 810–817, December 2015.
89. V. K. Sharma, M. Pattanaik, and B Raj, “ONOFIC approach: low power high speed nanoscale VLSI circuits design,” *International Journal of Electronics*, vol. 101, no. 1, pp. 61–73, 2014.
90. S. Khandelwal, Balwinder Raj, and R. D. Gupta, “FinFET based 6T SRAM cell design: analysis of performance metric, process variation and temperature effect,” *Journal of Computational and Theoretical Nanoscience*, vol. 12, pp. 2500–2506, 2015.
91. Sumit Singh; Shekhar Yadav, Jagdeep Rahul, Anurag Srivastava, and B. Raj, “Impact of HfO_2 in graded channel dual insulator double gate MOSFET,” *Journal of Computational and Theoretical Nanoscience*, vol. 12, no. 6, pp. 950–953, April 2015.
92. Vijay Kumar Sharma, Manisha Pattanaik, and B. Raj, “PVT variations aware low leakage INDEP approach for nanoscale CMOS Circuits,” *Microelectronics Reliability*, vol. 54, pp 90–99, 2014.
93. B. Raj, A. K. Saxena, and S. Dasgupta, “Quantum mechanical analytical modeling of nanoscale DG FinFET: evaluation of potential, threshold voltage and source/drain resistance,” *Journal of Material Science in Semiconductor Processing,* vol. 16, no. 4, pp. 1131–1137, 2013.
94. Maisagalla Gopal, Siva Sankar D. Prasad, and Balwinder Raj, “8T SRAM cell design for dynamic and leakage power reduction,” *International Journal of Computer Applications*, vol. 71, no. 9, pp. 43–48, June 2013.
95. Manisha Pattanaik, B. Raj, Shashikant Sharma, and Anjan Kumar, “Diode based trimode multi-threshold CMOS technique for ground bounce noise reduction in static CMOS adders,” *Advanced Materials Research*, vol. 548, pp 885–889, 2012.
96. Balwinder Raj, A. K. Saxena, and S. Dasgupta, “Nanoscale FinFET based SRAM cell design: analysis of performance metric, process variation, underlapped FinFET and temperature effect,” *IEEE Circuits and System Magazine*, vol. 11, no. 2, pp. 38–50, 2011.

97. V. K. Sharma, M. Pattanaik, and Balwinder Raj, "Leakage current ONOFIC approach for deep submicron VLSI circuit design," *International Journal of Electrical, Computer, Electronics and Communication Engineering*, vol. 7, no. 4, pp. 239–244, 2013.
98. Parminder Kaur, Sandeep Singh Gill, and B Raj, "Comparative analysis of OFETs materials and devices for sensor applications," *Journal of Silicon*, vol 14, pp. 4463–4471, 2022.
99. Sanjeev Kumar Sharma, Parveen Kumar, Balwant Raj, and B. Raj, "$In_{1-x}Ga_xAs$ double metal gate-stacking cylindrical nanowire MOSFET for highly sensitive photo detector," *Journal of Silicon*, vol. 14, pp. 3535–3541, 2022.
100. B. Raj, A. K. Saxena, and S. Dasgupta, "Analytical modeling of quasi planar nanoscale double gate FinFET with source/drain resistance and field dependent carrier mobility: a quantum mechanical study," *Journal of Computer*, vol. 4, no. 9, pp.1–8, 2009.
101. S. Bhushan, S. Khandelwal, and B. Raj, "Analyzing different mode FinFET based memory cell at different power supply for leakage reduction," in *Seventh International Conference on Bio-Inspired Computing: Theories and Application (BIC-TA 2012) Advances in Intelligent Systems and Computing*, vol. 202, pp. 89–100, 2013.
102. Jeetendra Singh, and B. Raj, "Temperature dependent analytical modeling and simulations of nanoscale memristor," *Engineering Science and Technology, an International Journal*, vol. 21, pp. 862–868, October 2018.
103. Shradhya Singh, Shashi Bala, Balwant Raj, and B. Raj, "Improved sensitivity of dielectric modulated junctionless transistor for nanoscale biosensor design," *Sensor Letter*, vol. 18, pp. 328–333, April 2020.
104. Vivek Kumar, Santosh Kumar Vishvakarma, and B. Raj, "Design and performance analysis of ASIC for IoT applications," *Sensor Letter*, vol. 18, pp. 31–38, January 2020.
105. Akanksha Jaiswal, R. K. Sarin, B. Raj, and Shikha Sukhija, "A novel circular slotted microstrip-fed patch antenna with three triangle shape defected ground structure for multiband applications," *Advanced Electromagnetic*, vol. 7, no. 3, pp. 56–63, August 2018.
106. Girish Wadhwa, and B. Raj, "Label free detection of biomolecules using charge–plasma-based gate underlap dielectric modulated junctionless TFET," *Journal of Electronic Materials*, vol. 47, no. 8, pp 4683–4693, August 2018.
107. Gurmohan Singh, R. K. Sarin, and B Raj, "Design and performance analysis of a new efficient coplanar quantum-dot cellular automata adder," *Indian Journal of Pure & Applied Physics (IJPAP)*, vol. 55, pp. 97–103, February 2017.
108. Parminder Kaur, Vikas Pandey, and B Raj, "Comparative study of efficient design, control and monitoring of solar power using IoT," *Sensor Letter*, vol. 18, pp. 419–426, May 2020.
109. Anil Kumar Bhardwaj, Sumeet Gupta, and B. Raj, "Development & analysis of compact model for double gate Schottky barrier CNTFET," *Journal of Nanoelectronics and Optoelectronics*, vol. 15, pp. 1199–1208, August 2020.

110. Girish Wadhwa, Priyanka Kamboj, Jeetendra Singh, and B. Raj, "Design and investigation of junctionless DGTFET for biological molecule recognition," *Transactions on Electrical and Electronic Materials*, vol. 22, pp. 282–289, 2021.
111. Sachin Kumar Verma, Shailendra Singh, Girish Wadhwa, and B. Raj, "Detection of biomolecules using charge-plasma based gate underlap dielectric modulated dopingless TFET," *Transactions on Electrical and Electronic Materials,* vol. 21, pp. 528–535, June 2020.
112. Neeraj Jain, and B. Raj, "Impact of underlap spacer region variation on electrostatic and analog/RF performance of symmetrical high-k SOI FinFET at 20 nm channel length," *Journal of Semiconductors*, vol. 38, no. 12, pp. 122002, December 2017.
113. Shailendra Singh, and B. Raj," Analytical modeling and simulation analysis of T-shaped III-V heterojunction vertical T-FET," *Superlattices and Microstructures*, vol. 147, pp. 106717, November 2020.
114. Gurmohan Singh, R. K. Sarin, and B. Raj, "Design and analysis of area efficient QCA based reversible logic gates," *Journal of Microprocessors and Microsystems*, vol. 52, pp. 59–68, May 2017.
115. Shailendra Singh, and B. Raj, "Parametric variation analysis on hetero-junction vertical t-shape TFET for suppressing ambipolar conduction," *Indian Journal of Pure and Applied Physics*, vol. 58, pp. 478–485, June 2020.
116. Shailendra Singh Soniya, Girish Wadhwa, and Balwinder Raj, "Design and analysis of dual source vertical tunnel field effect transistor for high performance," *Transactions on Electrical and Electronics Materials*, vol. 21, pp. 74–82, October 2019.
117. Manjit Kaur, Neena Gupta, Sanjeev Kumar, B. Raj, and Arun Kumar Singh, "RF performance analysis of intercalated graphene nanoribbon based global level interconnects," *Journal of Computational Electronics*, vol. 19, pp. 1002–1013, June 2020.
118. Girish Wadhwa, and B. Raj, "Design and performance analysis of junctionless TFET biosensor for high sensitivity," *IEEE Nanotechnology*, vol. 18, pp. 567–574, 2019.
119. Jeetendra Singh, and B. Raj, "Enhanced nonlinear memristor model encapsulating stochastic dopant drift," *Journal of Nanoelectronics and Optoelectronics*, vol. 14, pp. 958–963, 2019.
120. Anjana Bhardwaj, Pradeep Kumar, B. Raj, and Sunny Anand, "Design and performance optimization of doping-less vertical nanowire TFET using gate stack technique," *Journal of Electronic Materials*, vol. 41, no. 7, pp. 4005–4013, 2022.
121. Jeetendra Singh, and B. Raj, "Tunnel current model of asymmetric MIM structure levying various image forces to analyze the characteristics of filamentary memristor," *Applied Physics A*, vol. 125, no. 3, pp. 203.1–203.11, February 2019.
122. Candy Goyal, Jagpal Singh Ubhi, and B. Raj, "Low leakage zero ground noise nanoscale full adder using source biasing technique," *Journal of Nanoelectronics and Optoelectronics*, vol. 14, pp. 360–370, March 2019.

123. Gurmohan Singh, R. K. Sarin, and B. Raj, "A novel robust exclusive-OR function implementation in QCA nanotechnology with energy dissipation analysis," *Journal of Computational Electronics*, vol. 15, no. 2, pp. 455–465, June 2016.
124. Girish Wadhwa, Priyanka Kamboj, and Balwinder Raj, "Design optimisation of junctionless TFET biosensor for high sensitivity," *Advances in Natural Sciences: Nanoscience and Nanotechnology*, vol. 10, no. 7, p. 045001, 2019.
125. Priya Bansal, and B. Raj, "Memristor modeling and analysis for linear dopant drift kinetics," *Journal of Nanoengineering and Nanomanufacturing*, vol. 6, pp. 1–7, 2016.
126. D. Vaithiyanathan Ashima, and B. Raj, "Performance analysis of charge plasma induced graded channel Si nanotube," *Journal of Engineering Research*, special issue, pp. 146–154 August 2021.
127. Abhishek Singh Tomar, Vijay Kumar Magraiya, and B. Raj, "Scaling of access and data transistor for high performance DRAM cell design," *Quantum Matter*, vol. 2, pp. 412–416, October 2013.
128. Neeraj Jain, and B. Raj, "Parasitic capacitance and resistance model development and optimization of raised source/drain SOI FinFET structure for analog circuit applications," *Journal of Nanoelectronics and Optoelectronics*, vol. 13, pp. 531–539, April 2018.

Chapter 3

Future directions and roadmap for nanoelectronics

Shilpa Sonali, Yogesh Kumar Yadav, Mandeep Singh, Tarun Chaudhary, and Balwinder Raj

3.1 INTRODUCTION

Since the emergence of integrated circuits in the late 1950s, advancements in semiconductor technology have swiftly evolved toward increased packing density, enhanced functionalities, accelerated circuit speed, and reduced power dissipation. Moore's initial law posits that every 18 months witnesses a doubling of functionality (or a halving of feature size by a factor of $\sqrt{2}$) in the predominant silicon CMOS (complementary metal–oxide–semiconductor) technology. In the foreseeable future, CMOS is poised to maintain its prominent position in Very Large-Scale Integration (VLSI) for several compelling reasons. Firstly, silicon (Si) stands out as one of the most abundant elements on Earth. Secondly, owing to its complementary nature, CMOS exhibits minimal power consumption during standby [1–5]. Thirdly, metal–oxide–semiconductor (MOS) ranks among the simplest devices, making CMOS design relatively straightforward. Lastly, the performance of CMOS circuits continues to advance with the reduction of feature size, simultaneously achieving improved efficiency and cost-effectiveness. CMOS will push the boundaries of feature size scaling as long as it retains its distinctive characteristics, including saturation, minimal or no standby power, and the cost advantages associated with performance. Concurrently, the limiting factor will shift toward the interconnect. The critical question emerges: What shape will nanoelectronics ultimately adopt concerning the ongoing scaled silicon technology? To explore this inquiry, the devices' scaling in Moore's law is recalibrated in relation to the number of electrons. Figure 3.1 depicts a roadmap for nanoelectronics, illustrating a decrease in the number of electrons, reaching a point where it approaches a few electrons and eventually just a single electron [6–11]. Presently, there is evidence of single-electron memory functioning at room temperature. With the reduction of device size, the Coulomb self-energy will significantly surpass kT at room temperature (25 meV), implying the potential for room temperature operation of logic or signal processing devices. Additionally, other options such as carbon nanotubes and various quantum effect devices could be considered. The nanoelectronics discussed necessitate the adoption

 DOI: 10.1201/9781003612797-3

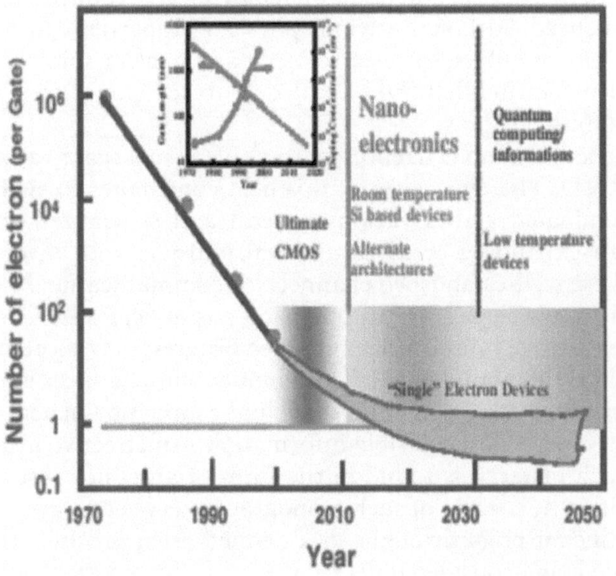

Figure 3.1 The nanoelectronics roadmap illustrates a decrease in the electron count to a fractional or single electron. The notion of fractional electrons can be accepted because quantum information systems and nanocellular automata offer the flexibility to go from discrete electrons to continuous waves [2].

of diverse architectures to capitalize on the distinctive features of emerging devices and overcome constraints stemming from numerous interconnects and fanouts [12–18].

These alternative architectures should minimize interconnects and fanouts for nanodevices characterized by small electron counts or low current drive. Moreover, these novel architectures must embrace massive parallelism to further amplify integrated circuit performance. Possible architectures might involve those with minimal global interconnects, like the cellular automata concepts explored in the past, albeit adapted for new nanodevice implementation [19–25]. Additionally, as previously indicated, it is imperative to integrate an input–output interface device into nanoelectronic architectures. Quantum coherent waves have the potential to effectively substitute explicit interconnects with implicit wave "interconnects". Quantum computing leverages a unique aspect of quantum mechanics, known as quantum entanglement, as a means of achieving interconnects and facilitating quantum parallelism in computation. This parallelism is accomplished through intricate entanglement, eliminating the necessity for physical interconnection wires due to the nonlocality of interaction in quantum mechanics. Consequently, this approach holds promise for alleviating the most critical bottleneck of integrated circuits – interconnects [26–31].

By doing so, there is the potential for achieving an exponential increase in computation speed. Although various physical properties can be exploited as quantum bits or qubits, the electron spin and nuclear spin present a convenient implementation that can be scaled up to accommodate a substantial number of qubits [1, 2].

Nanoscience and nanotech represent some of the most captivating domains in current R&D. The dynamism of this field contributes to swift advancements, disseminated across various general and specialized conferences, as well as an expanding array of archival journals with diverse focuses. Alongside these well-established channels for communicating research findings, there is a noticeable rise in open-access publications. Consequently, staying abreast of key discoveries becomes progressively more challenging [32–38]. While the abundance of communication channels is not unique to nanoscience, the increasing industrial and economic impacts of the field underscore the need for accessible information. An effective means of consolidating such material is found in the form of an edited research monograph. Specifically, a series of such monographs serves as a great vehicle for sharing significant breakthroughs in a defined area, precisely the purpose of the volume being considered [7].

3.2 NANOELECTRONIC DEVICES

3.2.1 Tunneling devices

The resonant tunneling diode (RTD) has attracted a lot of interest for use in nanoelectronic circuits because of its small size, quick operating speed, flexible design, and negative differential resistance [39–46]. Through the use of nanofabrication techniques such as metalorganic chemical vapor deposition (MOCVD), atomic layer deposition (ALD), and molecular beam epitaxy (MBE), researchers have effectively implemented the RTD (Figure 3.2).

A potential device for low-power applications like the Internet of Things (IoT) is the tunnel field-effect transistor (TFET) [47–55]. TFETs work on the basis of band-to-band tunneling (BTBT) at the source–channel interface, in contrast to traditional metal–oxide–semiconductor field-effect transistors (MOSFETs), which are unable to produce sub-60-mV/decade sub-threshold swing (SS) at room temperature because of non-scaled thermal voltage. With the use of this special property, TFET is able to attain SS values at 300 K that are lower than 60 mV/decade, exceeding the physical limit of SS [28].

Conventional TFETs, specifically p-i-n TFETs, have the advantage of having an average SS of less than 60 mV/decade, but they also have limited device performance due to insufficient on-state drive current and other factors [56–64]. They also have unidirectional current flow, which is caused by the asymmetry of their device structure. TFET performance measures

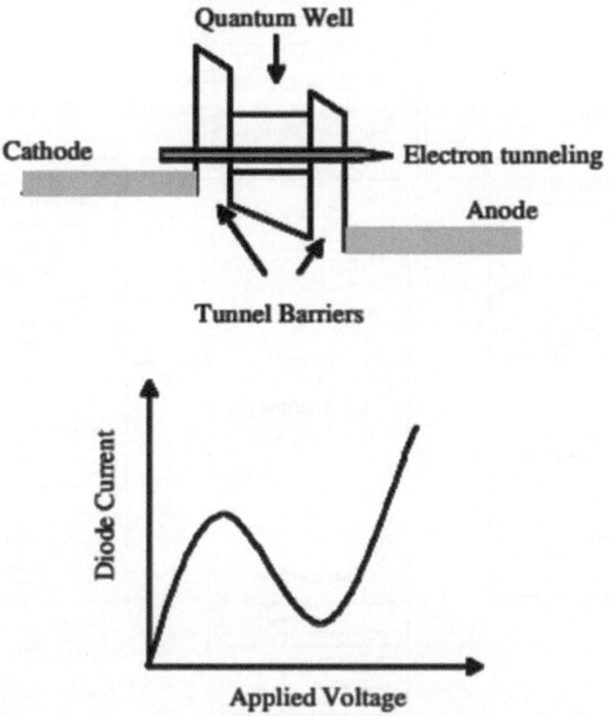

Figure 3.2 A standard band diagram illustrating the quantized energy levels of the resonant tunneling diode (RTD) and its corresponding current–voltage (I–V) characteristic [2].

such as the ION/IOFF ratio and SS still need to be improved, even if earlier research has suggested improvements like using tiny band-gap materials in the source region and utilizing a symmetrical device structure for bidirectional current flow [28].

3.2.2 Single electron transistors (SETs)

The number of electrons impacted by the gate diminishes as the size gets smaller until it reaches the nanoscale, at which point it influences just one electron. The fundamental idea underlying the "single-electron transistor" (SET), which allows devices to work, is the manipulation of a single electron [65–71]. This device uses a tunneling barrier to release an electron onto a small silicon island that is coupled to external reservoirs (source and drain). As seen in Figure 3.3a, the gate, or many gates, modulates the island's potential barrier. Owing to the island's nanoscale size, the charging energy ($e^2/2C$) surpasses tens of milli-electron-volts (meV), and the capacitance may be on the order of atto-farads (aF). This leads to the observation

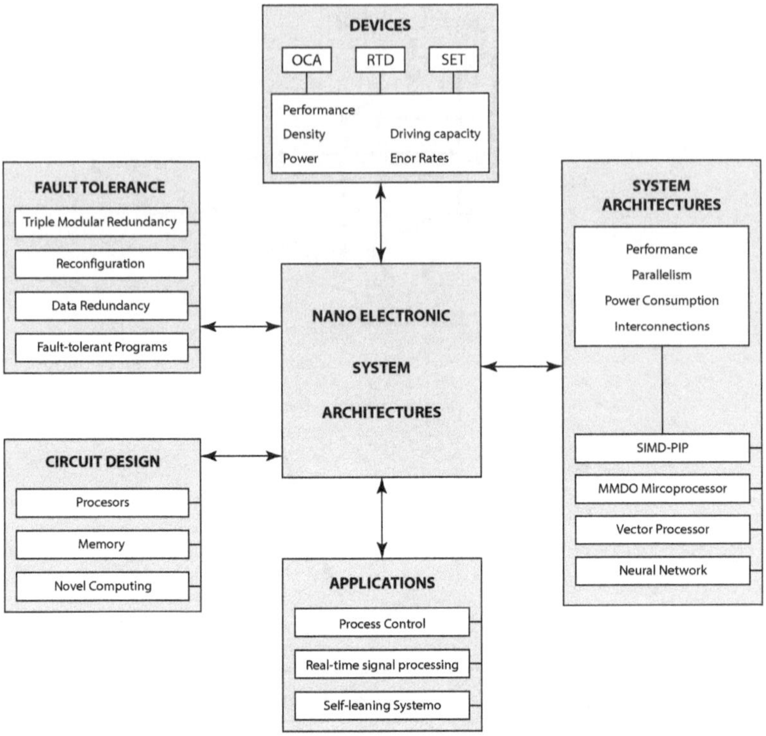

Figure 3.3 (a) A single-electron transistor's equivalent circuit. (b) The characteristic of current–voltage (I–V). The number of electrons in the silicon island determines the drain current (Id) at a low drain voltage. When there are an integer (or half-integer) number of electrons on the island, Id achieves a minimum (or maximum) [2].

of Coulomb blockage even at room temperature. The drain current, which is regulated by the gate voltage, displays what are known as "Coulomb oscillations", which are periodic oscillations with a period of (e/Cg) and are seen in Figure 3.3b. For some circuit applications, single-electron transistors offer the following advantages because of their unique properties, which were previously described: (1) remarkable low-power consumption due to the small number of electrons used and lack of significant standby power dissipation; (2) effective scalability made possible by the Coulomb blockade principle, which enables these devices to function at incredibly small physical dimensions, down to the atomic scale, facilitating the potential for ultra-large-scale integration; and (3) improved operational speed due to a reduced capacitance despite the low current [72–77].

3.2.3 Nanoelectronic tubes

Nanoelectronic tubes, also known as nanotubes or nanowires, signify a groundbreaking progression in the realm of nanotechnology. These structures, typically with diameters in the nanometer range, have attracted considerable attention due to their distinctive electrical, mechanical, and thermal attributes. This concise overview delves into the foundational principles of nanoelectronic tubes, elucidating their properties, synthesis methodologies, and potential applications. In the realm of nanostructures, nanotubes (NTs) represent another class [78–83]. Diodes, a straightforward category of devices, exemplify this class. Individual Y-junction carbon nanotube diodes have recently been claimed to have achieved their first experimental success. I–V measurements were used to demonstrate rectifying behavior at ambient temperature and to indicate future applications. The references provide previous research on nanotube p–n junction diodes that illustrate the basic idea of rectifying functioning. In addition to rectifying behavior, tunneling across a potential barrier in two-terminal NT devices can result in negative differential resistance (NDR), as noted by Leonard and Tersoff. Even at ambient temperature, theoretical treatment of both nanotube p–n junctions and undoped metal–nanotube–metal junctions revealed notable peak-to-valley current ratios that exceeded those observed in present devices. We successfully constructed and showed molecular field-effect transistors (FETs) utilizing single-wall and multi-wall carbon nanotubes (NTs) as three-terminal switching devices [84–90].

The nanotube changed from a conducting to an insulating state by applying voltage to a gate electrode. Certain devices behaved like transistors at ambient temperature, whereas other devices showed gate-voltage-independent wire behavior. At lower temperatures, Coulomb blockade usually dominated transport; at higher temperatures, power law characteristics were noted. An NT-FET with a single wall might have its conductance modulated by the gate electrode by a factor greater than 5. Large-diameter multi-wall NTs, on the other hand, usually did not display any gate effect; nonetheless, structural deformation might alter their electrical structure to the point where they displayed FET activity [91–96]. Notably, NT-FETs with hole transport capabilities and usually high contact resistance were produced on high-work-function metal electrodes, like platinum or gold. In order to study the p-type properties of semiconducting NTs, Martel et al. created carbon NT-FETs with a homogeneous SiO_2 layer and titanium carbide contacts. Unlike earlier devices, the titanium-carbide-contacted FET demonstrated that the gate field could modify the apparent barrier height for carrier injection, enabling the FETs to be ambipolar with low contact resistance for both n- and p-type conduction. This result implied that the typical p-type feature of NT-FETs might be a feature of the metal junction contacts between the nanotube and the metal rather than an intrinsic characteristic of the nanotube. Moreover, complementary logics could be implemented by utilizing the ambipolar features of NT-FETs [97–102].

Single-electron transistors have recently been realized within individual metallic single-wall carbon nanotube molecules, featuring a short nanotube section created by inducing local barriers. Coulomb charging was observed at room temperature. Although these devices show promise, significant work is still required before they can find practical circuit applications. It remains uncertain whether they will or should replace today's scaled CMOS, as entirely new devices and circuits for new architectures may be more suitable for these molecular-based structures [103–107].

3.2.4 Quantum dots

In recent times, there has been a significant surge in interest in the self-assembled growth of quantum dots due to several inherent advantages. These include the ability to utilize lattice-mismatched materials without encountering misfit dislocations, the absence of surface damage resulting from physical/chemical patterning, and, importantly, the potential for cost-effective production. However, owing to the spontaneous growth nature, controlling the formation process, especially in terms of size and precise dot placement, has proven to be challenging. Despite these challenges, considerable efforts have been dedicated to achieving progress in effectively controlling both uniformity and spatial arrangement. An instance of self-assembled growth involves depositing one material with lattice mismatch (Ge) onto another (Si) using epitaxy. In the initial growth stage, it follows a layer-by-layer mode. Once surpassing the critical thickness, three-dimensional growth, known as Stranki–Krastanov growth, occurs to alleviate strain within the film [108–113]. These three-dimensional structures evolve into quantum dots, showcasing quantum confinement properties due to their dimensions. Numerous publications have detailed the growth and properties of self-assembled Ge dots, with recent advancements achieving uniformity in Ge dots. Figure 3.4 illustrates self-assembled Ge quantum dots on a Si substrate, where the dots have a base and height of 70 nm and 15 nm, respectively. The resulting dot height uniformity is approximately 3%. A more complex challenge for electronic applications involves placing dots in a controlled, organized pattern. Addressing this challenge, Ge quantum dots with a one-dimensional (1D) arrangement have been successfully demonstrated on Si material through selective growth on patterned Si substrates [114–117] (Table 3.1).

3.2.5 Nanosheets

The horizontally stacked gate-all-around (GAA) nanosheet structure has been shown for the first time to be a viable alternative to fin field-effect transistor (FinFET) in the 5 nm technology node and beyond [118]. This configuration not only offers a higher effective width per active footprint than FinFET, but it also performs better. In addition, it uses extreme ultraviolet (EUV) lithography, a less complex patterning technique.

Approximately at L_g = 12 nm, favorable electrostatics with 44/48nm CPP (Contacted Poly Pitch) ground rules are detected. In the study, several

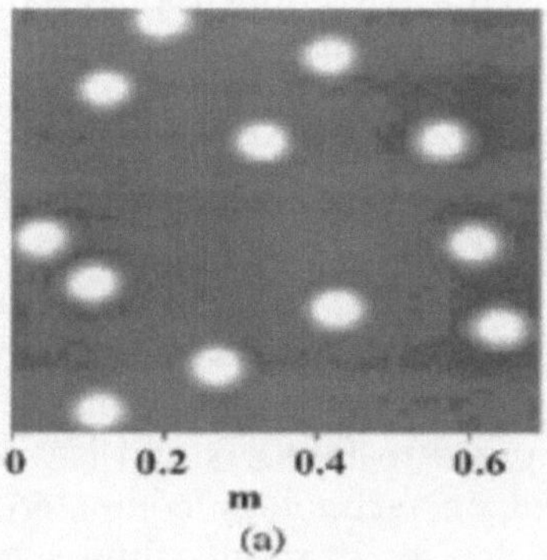

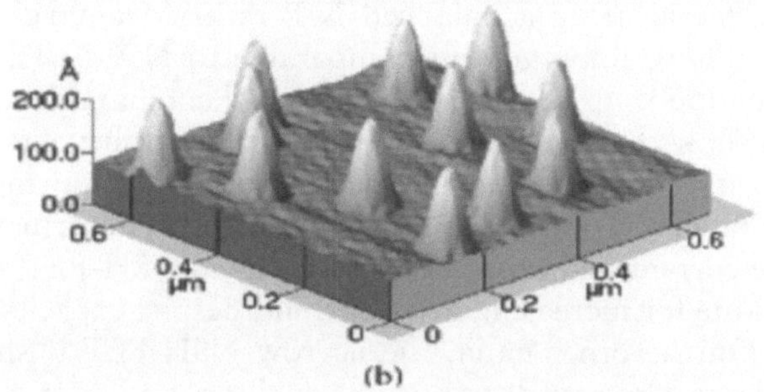

Figure 3.4 (a) Atomic force microscopic images in two dimensions and (b) top-down views depict uniform self-assembled Ge dots. The base dimensions and height of these dots measure 70 nm and 15 nm, respectively [2].

Table 3.1 Characteristics of quantum dots versus the size for a given Si sphere embedded in oxide [30]

Quantum-dot diameter (nm)	*Self-capacitance (aF)*	*Charging energy,* $e^2/2C$ *(eV)*	*Ground-state eigen-energy (eV)*	*Control gate capacitance,* C_{CG} *(aF)*	*Single electron* $\Delta V_T, e/C_{CG}(V)$
30	6.51	0.012	≈0.003	4.10	0.039
20	4.34	0.018	≈0.007	2.11	0.076
10	2.17	0.037	≈0.030	0.63	0.255
5	1.08	0.074	≈0.104	0.17	0.924
3	0.65	0.123	≈0.290	0.07	2.460
2	0.43	0.185	≈0.600	0.03	≥5
1	0.22	0.369	≈2.600	0.01	≥20

threshold voltages are applied in accordance with narrow sheet-to-sheet spacing for stacked sheets, and the work function metal (WFM) is substituted. Stiction in long-channel devices is successfully handled. Moreover, sub-sheet leakage control via dielectric isolation is demonstrated on a typical bulk substrate. Additionally, the efficiency of wrap around contact (WAC) in lowering extrinsic resistance is evaluated [4].

Nanosheets, categorized as two-dimensional nanomaterials with nanometer-scale thickness and micrometer-scale lateral dimensions, stand out for their distinctive properties, rendering them promising candidates for a wide array of applications in nanotechnology, electronics, energy storage, and beyond. This in-depth analysis explores the essential properties, manufacturing processes, and possible uses of nanosheets [119, 120].

The inherent performance, DC characteristics, and RO benchmarking of nanosheet field-effect transistors (NSH-FETs) have been thoroughly examined. Additionally, many device design options have been taken into consideration and compared to FinFETs and NW-FETs for the sub-7-nm technology node. Based on technology computer-aided design (TCAD), their intrinsic performance shows results similar to FinFETs at the same cross-section, emphasizing a balanced N/P ratio advantage. NSH width narrowing results in improvement comparable to NW-FETs, while NSH thickness reduction leads to electrostatic control comparable to FinFETs.

The results show that NSH-FETs have a number of benefits with regard to drive current and device design freedom without requiring drastic fin pitch scaling; however, optimization is still necessary to achieve a balance between drive current and capacitance. The broad NSH-FET shows promise as a substitute for increasing drive current, despite capacitance limiting its RO gain. On the other hand, the narrow NSH-FET is similar to the NW-FET and has a greater drive current, making it a better choice for circuits with high loads [5].

3.2.6 HEMT

Gallium nitride (GaN) high-electron-mobility transistors (HEMTs) are integral to modern power devices, offering advantages such as a wide band-gap, high electron mobility, and low thermal impedance. In power conditioners of electronic systems, GaN HEMTs control and manage electric currents efficiently (Figure 3.5).

The review emphasizes the significance of achieving low resistance (RON) and minimal leakage current for optimal performance. Specifically, it focuses on normally off GaN HEMTs, detailing aspects like the insulator's role and metal performance on p-GaN. The technology addresses reliability concerns, including drain, gate lag, and current collapse. Overall, GaN HEMTs play a vital role in enhancing the efficiency and reliability of power electronics, contributing to the progress of electrical power applications [12].

Modulation-doped field-effect transistor (MODFET), or HEMT, is a fascinating semiconductor device. It utilizes heterojunctions in group III–V compounds for effective band-gap engineering. In the context of a MOSFET, the challenge lies in balancing high transconductance with the need for high channel conductivity. While increased doping can enhance conductivity, it often leads to mobility degradation due to impurity scattering.

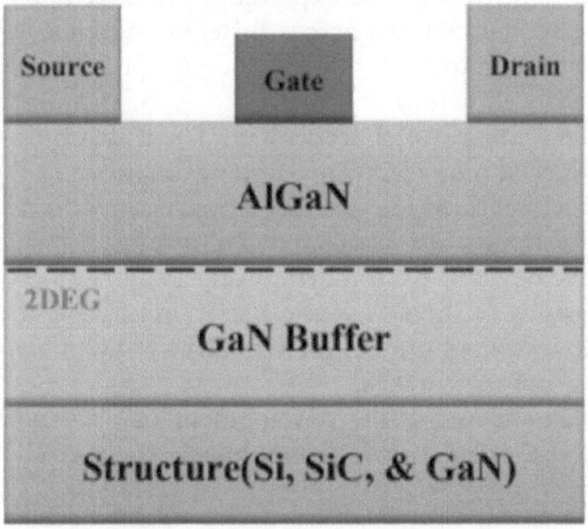

Figure 3.5 The dashed line displays the AlGaN/GaN HEMT arrangement and the 2DEG formed in the AlGaN/GaN contact [12].

The clever introduction of modulation doping in MODFET addresses this challenge. This technique involves incorporating an undoped well, such as GaAs, between wider band-gap-doped barriers (e.g., AlGaAs). Electrons from the doped barriers fall into the well, creating conductive GaAs with reduced impurity scattering. Higher mobility is facilitated by the creation of a 2D electron gas (2DEG).

The advantages of MODFET include lowering ionized impurity scattering, preserving high mobility, and reaching a high electron concentration. Under typical circumstances, the device behaves very much like a MOSFET since the AlGaAs layer is completely depleted. In comparison to Si MOSFETs, HEMT outperforms with higher mobility and maximum electron velocity in GaAs.

Additionally, there is potential for other materials like the InGaAsP/InP system, offering alternatives to address certain drawbacks associated with AlxGa1-xAs. Pseudomorphic HEMTs, featuring slight lattice mismatch, are capable of achieving high performance. The device is also known as a separately doped FET (SEDFET) or two-dimensional electron gas FET (2DEG FET), highlighting its special qualities and design concepts [29].

3.3 COMPARISON

Computational paradigms are classified based on key features crucial to nanoelectronics, including wiring, parallelism, and data transfer, as depicted in Figure 3.6. Notably, the distinction between physical wiring and non-physical data transfer is explored, with an emphasis on the efficiency of basic circuit functions across different systems. Artificial neural networks and cellular automata, which process input data entirely in parallel, are compared with traditional microprocessor-memory design, which places less emphasis on parallel processing. The domain of highly parallel and locally networked nanoelectronic systems is shown in Figure 3.6 [11].

Appropriate architectures are discussed indirectly within the framework of the MEL-ARI nanoscale integrated circuit cluster, which is centered around innovative devices. An important takeaway from the review is that every design depends on how well the basic circuit operations are implemented. Notable instances are Teramac and the RAW machine, which use programmable crossbar switches.

These examples encourage the development of prototype implementations that make use of nanoscale technologies, such as SET or RTD. We

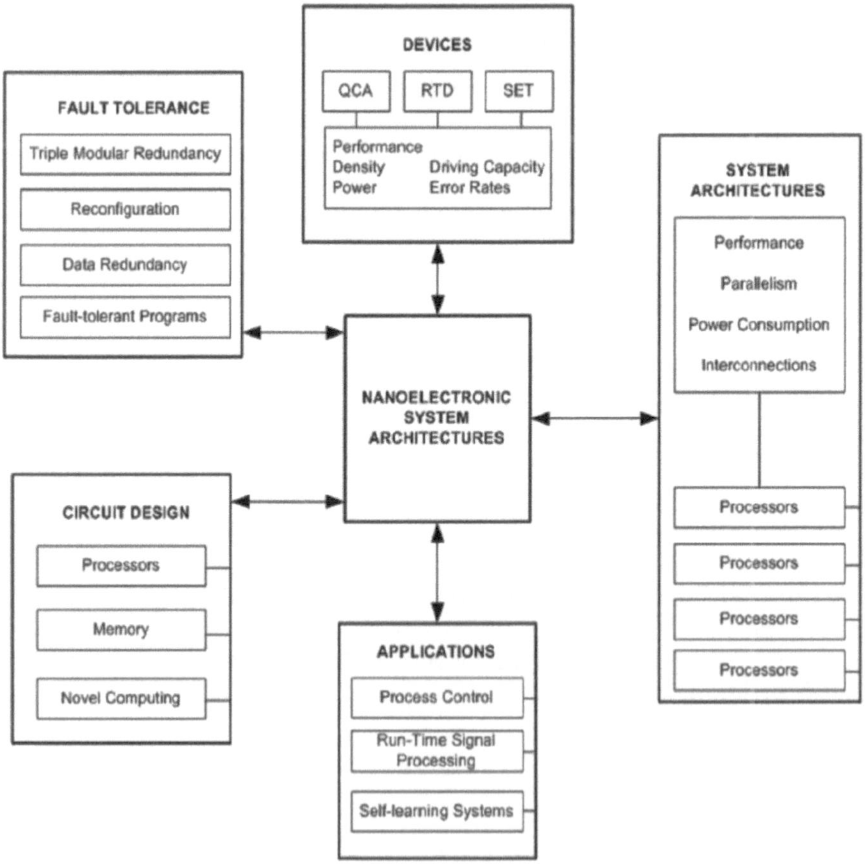

Figure 3.6 Aspects of different architectures for nanoelectronic systems [Fountain 1998b] [11].

are able to extrapolate the performance of larger circuits and systems by integrating the results of simulations and experiments. The review foresees the increasing significance of intelligent design tools within a hierarchical design flow that is defined by interfaces with design rules and standardized functional blocks as we face increased variety and design complexity. This investigation advances the field of nanoelectronic structures and paves the way for more developments (Table 3.2 and Figure 3.7).

3.4 FUTURE SCOPE

The trajectory of transistor scaling faces impending limitations, including rising power and capital costs. Nanoelectronics offer a promising avenue for integrated circuit (IC) miniaturization, with demonstrated capabilities in fabricating components like wires and molecular switches. However, whether they will replace or complement conventional ICs remains uncertain. The integration of nanoscale devices into architectures, while making progress, poses challenges in defect management and scalability, with array-based structures showing promise.

Research indicates that fault tolerance is crucial, demanding economical defect detection techniques. While nanoelectronics share similarities with current CAD flows, challenges include unique chip deployment models. Progress in fabricating molecules, silicon nanowires, and carbon nanotubes is notable, but reliability and ongoing enhancements are vital. The prospect

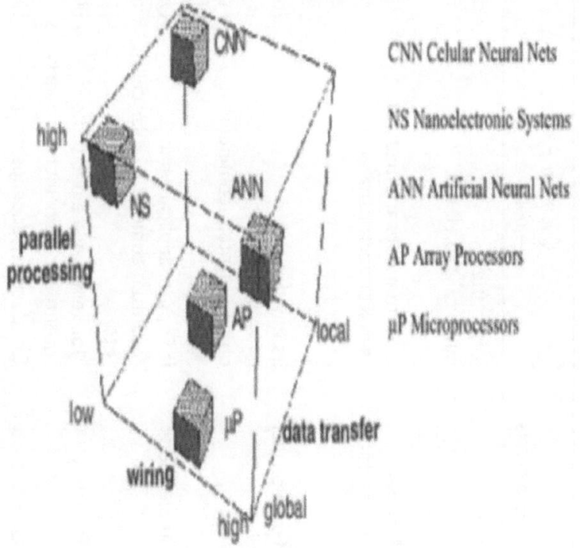

Figure 3.7 Comparison between different architectures with respect to data transfer, wiring, and parallel processing [11].

Table 3.2 Comparison between quantum dots, tunneling devices, nanosheets, and nanoelectronic tubes on various parameters

S. No.	*Parameters*	*Quantum dots*	*Tunneling devices*	*Nanosheets*	*Nanoelectronic tubes*
1.	Size	It is in the range of 2 to 10 nm.	Its range is from nanoscale to microscale.	It has size that of nanoscale.	Its range lies from nanoscale to that of microscale.
2.	Material composition	Quantum dots are made of semiconductor materials.	They are made of various materials which include semiconductors and insulators.	They are made of carbon-based materials.	It is made of semiconductors or carbon-based materials.
3.	Functionality	Its functional ity is based on the quantum confinement effect.	Its functionality is based on the electron tunneling effect.	It exhibits 2D electronic structure.	It promotes electron transport within tube.
4.	Applications	It finds its application in quantum computing, solar cells, etc.	Used in tunnel diodes, memory devices, etc.	Used in flexible electronic sensors, energy storage devices.	Find its application in nanoelectronics, sensors, etc.
5.	Flexibility	Not very flexible due to its quantum confinement property.	Flexible	Flexible	Flexible
6.	Conductivity	Conductivity depends on its size and the material it is made of.	Dependent on the tunneling effect.	Depends on the material it is made of.	It can exhibit high conductivity.

of integrating 1012 devices per chip fuels optimism, but challenges persist, including reducing defect levels and developing complementary software tools. With the anticipated end of scaling for lithography-based circuits in just over a decade, addressing these challenges within the next ten years is imperative for the realization of nanoelectronics as a transformative technology in the realm of integrated circuits [5].

The introduction of new interconnect technologies heralds an extension of Moore's law, ensuring the effective utilization of forthcoming transistor-scaling advancements rather than allowing them to diminish due to interconnect limitations.

Future generations of silicon technology are expected to be significantly shaped by these creative connectivity options. When copper interconnects replaced aluminum interconnects in advanced silicon chips in the late 1990s, silicon technology began to pivot from a transistor-centric to an interconnect-centric orientation [9, 10]. The chip industry is quite interested in the length of this transitional phase, which is effectively a forerunner to the interconnect age [8].

3.4.1 Opportunities

Smaller footprint: When fabrication is scaled down to nanodimensions, nanodevices exhibit a smaller footprint, allowing for higher density information integration in a smaller integrated circuit. This property is advantageous for securing resource-constrained computing platforms like RFID tags.

Low-cost fabrication: Numerous nanoscale devices have low fabrication complexity and are compatible with current CMOS fabrication techniques. Lower fabrication costs are anticipated in the long run when these devices are manufactured in large quantities.

C2C variations: By utilizing the distinct C2C variations found in memristors, reconfigurable PUFs (rPUFs) can be created without the need for additional circuitry. Refreshable control and relay panels (CRPs) maintain their initial level of protection, and the reconfiguration operation is irreversible, a key requirement for secure rPUF design [14, 15].

Programming sensitivity: Mobility load balancing (MLB) capability and programming sensitivity in phase-change memory (PCM) and spin-transfer torque magnetic random access memory (STT-MRAM) provide opportunities for building rPUFs with enhanced security properties [16–18].

Nonvolatility: Nonvolatility enables reliable regeneration of physical unclonable function (PUF) responses and functions as a key generator in addition to buffer storage, making it a more affordable option than traditional memory-based PUFs.

Bidirectionality: Two-terminal nanodevices are bidirectional, facilitating the design of novel bidirectional nano-PPUFs. [19–21].

Formation process: The one-time irreversible formation process in memristors contributes to PUF security by eliminating issues related to untrusted manufacturers, ensuring tamper detection [22].

Security enhancement: PUF structures built on cutting-edge nanoelectronic devices could solve security issues with PUFs that are now silicon-based. For example, problems with SRAM PUFs have been shown, and new nanodevices may provide better protection against invasive techniques and fault injection assaults [24].

3.4.2 Challenges

Experimental validation: Most PUF structures lack experimental validation; limited studies provide partial experimental results [14, 15].

PUF performance evaluation: A comprehensive evaluation, considering uniqueness, randomness, error rates, and resilience, is needed. Some evaluations rely on simplistic models, neglecting factors like C2C variations [25, 26].

Standard models: The absence of industry-standard models makes simulation results speculative. Collaboration between security designers and device engineers is vital for practical applications.

Reliability: Memristor-based PPUFs face reliability issues; responses may differ between physical devices and simulators. Traditional reliability techniques may not apply, and repeated programming can affect reliability. The lack of temperature-dependent memristor models adds uncertainty to reliability results. Addressing these challenges is crucial for practical PUF applications [27].

3.5 CONCLUSION

It has been stated that any "new switch's" ability to achieve lower dissipation in device switching is essential to its success. To tackle this problem, two different strategies are being used. Less energy is stored in the device in the first technique, which is especially important for conventional FET logic, where the rapid application of continuous power supply potentials causes resistive energy loss equivalent to the stored energy at each switching event. The fact that the stored energy is $(1/2)CV^2$ makes lowering the power source voltage extremely desirable. However, in the sub-threshold current domain of common FETs, minimum supply voltages are limited by kT/e scaling. Consequently, improving the turn-on nonlinearity necessitates changing the device physics in order to achieve quick logic at very low voltages. A few distinct physical representations have been identified: the tunnel-FET symbolizes internal transduction, the spin FET depicts energy filtering, and the ferroelectric-gate FET indicates internal voltage step-up.

Using energy-conserving logic circuits and near-adiabatic switching protocols is the second method of lowering power dissipation. Although these circuits and protocols can be used with traditional FETs, they have not been able to compete with conventional methods for low-power FET logic. Novel apparatuses could provide essential benefits for energy-efficient computing, such as quicker transitioning near the adiabatic limit. While conceptual approaches for spin-based logic are promising, efforts to find devices suitable for energy-conserving logic are still in their infancy [13].

REFERENCES

1. G. Pandey, D. Rawtani and Y. K. Agrawal, "Aspects of nanoelectronics in materials development", *InTech eBooks*, 27 July 2016. https://doi.org/10.5772/64414.
2. Wang, K. L. "Issues of nanoelectronics: a possible roadmap," *Journal of Nanoscience and Nanotechnology*, vol. 2, 1 July 2002. https://doi.org/10.1166/jnn.2002.115.
3. L. Esaki and R. Tsu, "Superlattice and negative differential conductivity in semiconductors", *IBM Journal of Research and Development*, vol. 14, no. 1, pp. 61–65, Jan. 1970, https://doi.org/10.1147/rd.141.0061.
4. N. Loubet et al., "Stacked nanosheet gate-all-around transistor to enable scaling beyond FinFET", *2017 Symposium on VLSI Technology*, Kyoto, Japan, 2017, pp. T230–T231. https://doi.org/10.23919/VLSIT.2017.7998183.
5. D. Jang et al., "Device exploration of nanosheet transistors for sub-7-nm technology node", *IEEE Transactions on Electron Devices*, vol. 64, no. 6, pp. 2707–2713, June 2017. https://doi.org/10.1109/TED.2017.2695455.
6. Y. Yorozu, M. Hirano, K. Oka and Y. Tagawa, "Electron spectroscopy studies on magneto-optical media and plastic substrate interface", *IEEE Translation Journal on Magnetics in Japan*, vol. 2, pp. 740–741, Aug. 1987 [*Digests 9th Annual Conf. Magnetics Japan*, p. 301, 1982].
7. Shore, K. A. "Introduction to nanoelectronics, by V.V. Mitin, V.A. Kochelap and M.A. Stroscio", *Contemporary Physics*, vol. 51, no. 3, pp. 285–286, May 2010. https://doi.org/10.1080/00107510903198273.
8. J. D. Meindl, "Beyond Moore's law: the interconnect era", *Computing in Science and Engineering*, vol. 5, no. 1, pp. 20–24, Jan.–Feb. 2003. https://doi.org/10.1109/MCISE.2003.1166548.
9. J. D. Meindl and J. A. Davis, "The fundamental limit on binary switching energy for terascale integration (TSI)", *IEEE Journal of Solid-State Circuits*, vol. 35, no. 10, pp. 1515–1516, Oct. 2000. https://doi.org/10.1109/4.871332.
10. Qiang Chen, B. Agrawal and J. D. Meindl, "A comprehensive analytical subthreshold swing (S) model for double-gate MOSFETs", *IEEE Transactions on Electron Devices*, vol. 49, no. 6, pp. 1086–1090, June 2002. https://doi.org/10.1109/TED.2002.1003757.
11. MELARI NANO, "Technology roadmap for nanoelectronics" [Online]. Available at: https://apps.dtic.mil/sti/tr/pdf/ADA572139.pdf, accessed 2004.

12. N. Islam, M. F. P. Mohamed, M. F. A. J. Khan, S. Falina, H. Kawarada and M. Syamsul, "Reliability, applications and challenges of GaN HEMT technology for modern power devices: a review", *Crystals*, vol. 12, no. 11, p. 1581, 2022. https://doi.org/10.3390/cryst12111581.
13. T. N. Theis and P. M. Solomon, "In quest of the "Next Switch": prospects for greatly reduced power dissipation in a successor to the silicon field-effect transistor", *Proceedings of the IEEE*, vol. 98, no. 12, pp. 2005–2014, Dec. 2010. https://doi.org/10.1109/JPROC.2010.2066531.
14. A. Chen, "Utilizing the variability of resistive random access memory to implement reconfigurable physical unclonable functions", *IEEE Electron Device Letters*, vol. 36, no. 2, pp. 138–140, Feb. 2015.
15. Y. Gao, D. C. Ranasinghe, S. F. Al-Sarawi, O. Kavehei and D. Abbott, "Memristive crypto primitive for building highly secure physical unclonable functions", *Scientific Reports*, vol. 5, pp. 26239669–26239675, Aug. 2015.
16. L. Zhang, X. Fong, C.-H. Chang, Z. H. Kong and K. Roy, "Highly reliable memory-based physical unclonable function using spin-transfer torque MRAM", *Proc. IEEE Int. Symp. Circuits Syst. (ISCAS)*, pp. 2169–2172, Jun. 2014.
17. L. Zhang, Z. H. Kong, C.-H. Chang, A. Cabrini and G. Torelli, "Exploiting process variations and programming sensitivity of phase change memory for reconfigurable physical unclonable functions", *IEEE Transactions on Information Forensics and Security*, vol. 9, no. 6, pp. 921–932, Jun. 2014.
18. J. Rajendran, G. S. Rose, R. Karri and M. Potkonjak, "Nano-PPUF: a memristor-based security primitive", *Proc. IEEE Comput. Soc. Annu. Symp. VLSI (ISVLSI)*, pp. 84–87, Aug. 2012.
19. J. B. Wendt and M. Potkonjak, "Nanotechnology-based trusted remote sensing", *Proc. IEEE Sensors*, pp. 1213–1216, Oct. 2011.
20. J. B. Wendt and M. Potkonjak, "The bidirectional polyomino partitioned PPUF as a hardware security primitive", *Proc. IEEE Global Conf. Signal Inf. Process. (GlobalSIP)*, pp. 257–260, Dec. 2013.
21. J. P. Strachan, D. B. Strukov, J. Borghetti, J. J. Yang, G. Medeiros Ribeir and R. S. Williams, "The switching location of a bipolar memristor: chemical thermal and structural mapping", *Nanotechnology*, vol. 22, no. 25, pp.1–8, May 2011.
22. Y. Oren, A. -R. Sadeghi and C. Wachsmann, "On the effectiveness of the remanence decay side-channel to clone memory-based PUFs", in *Cryptographic Hardware and Embedded Systems (CHES)*, Berlin, Germany: Springer, pp. 107–125, 2013.
23. C. Helfmeier, C. Boit, D. Nedospasov and J.-P. Seifert, "Cloning physically unclonable functions", *Proc. IEEE Int. Symp. Hardw.-Oriented Secur. Trust (HOST)*, pp.1–6, Jun. 2013.
24. Y. Gao, D. C. Ranasinghe, S. F. Al-Sarawi, O. Kavehei and D. Abbott, "Emerging physical unclonable functions with nanotechnology", *IEEE Access*, vol. 4, pp. 61–80, 2016. https://doi.org/10.1109/ACCESS.2015.2503432.
25. G. S. Rose, N. McDonald, L. -K. Yan and B. Wysocki, "A writetime based memristive PUF for hardware security applications", *Proc. IEEE/ACM Int. Conf. Comput.-Aided Design (ICCAD)*, pp. 830–833, Nov. 2013.

26. P. Koeberl, U. Kocabas and A.-R. Sadeghi, "Memristor PUFs: a new generation of memory-based physically unclonable functions", *Proc. Conf. Design Autom. Test Eur.*, pp. 428–431, Mar. 2013.
27. G. S. Rose and C. A. Meade, "Performance analysis of a memristive crossbar PUF design", *Proc. 52nd ACM/EDAC/IEEE Design Autom. Conf. (DAC)*, pp. 1–6, Jun. 2015.
28. J. Park and C. Shin, "Tunnel field-effect transistor with segmented channel", *IEEE Journal of the Electron Devices Society*, vol. 7, pp. 621–625, 2019. https://doi.org/10.1109/JEDS.2019.2919331.
29. B. G. Streetman and S. Banerjee, *Solid State Electronic Devices*, Prentice Hall, 2015. [Online]. Available at http://books.google.ie/books?id=-nnzn-QEACAAJ&dq=ben+g+streetman&hl=&cd=1&source=gbs_api
30. J. J. Welser, S. Tiwari, S. Rishton, K. Y. Lee and Y. Lee, "Room temperature operation of a quantum-dot flash memory," *IEEE Electron Device Letters*, vol. 18, p. 278, 1997.
31. Anil Kumar Bhardwaj, Sumeet Gupta, B. Raj and Amandeep Singh, "Impact of double gate geometry on the performance of carbon nanotube field effect transistor structures for low power digital design", *Computational and Theoretical Nanoscience*, vol. 16, pp. 1813–1820, 2019.
32. Neeraj Jain and B. Raj, "Thermal stability analysis and performance exploration of asymmetrical dual-k underlap spacer (ADKUS) SOI FinFET for security and privacy applications", *Indian Journal of Pure & Applied Physics (IJPAP)*, vol. 57, pp. 352–360, May 2019.
33. Neeraj Jain and Balwinder Raj, "Dual-k spacer region variation at the drain side of asymmetric SOI FinFET structure: performance analysis towards the analog/RF design applications", *Journal of Nanoelectronics and Optoelectronics*, vol. 14, pp. 349–359, March 2019.
34. Jeetendra Singh, Sanjeev Sharma, and B. Raj, "Mamta Khosla analysis of barrier layer thickness on performance of In1-xGaxAs based gate stack cylindrical gate nanowire MOSFET", *Journal of Nanoelectronics and Optoelectronics*, vol. 13, PP 1473–1477, Oct. 2018.
35. Neeraj Jain and B. Raj, "Analysis and performance exploration of high-k SOI FinFETs over the conventional low-k SOI FinFET toward analog/RF design", *Journal of Semiconductors*, vol. 39, no. 12, pp. 124002-1–7, Dec. 2018.
36. Candy Goyal, Jagpal Singh Ubhi and B. Raj, "A reliable leakage reduction technique for approximate full adder with reduced ground bounce noise", *Journal of Mathematical Problems in Engineering*, vol. 2018, Article ID 3501041, p. 16, 15 Oct. 2018.
37. Jeetendra Singh Anuradha, B. Raj and Mamta Khosla, "Design and performance analysis of nano-scale memristor-based nonvolatile SRAM", *Journal of Sensor Letter*, vol. 16, pp. 798–805, Oct. 2018.
38. Girish Wadhwa and B. Raj, "Parametric variation analysis of charge-plasma-based dielectric modulated JLTFET for biosensor application", *IEEE Sensor Journal*, vol. 18, no. 15, pp. 6070–6077, 1 Aug. 2018.
39. Jeetendra Singh and B. Raj, "Comparative analysis of memristor models for memories design", *Journal of Semiconductors*, vol. 39, no. 7, pp. 074006-1–12, July 2018.

40. Divya Yadav, Shailesh Singh Chouhan, Santosh Kumar Vishvakarma and B. Raj, "Application specific microcontroller design for IoT based WSN", *Sensor Letter*, vol. 16, pp. 374–385, May 2018.
41. Gurmohan Singh, R. K. Sarin and B. Raj, "Fault-tolerant design and analysis of quantum-dot cellular automata based circuits", *IEEE/IET Circuits, Devices & Systems*, vol. 12, pp. 638–64, 2018.
42. Jeetendra Singh and B. Raj, "Modeling of mean barrier height levying various image forces of metal insulator metal structure to enhance the performance of conductive filament based memristor model", *IEEE Nanotechnology*, vol. 17, no. 2, pp. 268–267, March 2018.
43. Aakash Jain, Sanjeev Sharma and B. Raj, "Analysis of triple metal surrounding gate (TM-SG) III-V nanowire MOSFET for photosensing application", *Opto-Electronics Journal*, vol. 26, Issue 2, pp. 141–148, May 2018.
44. Aakash Jain, Sanjeev Sharma and B. Raj, "Design and analysis of high sensitivity photosensor using cylindrical surrounding gate MOSFET for low power sensor applications", *Engineering Science and Technology, an International Journal*, vol. 19, no. 4, pp. 1864–1870, Dec. 2016.
45. G. Saiphani Kumar, Amandeep Singh and B. Raj, "Design and analysis of gate all around CNTFET based SRAM cell design", *Journal of Computational Electronics*, vol. 17, no. 1, pp 138–145, March 2018.
46. Gurinder pal Singh, B. S. Sohi and Balwinder Raj, "Material properties analysis of graphene base transistor (GBT) for VLSI analog circuits", *Indian Journal of Pure & Applied Physics (IJPAP)*, vol. 55, pp. 896–902, Dec. 2017.
47. Sunil Kumar and B. Raj, "Estimation of stability and performance metric for inward access transistor based 6T SRAM cell design using n-type/p-type DMDG-GDOV TFET", *IEEE VLSI Circuits and Systems Letter*, vol. 3, no. 2, pp. 25–39, June 2017.
48. Shashikant Sharma, Anjan Kumar, Manisha Pattanaik and B. Raj, "Forward body biased multimode multi-threshold CMOS technique for ground bounce noise reduction in static CMOS adders", *International Journal of Information and Electronics Engineering*, vol. 3, no. 3, pp. 567–572, 2013.
49. Hamendra Singh, Pankaj Kumar and Balwinder Raj, "Performance analysis of majority gate SET based 1-bit full adder", *International Journal of Computer and Communication Engineering* (IJCCE), vol. 2, no. 4, pp. 1–8, 2013.
50. Anil Kumar Bhardwaj, Sumeet Gupta and B. Raj, "Investigation of parameters for schottky barrier (SB) height for schottky barrier based carbon nanotube field effect transistor device", *Journal of Nanoelectronics and Optoelectronics*, vol. 15, pp. 783–791, July 2020.
51. Priya Bansal and B. Raj, "Memristor: a versatile nonlinear model for dopant drift and boundary issues", *Journal of Computational and Theoretical Nanoscience*, vol. 14, no. 5, pp. 2319–2325, May 2017.
52. Neeraj Jain and B. Raj, "An analog and digital design perspective comprehensive approach on Fin-FET (Fin-Field Effect transistor) technology: a review", *Reviews in Advanced Sciences and Engineering*, vol. 5, pp. 1–14, 2016.
53. Balwinder Raj, A. K. Saxena and S. Dasgupta, "Analytical modeling for the estimation of leakage current and subthreshold swing factor of nanoscale double gate FinFET device", *Microelectronics International*, vol. 26, pp. 53–63, 2009.

54. Shailendra Singh Soniya, Girish Wadhwa and Balwinder Raj, "An analytical modeling for dual source vertical tunnel field effect transistor", *International Journal of Recent Technology and Engineering*, vol. 8, no. 2, pp. 603–608, July 2019.
55. Shailendra Singh and B. Raj, "Design and analysis of hetrojunction vertical T-shaped tunnel field effect transistor", *Journal of Electronics Material*, vol. 48, no. 10, pp 6253–6260, Oct. 2019.
56. Candy Goyal, Jagpal Singh Ubhi and B. Raj, "A low leakage CNTFET based inexact full adder for low power image processing applications", *International Journal of Circuit Theory and Applications*, vol. 47, no. 9, pp. 1446–1458, Sep. 2019.
57. B. Raj, A. K. Saxena and S. Dasgupta, "A compact drain current and threshold voltage quantum mechanical analytical modeling for FinFETs", *Journal of Nanoelectronics and Optoelectronics (JNO)*, vol. 3, no. 2, pp. 163–170, 2008.
58. Girish Wadhwa and B. Raj, "An analytical modeling of charge plasma based tunnel field effect transistor with impacts of gate underlap region", *Superlattices and Microstructures*, vol. 142, pp. 106512, June 2020.
59. Shailendra Singh and B. Raj "Modeling and simulation analysis of SiGe hetrojunction double gatevertical t-shaped tunnel FET", *Superlattices and Microstructures*, vol. 142, pp. 106496, June 2020.
60. Neeraj Jain and Balwinder Raj, "Device and circuit co-design perspective comprehensive approach on FinFET technology - a review", *Journal of Electron Devices*, vol. 23, no. 1, pp 1890–1901, 2016.
61. Sunil Kumar and B. Raj, "Analysis of I_{ON} and Ambipolar current for dual-material gate-drain overlapped DG-TFET", *Journal of Nanoelectronics and Optoelectronics*, vol. 11, pp. 323–333, June 2016.
62. Naveed Anjum, Tarun Bali and B. Raj, "Design and simulation of handwritten multiscript character recognition", *International Journal of Advanced Research in Computer and Communication Engineering*, vol. 2, no. 7, pp. 2544–2549, July 2013.
63. Karmjit Singh and B. Raj, "Performance and analysis of temperature dependent multi-walled carbon nanotubes as global Interconnects at different technology nodes", *Journal of Computational Electronics*, vol. 14 no. 2, pp. 469–476, June 2015.
64. Sunil Kumar and B. Raj, "Compact channel potential analytical Modeling of DG-TFET based on evanescent–mode Approach", *Journal of Computational Electronics*, vol. 14, no. 2, pp. 820–827, July 2015.
65. Karmjit Singh and B. Raj, "Temperature dependent modeling and performance evaluation of multi-walled CNT and single-walled CNT as global interconnects", *Journal of Electronic Materials*, vol. 44, no. 12, pp 4825–4835, Dec. 2015.
66. V. K. Sharma, M. Pattanaik and B. Raj, "INDEP approach for leakage reduction in nanoscale CMOS circuits", *International Journal of Electronics*, vol. 102, no. 2, pp. 200–215, 2014.

67. Karmjit Singh and B. Raj, "Influence of temperature on MWCNT bundle, SWCNT bundle and copper interconnects for nanoscaled technology nodes", *Journal of Materials Science: Materials in Electronics*, vol. 26, no. 8, pp. 6134–6142, 2015.
68. Naveed Anjum, Tarun Bali and B. Raj, "Design and simulation of handwritten Gurumukhi and Devanagri numerical recognition", *International Journal of Computer Applications*, vol. 73, no. 12, pp. 16–21, 2013.
69. Khandelwal, S.,V. Gupta, B. Raj and R. D. Gupta, "Process variability aware low leakage reliable nano scale DG-FinFET SRAM cell design technique", *Journal of Nanoelectronics and Optoelectronics*, vol. 10, no. 6, pp. 810–817, Dec. 2015.
70. V. K. Sharma, M. Pattanaik and B. Raj, "ONOFIC approach: low power high speed nanoscale VLSI circuits design", *International Journal of Electronics*, vol. 101, no. 1, pp. 61–73, 2014.
71. S. Khandelwal, Balwinder Raj and R. D. Gupta, "FinFET based 6T SRAM cell design: analysis of performance metric, process variation and temperature effect", *Journal of Computational and Theoretical Nanoscience*, vol. 12, pp. 2500–2506, 2015.
72. Sumit Singh, Shekhar Yadav, Jagdeep Rahul, Anurag Srivastava, and B. Raj, "Impact of HfO_2 in graded channel dual insulator double gate MOSFET", *Journal of Computational and Theoretical Nanoscience*, vol. 12, no. 6, pp. 950–953, April 2015.
73. Vijay Kumar Sharma, Manisha Pattanaik and B. Raj, "PVT variations aware low leakage INDEP approach for nanoscale CMOS circuits", *Microelectronics Reliability*, vol. 54, pp 90–99, 2014.
74. B. Raj, A. K. Saxena and S. Dasgupta, "Quantum mechanical analytical modeling of nanoscale DG FinFET: evaluation of potential, threshold voltage and source/drain resistance ", *Elsevier's Journal of Material Science in Semiconductor Processing*, vol. 16, no. 4, pp. 1131–1137, 2013.
75. Maisagalla Gopal, Siva Sankar D. Prasad and Balwinder Raj, "8T SRAM cell design for dynamic and leakage power reduction", *International Journal of Computer Applications*, vol. 71, no. 9, pp. 43–48, June 2013.
76. Manisha Pattanaik, B. Raj, Shashikant Sharma and Anjan Kumar, "Diode based trimode multi-threshold CMOS technique for ground bounce noise reduction in static CMOS adders", *Advanced Materials Research*, vol. 548, pp 885–889, 2012.
77. Balwinder Raj, A. K. Saxena and S. Dasgupta, "Nanoscale FinFET based SRAM cell design: analysis of performance metric, process variation, underlapped FinFET and temperature effect", *IEEE Circuits and System Magazine*, vol. 11, no. 2, pp. 38–50, 2011.
78. V. K. Sharma, M. Pattanaik and Balwinder Raj, "Leakage current ONOFIC approach for deep submicron VLSI circuit design", *International Journal of Electrical, Computer, Electronics and Communication Engineering*, vol. 7, no. 4, pp. 239–244, 2013.
79. Parminder Kaur, Sandeep Singh Gill and B. Raj, "Comparative analysis of OFETs materials and devices for sensor applications", *Journal of Silicon*, vol. 14, pp. 4463–4471, 2022.

80. Sanjeev Kumar Sharma, Parveen Kumar, Balwant Raj and B. Raj, "$In_{1-x}Ga_x$as double metal gate-stacking cylindrical nanowire MOSFET for highly sensitive photo detector", *Journal of Silicon*, vol. 14, pp. 3535–3541, 2022.
81. B. Raj, A. K. Saxena and S. Dasgupta, "Analytical modeling of quasi planar nanoscale double gate FinFET with source/drain resistance and field dependent carrier mobility: a quantum mechanical study", *Journal of Computer*, vol. 4, no. 9, pp. 1–8, 2009.
82. S. Bhushan, S. Khandelwal and B. Raj, "Analyzing different mode FinFET based memory cell at different power supply for leakage reduction", *Seventh International Conference on Bio-Inspired Computing: Theories and Application, (BIC-TA 2012)*, 2012 and *Advances in Intelligent Systems and Computing*, vol. 202, pp. 89–100, 2013.
83. Jeetendra Singh and B. Raj, "Temperature dependent analytical modeling and simulations of nanoscale memristor", *Engineering Science and Technology, an International Journal*, vol. 21, pp. 862–868, Oct. 2018.
84. Shradhya Singh, Shashi Bala, Balwant Raj and B. Raj, "Improved sensitivity of dielectric modulated junctionless transistor for nanoscale biosensor design", *Sensor Letter*, vol. 18, pp. 328–333, April 2020.
85. Vivek Kumar, Santosh Kumar Vishvakarma and B. Raj, "Design and performance analysis of ASIC for IoT applications", *Sensor Letter ASP*, vol. 18, pp. 31–38, Jan. 2020.
86. Akanksha Jaiswal, R. K. Sarin, B. Raj and Shikha Sukhija, "A novel circular slotted microstrip-fed patch antenna with three triangle shape defected ground structure for multiband applications", *Advanced Electromagnetic*, vol. 7, no. 3, pp. 56–63, Aug. 2018.
87. Girish Wadhwa and B. Raj, "Label free detection of biomolecules using charge-plasma-based gate underlap dielectric modulated junctionless TFET", *Journal of Electronic Materials*, vol. 47, no. 8, pp. 4683–4693, Aug. 2018.
88. Gurmohan Singh, R. K. Sarin and B. Raj, "Design and performance analysis of a new efficient coplanar quantum-dot cellular automata adder", *Indian Journal of Pure & Applied Physics*, vol. 55, pp. 97–103, Feb. 2017.
89. Parminder Kaur, Vikas Pandey and B. Raj, "Comparative study of efficient design, control and monitoring of solar power using IoT", *Sensor Letter*, vol. 18, pp. 419–426, May 2020.
90. Anil Kumar Bhardwaj, Sumeet Gupta and B. Raj, "Development & analysis of compact model for double gate Schottky barrier CNTFET", *Journal of Nanoelectronics and Optoelectronics*, vol. 15, pp. 1199–1208, Aug. 2020.
91. Girish Wadhwa, Priyanka Kamboj, Jeetendra Singh and B. Raj, "Design and investigation of junctionless DGTFET for biological molecule recognition", *Transactions on Electrical and Electronic Materials*, vol. 22, pp. 282–289, 2021.
92. Sachin Kumar Verma, Shailendra Singh, Girish Wadhwa and B. Raj, "Detection of biomolecules using charge-plasma based gate underlap dielectric modulated dopingless TFET", *Transactions on Electrical and Electronic Materials*, vol. 21, pp. 528–535, June 2020.

93. Neeraj Jain and B. Raj, "Impact of underlap spacer region variation on Electrostatic and analog/RF performance of symmetrical high-k SOI FinFET at 20 nm channel length", *Journal of Semiconductors*, vol. 38, no. 12, pp. 122002, Dec. 2017.
94. Shailendra Singh and B. Raj, "Analytical modeling and simulation analysis of T-shaped III-V heterojunction vertical T-FET", *Superlattices and Microstructures*, vol. 147, pp. 106717, Nov. 2020.
95. Gurmohan Singh, R. K. Sarin and B. Raj, "Design and analysis of area efficient QCA based reversible logic gates", *Journal of Microprocessors and Microsystems*, vol. 52, pp. 59–68, May 2017.
96. Shailendra Singh and B. Raj, "Parametric variation analysis on hetero-junction vertical t-shape TFET for supressing ambipolar conduction", *Indian Journal of Pure and Applied Physics*, vol. 58, pp. 478–485, June 2020.
97. Shailendra Singh Soniya, Girish Wadhwa and Balwinder Raj, "Design and analysis of dual source vertical tunnel field effect transistor for high performance", *Transactions on Electrical and Electronics Materials*, vol. 21, pp. 74–82, Oct. 2019.
98. Manjit Kaur, Neena Gupta, Sanjeev Kumar, B. Raj and Arun Kumar Singh, "RF performance analysis of intercalated graphene nanoribbon based global level interconnects", *Journal of Computational Electronics*, vol. 19, pp. 1002–1013, June 2020.
99. Girish Wadhwa and B. Raj, "Design and performance analysis of junctionless TFET biosensor for high sensitivity", *IEEE Nanotechnology*, vol. 18, pp. 567–574, 2019.
100. Jeetendra Singh and B. Raj, "Enhanced nonlinear memristor model encapsulating stochastic dopant drift", *Journal of Nanoelectronics and Optoelectronics*, vol. 14, pp. 958–963, 2019.
101. Anjana Bhardwaj, Pradeep Kumar, B. Raj and Sunny Anand, "Design and performance optimization of doping-less vertical nanowire TFET using gate stack technique", *Journal of Electronic Materials*, vol. 41, no. 7, pp. 4005–4013, 2022.
102. Jeetendra Singh and B. Raj, "Tunnel current model of asymmetric MIM structure levying various image forces to analyze the characteristics of filamentary memristor", *Applied Physics A*, vol. 125, no. 3, pp. 203.1–203.11, Feb. 2019.
103. Candy Goyal, Jagpal Singh Ubhi and B. Raj, "Low leakage zero ground noise nanoscale full adder using source biasing technique", *Journal of Nanoelectronics and Optoelectronics*, vol. 14, pp. 360–370, March 2019.
104. Gurmohan Singh, R. K. Sarin and B. Raj, "A novel robust exclusive-OR function implementation in QCA nanotechnology with energy dissipation analysis", *Journal of Computational Electronics*, vol. 15, no. 2, pp. 455–465, June 2016.
105. Girish Wadhwa, Priyanka Kamboj and Balwinder Raj, "Design optimisation of junctionless TFET biosensor for high sensitivity", *Advances in Natural Sciences: Nanoscience and Nanotechnology*, vol. 10, no. 7, pp. 045001, 2019.
106. Priya Bansal and B. Raj, "Memristor modeling and analysis for linear dopant drift kinetics", *Journal of Nanoengineering and Nanomanufacturing*, vol. 6, pp. 1–7, 2016.

107. D. Vaithiyanathan Ashima and B. Raj, "Performance analysis of charge plasma induced graded channel Si nanotube", *Journal of Engineering Research*, Special Issue, pp. 146–154, Aug. 2021.
108. Abhishek Singh Tomar, Vijay Kumar Magraiya and B, Raj, "Scaling of access and data transistor for high performance DRAM cell design", *Quantum Matter*, vol. 2, pp. 412–416, Oct. 2013.
109. Neeraj Jain and B. Raj, "Parasitic capacitance and resistance model development and optimization of raised source/drain SOI FinFET structure for analog circuit applications", *Journal of Nanoelectronics and Optoelectronics*, vol. 13, pp. 531–539, April 2018.
110. Shradhya Singh, S. K. Vishvakarma and B, Raj, "Analytical modeling of split-gate junction-less transistor for a biosensor application", *Sensing and Bio-Sensing*, vol. 18, pp. 31–36, April 2018.
111. Maisagalla Gopal and Balwinder Raj, "Low power 8T SRAM cell design for high stability video applications", *ITSI Transaction on Electrical and Electronics Engineering*, vol. 1, no. 5, pp. 91–97, 2013.
112. Balwinder Raj, Jatin Mitra, Deepak Kumar Bihani, Rangharajan V, A.K. Saxena and S. Dasgupta, "Analysis of noise margin, power and process variation for 32 nm FinFET based 6T SRAM cell", *Journal of Computer*, vol. 5, no. 6, pp. 1–8, 2010.
113. Divya Sharma, Rajesh Mehra and B. Raj, "Comparative analysis of photovoltaic technologies for high efficiency solar cell design", *Superlattices and Microstructures*, vol. 153, pp. 106861, May 2021.
114. Pawandeep Kaur, Avtar Singh Buttar and B. Raj, "A comprehensive analysis of nanoscale transistor based biosensor: a review", *Indian Journal of Pure and Applied Physics*, vol. 59, pp. 304–318, April 2021.
115. Divya Yadav, Balwant Raj and B. Raj, "Design and simulation of low power microcontroller for IoT applications", *Journal of Sensor Letters*, vol. 18, pp. 401–409, May 2020.
116. Shailendra Singh and B. Raj, "A 2-D analytical surface potential and drain current modeling of double-gate vertical t-shaped tunnel FET", *Journal of Computational Electronics*, vol. 19, pp. 1154–1163, April 2020.
117. Jeetendra Singh, and B. Raj, "An accurate and generic window function for non-linear memristor model", *Journal of Computational Electronics*, vol. 18, no. 2, pp. 640–647, June 2019.
118. Manjit Kaur, Neena Gupta, Sanjeev Kumar, B. Raj, and Arun Kumar Singh, "Comparative RF and crosstalk analysis of carbon based nano interconnects", *IET Circuits, Devices & Systems*, vol. 15, no. 6, pp. 493–503, Feb. 2021.
119. Nehru Kandasamy, Firdous Ahmad, D. Ajitha, B. Raj and Nagarjuna Telagam "Quantum dot cellular automata based scan flip flop and boundary scan register", *IETE Journal of Research*, vol. 66, pp. 535–548, 2020.
120. Jeetendra Singh and B. Raj, "Design and investigation of 7T2M NVSARM with enhanced stability and temperature impact on store/restore energy", *IEEE Transactions on Very Large Scale Integration Systems*, vol. 27, no. 6, pp. 1322–1328, June 2019.

Chapter 4

Analysis of HEMT and its applications

Shubham Kumar Gupta, Sourabh Khemot, Deepak Kumar, Mandeep Singh, Tarun Chaudhary, and Balwinder Raj

4.1 INTRODUCTION

Transistors had to come up with high electron mobility and better transport properties due to the necessity for high switching speed, which is required in the fields of microwave communications and RF technologies. T. Mimura, who worked at Fujitsu Laboratories Ltd. in Kobe, Japan, researching high-frequency, high-speed III–V compound semiconductor devices, is regarded as the inventor of HEMT devices. Afterward, HEMT was initially used in commerce. After that, HEMT was first commercially utilized in 1985 at the Nobeyama Radio Observatory (NRO) in Nagano, Japan, as a cryogenic low-noise amplifier [1].

In order to satisfy the demands of high-frequency, low-noise, and high-power density applications, traditional metal–oxide–semiconductor field-effect transistors (MOSFETs) and metal semiconductor field effect transistor (MESFETs) must be constructed with extremely small channel lengths. This reduces impurity scattering and performance degradation for the majority of carriers. These applications also suggest performance and design constraints that call for substantial transconductance and high saturation current, both of which can be obtained by strong doping.

Heterojunctions, which are created between two distinct bandgap materials and contain electrons trapped in a quantum well to prevent impurity scattering, are used in HEMT devices to get around these restrictions. Due to its increased electron mobility and dielectric constant, the direct bandgap semiconductor GaAs has been used in both optoelectronic integrated circuits and high-frequency operation. With a bandgap greater than that of GaAs and almost the same lattice constant, AlGaAs is the best option for a barrier material. GaAs/AlGaAs heterostructure is therefore thought to be the most often used option for incorporation into HEMTs. But another great device that has received a lot of attention recently is the AlGaN/GaN HEMT. It has high breakdown strength and high electron velocity in saturation, and it can perform satisfactorily at very high frequencies [2].

Strong piezoelectric polarization in GaN facilitates the buildup of massive carriers at the AlGaN/GaN contact. Device performance in these kinds

 DOI: 10.1201/9781003612797-4

of HEMTs is contingent upon the types of material layers, layer thicknesses, and AlGaN layer doping concentrations, which allow for design flexibility. For these reasons, AlGaN/GaN HEMT has been chosen as an example due to its superiority over HEMT devices made of other materials [3–7].

4.2 VARIOUS STRUCTURES

4.2.1 GaAs-based HEMTs

Figure 4.1 illustrates a typical GaAs-based HEMT structure. An abrupt heterostructure is formed between the majority carriers and ionized impurities in order to separate them. Figure 4.1 depicts a typical GaAs-based HEMT structure. An abrupt heterostructure is formed between the lower bandgap material GaAs and the wide bandgap material AlGaAs, while the wide bandgap material is doped (e.g., doping density, n = 7×10^{17} cm^{-3}) in order to separate the majority carriers from ionized impurities. As a result, a channel forms at the GaAs/AlGaAs heterojunction interface. One way to lessen coulombic scattering is by using an undoped AlGaAs spacer layer, which is thin. The Si or GaAs layer acts as a substrate at the bottom [8–12].

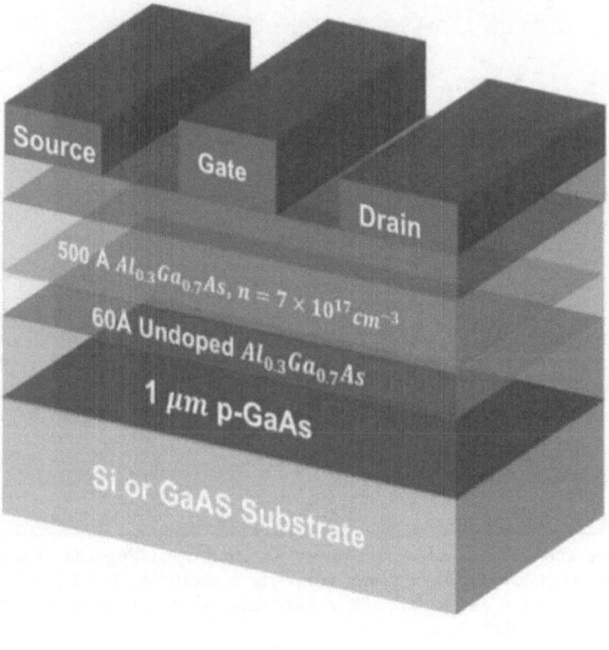

Figure 4.1 Structure of GaAs-based HEMTs [23].

4.2.2 GaN-based HEMTs

Figure 4.2 illustrates how the layered structure of GaN-based HEMTs is comparable to that of traditional GaAs-based HEMTs. But in AlGaN/GaN HEMTs, no deliberate doping is necessary. Instead, because of the spontaneous polarization present in wurtzite-structured GaN, electrons originate from surface states [13–18]. A 2DEG channel results from the buildup of free carriers creating a high carrier concentration at the interface. Additionally, donor-like surface traps (empty) are shown in the figure, which illustrates the positively polarized charge at the AlGaN/GaN interface. The surface barrier, AlGaN thickness, and bound positive charge at the interface are the explicit functions that make up the 2DEG [19–26] (Table 4.1).

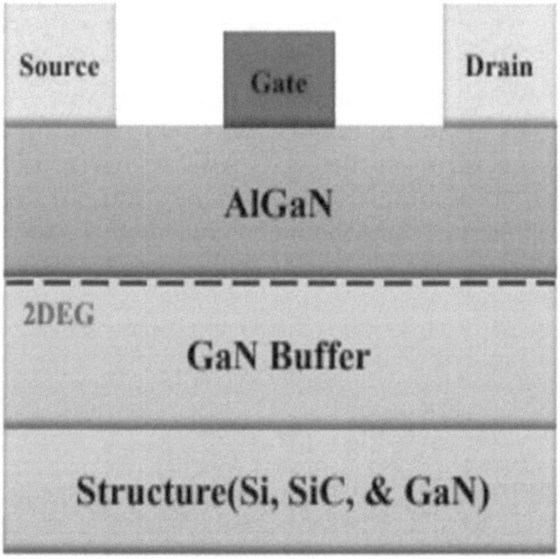

Figure 4.2 AlGaN/GaN HEMT layout and 2DEG generated in AlGaN/GaN contact are shown by the dashed line [24].

Table 4.1 Material properties of SiC and GaN in comparison

S. No.	*Parameter*	*Silicon-MOSFET*	*GaN HEMT*	*Reference*
1.	Threshold voltage (V_{th})	2 V	1.68 V	[20]
2.	On-state resistance (R_{on})	10.8 mOhm	0.1 uOhm	[20]
3.	Breakdown voltage (VBR)	50 V	550 V	[20]
4.	Relative permittivity	11.8	9.0	[21]
5.	Relative permeability	1400	1700	[21]

[20] Newar, Tannu and Veera Venkata Subrahmanya Kumar Bhajana. "Modelling and comparison of Si-MOSFET and eGaN-HEMT for power converter applications using TCAD". *2017 2nd IEEE International Conference on Recent Trends in Electronics, Information & Communication Technology (RTEICT)* (2017): 430–434.

[21] Yuanhang Zhang 2020 *IOP Conf. Ser.: Mater. Sci. Eng.* **738** 012004.

4.2.3 InP-based HEMTs

When compared to traditional GaAs-based HEMTs, InP HEMTs produce a lower electron-effective mass in the InGaAs channel layer. These HEMTs have a somewhat wide conduction band offset (about 0.5 eV) between the InAlAs barrier layer and the channel layer [3]. As a result, InP-based HEMTs exhibit high electron concentration, high electron mobility, and high electron saturation velocity [27–34]. As shown in Figure 4.3, the device typically consists of an undoped InAlAs Schottky barrier, an InGaAs/InAs composite channel for superior electron transport properties, and an InGaAs/InAs composite cap layer for improved ohmic contact.

4.3 WORKING PRINCIPLE

Essentially, HEMTs are heterojunctions made of semiconductors with different bandgap sizes. The valence and conduction bands in the material must bend to create a continuous level when a heterojunction forms. Because the broad-band element is doped with donor atoms (or because of polarization charge in GaN-based HEMTs), it contains extra electrons in the conduction band. Lower energy conduction band states are present in the narrow-band material. Since the next lower bandgap material has

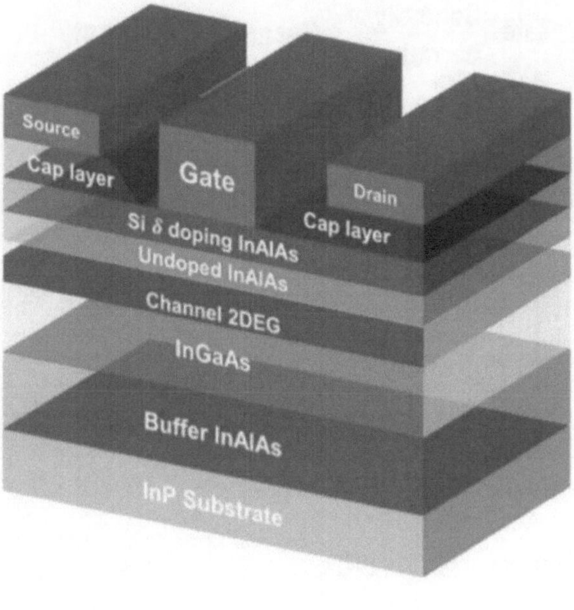

Figure 4.3 InP-based HEMTs [23].

lower energy levels, electrons from the broad-bandgap material will diffuse to it. As a result, the flow of electrons will cause a change in potential, and the materials will create an electric field. Electrons will return to the wide-bandgap element's conduction band as a result of the generated electric field. Drift and diffusion processes persist until they achieve equilibrium, resulting in an equilibrium junction akin to a p–n junction [35–41]. It should be noted that the undoped narrow-bandgap material now contains an abundance of majority charge carriers, resulting in fast switching. One noteworthy aspect is that the low-bandgap undoped semiconductor guarantees good mobility since it lacks donor atoms that could scatter light.

The ability to manipulate the kinds of carriers entering and leaving the device by engineering band discontinuities across the valence and conduction bands is another intriguing feature of HEMTs [42–47].

The concentration of electrons along the boundaries between the two zones within the narrow bandgap material is caused by the diffusion of carriers. In these devices, an accumulation of electrons may result in a very high current. Another name for the collected electrons is 2DEG [48–56]. The generalized band diagram for typical HEMTs generated at the heterojunction is displayed in Figure 4.4. A conducting channel finally forms when the conduction band (Ec) and valence band (Ev) bend with respect to the Fermi level (EF), creating a quantum well filled with 2DEG.

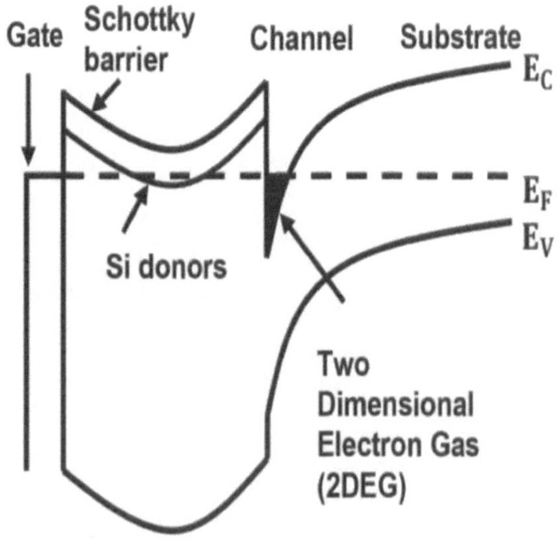

Figure 4.4 Band diagram of HEMTs [23].

4.4 PERFORMANCE ANALYSIS

4.4.1 Current–voltage characteristics using charge control model

Li et al. (2008) presented an enhanced charge control model for the I–V characteristics of AlGaN/GaN HEMTs [4]. This model incorporates adjustable eigenvalues for solving the 2D Poisson's equation and Robin boundary conditions for solving the 1D Schrödinger equation. This model incorporates both the parasitic resistance of the source and drain as well as non-linear polarization [57–63]. This model's device structure and the device structure of Figure 4.2 are nearly identical. The main distinction is that, in order to improve polarization, a doped AlGaN layer, measuring 22 nm and having a doping concentration of ND = 2 × 1018 cm^{-3}, is present above the undoped AlGaN layer. For various gate voltages, the I–V result shown using this analytical model is displayed in Figure 4.5 (Table 4.2).

4.4.2 Dependence of 2DEG charge density on gate bias

A physics-based analytical model for 2DEG density in AlGaN/GaN HEMTs was proposed by Khandelwal et al [5]. They demonstrate the dependency between 2DEG and Fermi levels using this model. There is no need for any fitting parameters using the suggested model. It models 2DEG by taking into account the charge concentration in two distinct areas [64–71]. In relation to

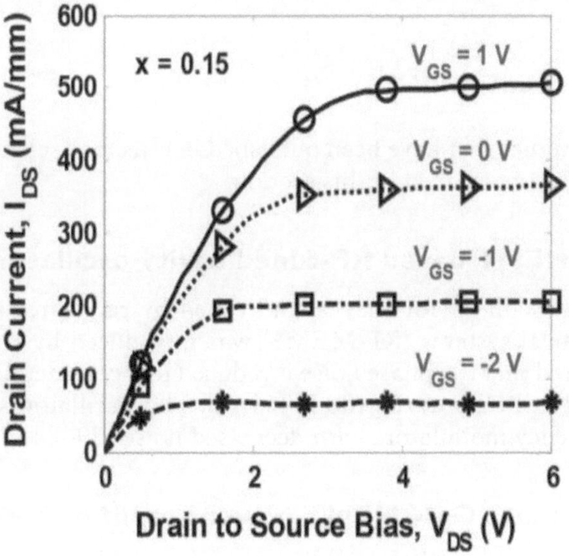

Figure 4.5 I–V characteristics for an AlGaN/GaN [23].

Table 4.2 Pros and cons of typical substrate materials for GaN-epitaxy (data relative to GaN properties) [22]

Substrate	*Bulk GaN*	*Silicon carbide*	*Sapphire*	*Silicon*
Lattice mismatch	None	+3.5%	−16%	−17%
Thermal mismatch	None	+33%	−25%	+116%
Electrical resistance	Low	Low	Infinite	Very low
Thermal resistance	Same	0.3×	3×	0.9×

[22] Kaminski N and Hilt O 2012 SiC and GaN Devices-Competition or Coexistence (ISBN 978-3-8007-3414-6) p. 11.

the Fermi level, one has a higher first subband energy and the other a lower one. Additionally, a combined model of these two locations is shown. It displays a 2DEG variation with gate bias voltage, as seen in the figure. Excellent agreement between the results and numerical calculations is shown.

4.4.3 Gate capacitance including parasitic components

An analytical approach based on surface potential was presented by Zhang et al. to calculate capacitance for AlGaN/GaN HEMTs that includes parasitic components [6]. Capacitance is computed using the idea of surface charge potential, which is consistent with the sheet charge density model, and the sheet charge density is modified by solving charge control equations. To offer a full model, the parasitic components are incorporated into the model further [72–78]. The developed model exhibits consistency with both experimental data and TCAD simulations.

4.5 LATEST RESEARCH

A few recent studies that have been published in esteemed scientific journals have been briefly mentioned in this section.

4.5.1 GaN HEMT-based RF-tuned cavity oscillator

A GaN-based oscillator for the X-band tuned by radio frequency microelectromechanical systems (RF-MEMS) was introduced by Hörberg et al. [7]. It is reported that the phase noise is reduced to a considerably low range of –140 to –129 dBc/Hz at the 100 kHz level. This oscillator is appropriate for high-frequency modulators with decreased noise [79–84].

4.5.2 A compact GaN HEMT power amplifier MMIC

Recently, a detailed performance investigation of a small GaN HEMT-based X-band power amplifier MMIC (monolithic microwave integrated circuit) was disclosed [8]. With this amplifier, an output power range of 47.5–48.7 dBm is achievable. Electronic systems requiring satellite transmitters or aerial phased radar arrays can be constructed using such amplifiers. Increased amplifier output power also boosts these electronic systems' performance, stability, and dependability. Figure 4.6 displays this power amplifier's output power performance with frequency variation in both pulse and continuous wave (CW) modes [85–91] (Figure 4.7).

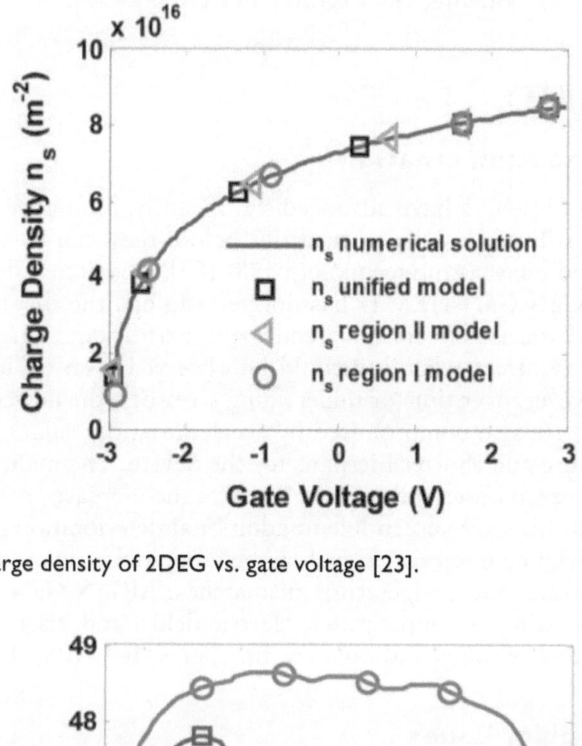

Figure 4.6 Charge density of 2DEG vs. gate voltage [23].

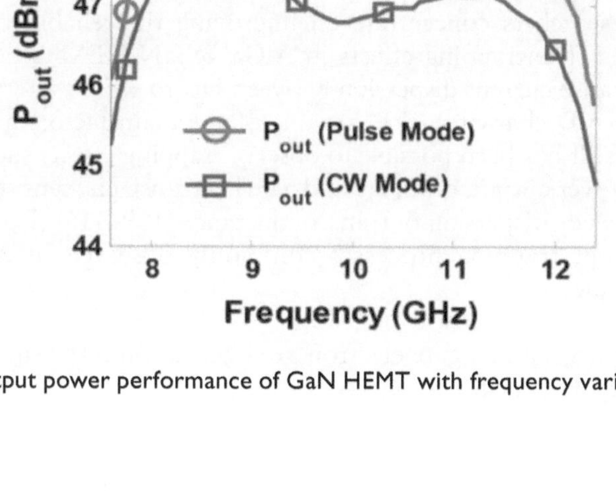

Figure 4.7 Output power performance of GaN HEMT with frequency variation [23].

4.5.3 Kink effect in GaN HEMT technology

Kink effect (KE) in advanced GaN HEMT technology was studied by Crupi et al. [9]. To enhance comprehension, KE is thoroughly examined under varying temperature and bias circumstances. It is demonstrated that device transconductance is primarily responsible for KE's dependency on operating conditions. Microwave engineers would benefit from the characterization of anomalous KE as they require this understanding of KE for constructing and modeling GaN HEMT devices [92–97].

4.6 RELIABILITY

4.6.1 Degradation creation

Although GaN HEMTs have advanced significantly, a number of material-related issues still need to be worked out before they can be considered a dependable and long-lasting technology [98–103]. The actual device performance of Al-GaN/GaN HEMTs has not yet attained the theoretically predicted levels. Parasitic effects that reduce the performance metrics but do not significantly affect reliability could be observed even on unaged GaN HEMTs. However, over time or under aging stressors, the device's resilience and reliability are also compromised by accelerating degradation processes, leading to failure and shorter lifespans for the device. The main goals of the GaN business are to lessen the parasitic impacts and increase product dependability. GaN devices are susceptible to additional deterioration issues due to inherent material characteristics such as heteroepitaxy, polarization effects with the substrate, and temperature mismatches. AlGaN/GaN HEMTs are capable of operating at temperatures, electric fields, and biases that are far higher than those of other traditional technologies [104–108] (Figure 4.8).

4.6.2 Reliability issues

These days, owing to the trapping effect during the AlGaN/GaN HEMT operation, researchers concentrate on improving the reliability of AlGaN/GaN HEMTs. The trapping effects in AlGaN/GaN HEMTs during operation, which cause current dispersion between DC to RF or, alternatively, DC to pulsed ID–VD characteristics, are one of the main factors affecting performance loss. It has been possible to observe trapping effects such as limited microwave power output, drain lag and gate lag transients, sensitivity to light, and the frequency dispersion of transconductance [109–116] (Figure 4.9).

The different trapping processes commonly observed in AlGaN/GaN HEMTs include:

- Barrier traps and a gate electron are injected onto the surface under a high negative VGS.

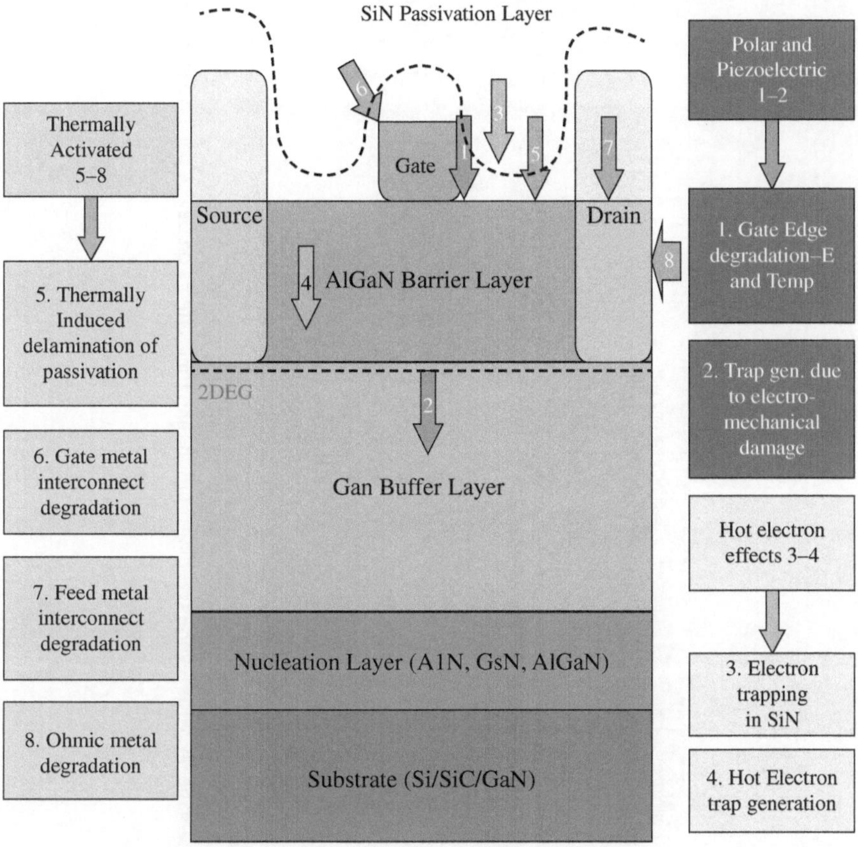

Figure 4.8 Several degrading problems affect AlGaN/GaN HEMTs with their origin [25].

- Hot electrons are used to inject the barrier/buffer traps under a high on-state VDS.
- The electrons that are absorbed by the deep layers are induced during the growth of the material.

4.7 APPLICATION

4.7.1 Broadband communication

The most significant non-military use of HEMT devices is in cellular communication, where they are replacing Si transistors. We receive numerous benefits from these broadband/multiband communication applications. One such is the rise in relative bandwidth at a particular power level. More efficient bandwidth is provided by certain novel circuit and system designs.

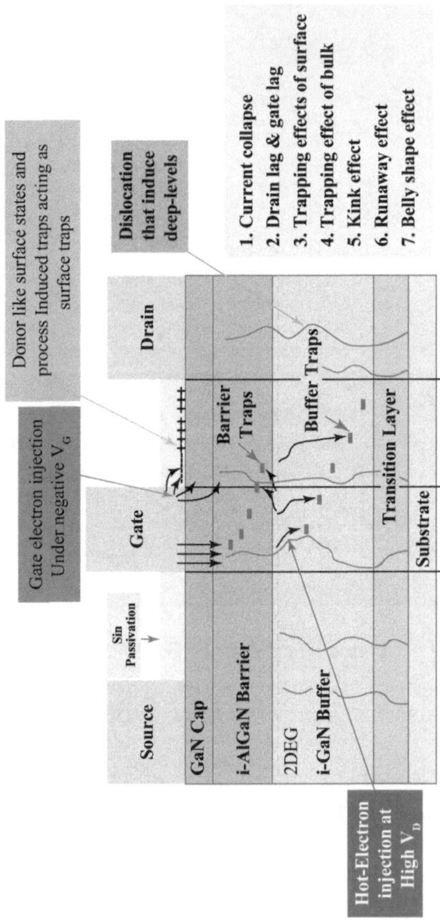

Figure 4.9 The AlGaN/GaN heterostructure is shown schematically, with several trapping methods and their effects on device performance [26].

It is now more linear at the same output power. Using GaN HEMT devices has also been shown to reduce memory effects [10].

4.7.2 Radar components and space applications

The key components of radar composite amplifiers are high gain and low noise. GaN HEMTs are among the top options for these kinds of parts. Ground-based and aerial radars use GaN-based HEMTs to construct active electronic sensor arrays, such as radars for air defense and the navy [11]. Literature is also debating 35 GHz Ka-band missile applications [12]. Because of its exceptional low-noise properties, discrete HEMTs are almost invariably employed as the preamplifier in a standard direct broadcast satellite (DBS) receiver, followed by one or more GaAs MESFET monolithic microwave integrated circuits (MMICs). The implementation of the low-noise HEMT preamplifier has led to notable enhancements in system efficiency at minimal extra expense.

4.7.3 Sensor applications

Chemical sensors have become more significant in the last ten years due to their use in food safety, medical and environmental monitoring, and homeland security. Acquiring the capacity to concurrently examine a broad range of environmental and biological materials has the ability to precisely and sensitively identify a target analyte in the presence of gases and liquids in the field. The HEMTs' conducting 2DEG channel is extremely sensitive to analyte adsorption and is located very near to the surface. As a result, HEMT sensors may be a useful substitute for detecting chemicals, ions, and gases [13].

4.7.4 DNA detection

The AlGaN surface can detect the hybridization of matching target DNAs through the binding layer of Au-gated AlGaN/GaN HEMTs functionalized in the gate region with label-free 3′-thiol-modified oligonucleotides. Thiol-modified DNA covalently bound to gold on the gated area is immobilized, as demonstrated by XPS. The reduction in drain–source current by 115 μA is evident upon the introduction of matched target DNA to the probe DNA on the surface, indicating the potential for DNA sequence detection in biological sensing [14].

4.7.5 Protein detection

Ungated AlGaN/GaN HEMT structures can be activated by adding aminopropyl silane to the gate region. This can operate as a binding layer to the AlGaN surface, allowing biotin to be attached. Streptavidin proteins are highly affinized for biotin. The charges on the chemicals that are linked to

AlGaN/GaN HEMTs have an impact on the device's current. When this protein is gathered at the surface, the device's current clearly decreases by 4 μA, indicating protein sensing [14].

4.7.6 pH detection

In the gate region, the usage of Sc_2O_3 gate dielectric yields better results than native oxide or UV ozone-induced oxide. The current of the ungated HEMTs with Sc_2O_3 in the gate region changes linearly between pH 3 and pH 10 at a rate of 37 μA/pH. Throughout the whole pH range, the HEMT pH sensors operate steadily with a resolution of <0.1 pH. The findings suggest that HEMTs could be useful for tracking pH solution variations between 7 and 8, which is the range relevant for analyzing human blood [13].

4.8 CHALLENGES FOR HEMTS DEVICE

High-electron-mobility transistors (HEMTs) have seen significant advancements in various fields, but they also face challenges and have room for future developments. Here are some of the challenges and potential areas for improvement in HEMT technology:

4.8.1 Complexity of manufacturing

The production of HEMTs is frequently associated with intricate procedures, particularly when combining materials and heterostructures with varying lattice constants. Higher production costs and problems with yield may result from this.

4.8.2 Reliability

It can be difficult to guarantee HEMT long-term durability, especially in demanding settings like space, military applications, or high-power operations. To increase their strength and durability, efforts must be made.

4.8.3 Heat dissipation

When HEMTs are used at increased frequencies and power levels, efficient heat dissipation becomes a major challenge. To keep a device from overheating and to preserve performance, thermal management is essential.

4.8.4 Integration challenges

Because of their distinct electrical properties, integrating HEMTs into intricate electronic systems can be difficult. Maintaining signal integrity and achieving compatibility with other components are constant challenges.

4.8.5 Linearity

Linearity is essential in communication systems, and HEMTs, like other high-frequency transistors, can exhibit nonlinear behavior at high power levels. Researchers are working on improving linearity for better performance in communication applications.

4.9 FUTURE TRENDS

Future HEMT devices in electronics, communications, physics, and other fields appear to be highly promising since they are based on two-dimensional carrier confinement. GaAs-, InP-, and GaN-based HEMTs will persist in their pursuit of more integration, increased frequency, increased power, increased efficiency, reduced noise, and decreased cost. In particular, GaN offers the high-power, high-frequency domain of vacuum tubes and contributes to communication systems that are lighter, more reliable, and more efficient.

In order to take advantage of the special qualities of 2DEG in a variety of material systems, HEMTs will keep evolving into new types of FETs. GaN-based HEMTs in power electronics can have a significant impact on military, consumer, industrial, transportation, and communication systems. Conversely, MOS-HEMT or MISFET structures are probably going to be used in enhancement mode, which has very low leakage current. For these kinds of devices, the idea of various physical and biosensor types is still relatively new. High-sensitivity micro-Hall sensors are made possible by the ultra-high mobility of InAlSb/InAsSb-based systems, which can be used for a variety of purposes such as biorecognition and scanning Hall probe microscopy [15]. Micromachined AlGaAs/GaAs-based HEMTs have been reported to contain three-axis Hall magnetic sensors [16]. In the future, electronic compasses and navigation may employ these gadgets. Devices based on 2DEG can also take advantage of THz detection, mixing, and frequency multiplication capabilities. Strong piezoelectric polarization and chemical stability characterize GaN and similar materials as semiconductors [17]. It is possible to create complex sensors by combining free-standing resonators with functionalized GaN-based 2DEG devices [18]. These can provide techniques for measuring a variety of parameters, including temperature, pH, and viscosity.

It is anticipated that the growth of this technology in the machine-to-machine (M2M) domain would be employed in cloud networking-based diverse sensing tasks, without citations. With its high speed, high mobility, and high sensitivity, newly developed sensing technology can be very helpful in a variety of applications, including environmental research, biotechnology, and structural analysis. At a device level, HEMT technology is anticipated to significantly alter the intelligent social infrastructure. Examples of areas where this technology can be an exception include smart city systems, transportation systems, the food industry, logistics, agriculture, health and welfare, and environmental science and education systems [19].

4.10 CONCLUSION

This chapter examined the device features and performance analysis of HEMTs by referencing the existing literature. The short-channel effects, thermal response, charge estimation, capacitance calculation, and I–V characteristics of HEMTs have all been covered.

These days, HEMT-based oscillators, amplifiers, switches, and diodes are becoming more and more common. Based on recently published research, they have been reviewed. Future research trends on HEMTs have also been reviewed based on these most recent studies. Finally, to highlight the enormous potential of HEMT devices, a number of significant applications of HEMTs have been described, including broadband and radar communications, space and sensor constituents, and DNA, protein, and pH detections.

REFERENCES

1. T. Mimura, "The early history of the high electron mobility transistor (HEMT)," *IEEE Transactions on Microwave Theory and Techniques*, vol. 50, no. 3, pp. 780–782, Mar. 2002.
2. P. Gangwani, S. Pandey, S. Haldar, M. Gupta, R. S. Gupta, "Polarization dependent analysis of AlGaN/GaN HEMT for high power applications," *Solid-State Electronics,* vol. 51 , no. 1, pp. 130–135, 31 Jan. 2007.
3. K. Shinohara and T. Matsui, "3–7 Nano-gate transistor-world's fastest InP-HEMT," *Journal of the National Institute of Information and Communications Technology,* vol. 51, pp. 95–102, 2004.
4. M. Li and Y. Wang, "2-D analytical model for current–voltage characteristics and transconductance of AlGaN/GaN MODFETs," *IEEE Transactions on Electron Devices,* vol. 55, no. 1, pp. 261–267, Jan. 2008.
5. S. Khandelwal, N. Goyal and T. A. Fjeldly, "A physics-based analytical model for 2DEG charge density in AlGaN/GaN HEMT devices," *IEEE Transactions on Electron Devices,* vol. 58 , no. 10, pp. 3622–3625, Oct. 2011.
6. A. Zhang, L. Zhang, Z. Tang, X. Cheng, Y. Wang, K. J. Chen and M. Chan, "Analytical modeling of capacitances for GaN HEMTs, including parasitic components," *IEEE Transactions on Electron Devices,* vol. 61, no. 3, pp. 755–761, Mar. 2014.
7. M. Hörberg, T. Emanuelsson, P. Ligander, S. Lai, H. Zirath and D. Kuylenstierna, "RF-MEMS tuned GaN HEMT based cavity oscillator for X-band," *IEEE Microwave and Wireless Components Letters,* vol. 27, no. 1, pp. 46–48, Jan. 2017.

8. H. Q. Tao, W. Hong, B. Zhang and X. M. Yu, "A compact 60W X-band GaN HEMT power amplifier MMIC," *IEEE Microwave and Wireless Components Letters*, vol. 27, no. 1, pp. 73–75, Jan. 2017.
9. G. Crupi, A. Rao, Z. Marinkovic, G. Avolio, A. Caddemi, V. Markovic, G. Vannini and D. M. Schreurs, "An extensive experimental analysis of the kink effects in S22 and h21 for a GaN HEMT," *IEEE Transactions on Microwave Theory and Techniques,* vol. 62, no. 3, pp. 513–520, Mar. 2014.
10. C. P. Wen, J. Wang and Y. Hao, "Current collapse, memory effect free GaN HEMT," *2010 IEEE MTT-S International Microwave Symposium Digest (MTT)*, Anaheim, CA, USA, 23 May 2010, pp. 149–152.
11. Y. Mancuso, P. Gremillet and P. Lacomme, "T/R-modules technological and technical trends for phased array antennas," *2005 European Microwave Conference*, Paris, France, 4–6 Oct. 2005, pp. 614–617.
12. H. Hommel and H. Feldle, "Current status of airborne active phased array (AESA) radar systems and future trends," *34th European Microwave Conference*, Amsterdam, Netherlands, 12–14 Oct. 2004, pp. 121–124.
13. B. H. Chu, B. S. Kang, H. T. Wang, C. Y. Chang, Y. Tseng, A. Goh, A. Sciullo, W. S. Wu, J. N. Lin, B. P. Gila and S. J. Pearton, "AlGaN/GaN HEMT and ZnO nanorod-based sensors for chemical and bio-applications," *SPIE OPTO: Integrated Optoelectronic Devices*, San Jose, CA, USA. International Society for Optics and Photonics, 12 Feb. 2009, p. 72162A.
14. B. S. Kang, *Fabrication and Characterization of Compound Semiconductor Sensors for Pressure, Gas, Chemical, and Biomaterial Sensing*, Doctoral dissertation, University of Florida.
15. M. Bando, T. Ohashi, M. Dede, R. Akram, A. Oral, S. Y. Park, I. Shibasaki, H. Handa and A. Sandhu, "High sensitivity and multifunctional micro-Hall sensors fabricated using InAlSb/InAsSb/InAlSb heterostructures," *Journal of Applied Physics*, vol. 105, no. 7, p. 07E909, 1 Apr. 2009.
16. A. Vorob'ev, A. Chesnitskiy, A. Toropov and V. Prinz, "Three-axis Hall transducer based on semiconductor microtubes," *Applied Physics Letters*, vol. 103, no. 17, p. 173513, 21 Oct. 2013.
17. M. Dyakonov and M. Shur, "Detection, mixing, and frequency multiplication of terahertz radiation by two-dimensional electronic fluid," *IEEE Transactions on Electron Devices*, vol. 43, no. 3, pp. 380–387, Mar. 1996.
18. K. Brueckner, F. Niebelschuetz, K. Tonisch, S. Michael, A. Dadgar, A. Krost, V. Cimalla, O. Ambacher, R. Stephan and M. A. Hein, "Two-dimensional electron gas based actuation of piezoelectric AlGaN/GaN microelectromechanical resonators," *Applied Physics Letters.*, vol. 93, no. 17, p. 173504, 27 Oct. 2008.
19. K. Joshin, T. Kikkawa, S. Masuda and K. Watanabe, "Outlook for GaN HEMT technology," *Fujitsu Scientific and Technical Journal*, vol. 50, no. 1, pp. 138–143, 1 Jan. 2014.
20. T. Newar and V. V. S. Kumar Bhajana, "Modelling and comparison of Si-MOSFET and eGaN-HEMT for power converter applications using TCAD," *2017 2nd IEEE International Conference on Recent Trends in Electronics, Information & Communication Technology (RTEICT)*, Bengaluru, India, 2017, pp. 430–434.

21. Z. Yuanhang, "Comparison between competing requirements of GaN and SiC family of power switching devices," *2020 IOP Conf. Ser. Mater. Sci. Eng.*, 738 012004.
22. N. Kaminski and O. Hilt, "SiC and GaN Devices-Competition or Coexistence," *Integrated Power Electronics Systems (CIPS), 7th International Conference*, Nuremberg, Germany, 2012. ISBN 978-3-8007-3414-6, p. 11.
23. Muhammad Navid Anjum Aadit, Sharadindu Gopal Kirtania, Farhana Afrin, Md. Kawsar Alam and Quazi Deen Mohd Khosru, "High electron mobility transistors: performance analysis, research trend and applications," researchgate.net/publication/315641333.
24. Naeemul Islam, Mohamed Fauzi Packeer Mohamed, Muhammad Firdaus Akbar Jalaludin Khan, Shaili Falina, Hiroshi Kawarada and Mohd Syamsul, "Reliability, applications and challenges of GaN HEMT technology for modern power devices: a review," *MDPI, Crystals*, vol. 12, pp. 1581–1589, 2022.
25. G. Meneghesso, M. Meneghini, A. Tazzoli, A. Stocco, A. Chini and E. Zanoni, "Reliability issues of gallium nitride high electron mobility transistors," *Int. J. Microw. Wirel. Technol*, vol. 2, pp. 39–50, 2010.
26. R. Ye, X. Cai, C. Du, H. Liu, Y. Zhang, X. Duan and J. Zhu, "An overview on analyses and suppression methods of trapping effects in AlGaN/GaN HEMTs," *IEEE Access*, vol. 10, pp. 21759–21773, 2021.
27. Priya Bansal and B. Raj, "Memristor modeling and analysis for linear dopant drift kinetics," *Journal of Nanoengineering and Nanomanufacturing*, vol. 6, pp. 1–7, 2016.
28. D. Vaithiyanathan Ashima and B. Raj, "Performance analysis of charge plasma induced graded channel Si nanotube," *Journal of Engineering Research*, Special Issue, pp. 146–154, Aug. 2021.
29. Abhishek Singh Tomar, Vijay Kumar Magraiya and B. Raj, "Scaling of access and data transistor for high performance DRAM cell design," *Quantum Matter*, vol. 2, pp. 412–416, Oct. 2013.
30. Neeraj Jain and B. Raj, "Parasitic capacitance and resistance model development and optimization of raised source/drain SOI FinFET structure for analog circuit applications," *Journal of Nanoelectronics and Optoelectronins*, vol. 13, pp. 531–539, April 2018.
31. Shradhya Singh, S. K. Vishvakarma and B Raj, "Analytical modeling of split-gate junction-less transistor for a biosensor application," *Sensing and Bio-Sensing*, vol. 18, pp. 31–36, April 2018.
32. Maisagalla Gopal and Balwinder Raj, "Low power 8T SRAM cell design for high stability video applications," *ITSI Transaction on Electrical and Electronics Engineering*, vol. 1, no. 5, pp. 91–97, 2013.
33. Balwinder Raj, Jatin Mitra, Deepak Kumar Bihani, Rangharajan V, A.K. Saxena and S. Dasgupta, "Analysis of noise margin, power and process variation for 32 nm FinFET based 6T SRAM cell," *Journal of Computer (JCP)*, vol. 5, no. 6, pp. 1–8, 2010.
34. Divya Sharma, Rajesh Mehra and B Raj, "Comparative analysis of photovoltaic technologies for high efficiency solar cell design," *Superlattices and Microstructures*, vol. 153, p. 106861, May 2021.

35. Pawandeep Kaur, Avtar Singh Buttar and B. Raj, "A comprehensive analysis of nanoscale transistor based biosensor: a review," *Indian Journal of Pure and Applied Physics*, vol. 59, pp. 304–318, April 2021.
36. Divya Yadav, Balwant Raj and B. Raj, "Design and simulation of low power microcontroller for IoT applications," *Journal of Sensor Letters*, vol. 18, pp. 401–409, May 2020.
37. Shailendra Singh and B. Raj, "A 2-D analytical surface potential and drain current modeling of double-gate vertical t-shaped tunnel FET," *Journal of Computational Electronics*, vol. 19, pp. 1154–1163, April 2020.
38. Jeetendra Singh and B. Raj, "An accurate and generic window function for non-linear memristor model," *Journal of Computational Electronics*, vol. 18, no. 2, pp. 640–647, June 2019.
39. Manjit Kaur, Neena Gupta, Sanjeev Kumar, B. Raj and Arun Kumar Singh, "Comparative RF and crosstalk analysis of carbon based nano interconnects," *IET Circuits, Devices & Systems*, vol. 15, no. 6, pp. 493–503, Feb. 2021.
40. Nehru Kandasamy, Firdous Ahmad, D. Ajitha, B. Raj and Nagarjuna Telagam, "Quantum dot cellular automata based scan flip flop and boundary scan register," *IETE Journal of Research*, vol. 66, pp. 535–548, 2020.
41. Jeetendra Singh and B. Raj, "Design and investigation of 7T2M NVSARM with enhanced stability and temperature impact on store/restore energy," *IEEE Transactions on Very Large Scale Integration Systems,* vol. 27, no. 6, pp. 1322–1328, June 2019.
42. Anil Kumar Bhardwaj, Sumeet Gupta, B. Raj and Amandeep Singh, "Impact of double gate geometry on the performance of carbon nanotube field effect transistor structures for low power digital design," *Computational and Theoretical Nanoscience*, vol. 16, pp. 1813–1820, 2019.
43. Neeraj Jain and B. Raj, "Thermal stability analysis and performance exploration of asymmetrical dual-k underlap spacer (ADKUS) SOI FinFET for security and privacy applications," *Indian Journal of Pure & Applied Physics (IJPAP)*, vol. 57, pp. 352–360, May 2019.
44. Neeraj Jain and Balwinder Raj, "Dual-k spacer region variation at the drain side of asymmetric SOI FinFET structure: performance analysis towards the analog/RF design applications," *Journal of Nanoelectronics and Optoelectronics*, vol. 14, pp. 349–359, March 2019.
45. Jeetendra Singh, Sanjeev Sharma and B. Raj, "Mamta Khosla analysis of barrier layer thickness on performance of In1-xGaxAs based gate stack cylindrical gate nanowire MOSFET," *Journal of Nanoelectronics and Optoelectronics*, vol. 13, pp. 1473–1477, Oct. 2018.
46. Neeraj Jain and B. Raj, "Analysis and performance exploration of high-k SOI FinFETs over the conventional low-k SOI FinFET toward analog/RF design," *Journal of Semiconductors*, vol. 39, no. 12, pp. 124002-1–7, Dec. 2018.
47. Candy Goyal, Jagpal Singh Ubhi and B. Raj, "A reliable leakage reduction technique for approximate full adder with reduced ground bounce noise," *Journal of Mathematical Problems in Engineering*, vol. 2018, Article ID 3501041, p. 16, 15 Oct. 2018.

48. Jeetendra Singh Anuradha, B. Raj and Mamta Khosla, "Design and performance analysis of nano-scale memristor-based nonvolatile SRAM," *Journal of Sensor Letter*, vol. 16, pp. 798–805, Oct. 2018.
49. Girish Wadhwa and B. Raj, "Parametric variation analysis of charge-plasma-based dielectric modulated JLTFET for biosensor application," *IEEE Sensor Journal*, vol. 18, no. 15, pp. 6070–6077, 1 Aug. 2018.
50. Jeetendra Singh and B. Raj, "comparative analysis of memristor models for memories design," *Journal of Semiconductors*, vol. 39, no. 7, pp. 074006-1–12, July 2018.
51. Divya Yadav, Shailesh Singh Chouhan, Santosh Kumar Vishvakarma and B. Raj, "Application specific microcontroller design for IoT based WSN," *Sensor Letter*, vol. 16, pp. 374–385, May 2018.
52. Gurmohan Singh, R. K. Sarin and B. Raj, "Fault-tolerant design and analysis of quantum-dot cellular automata based circuits," *IEEE/IET Circuits, Devices & Systems*, vol. 12, pp. 638–64, 2018.
53. Jeetendra Singh and B. Raj, "Modeling of mean barrier height levying various image forces of metal insulator metal structure to enhance the performance of conductive filament based memristor model," *IEEE Nanotechnology*, vol. 17, no. 2, pp. 268–267, Mar. 2018.
54. Aakash Jain, Sanjeev Sharma and B. Raj, "Analysis of triple metal surrounding gate (TM-SG) III–V nanowire MOSFET for photosensing application," *Opto-Electronics Journal,* vol. 26, no. 2, pp. 141–148, May 2018.
55. Aakash Jain, Sanjeev Sharma and B. Raj, "Design and analysis of high sensitivity photosensor using cylindrical surrounding gate MOSFET for low power sensor applications," *Engineering Science and Technology, an International Journal*, vol. 19, no. 4, pp. 1864–1870, Dec. 2016.
56. G. Saiphani Kumar, Amandeep Singh and B. Raj, "Design and analysis of gate all around CNTFET based SRAM cell design," *Journal of Computational Electronics*, vol. 17, no. 1, pp 138–145, Mar. 2018.
57. B. S. Sohi Gurinder pal Singh and Balwinder Raj, "Material properties analysis of graphene base transistor (GBT) for VLSI analog circuits," *Indian Journal of Pure & Applied Physics (IJPAP),* vol. 55, pp. 896–902, Dec. 2017.
58. Sunil Kumar and B. Raj, "Estimation of stability and performance metric for inward access transistor based 6T SRAM cell design using n-type/p-type DMDG-GDOV TFET," *IEEE VLSI Circuits and Systems Letter,* vol. 3, no. 2, pp. 25–39, June 2017.
59. Shashikant Sharma, Anjan Kumar, Manisha Pattanaik and B Raj, "Forward body biased multimode multi-threshold CMOS technique for ground bounce noise reduction in static CMOS adders," *International Journal of Information and Electronics Engineering*, vol. 3, no. 3, pp. 567–572, 2013.
60. Hamendra Singh, Pankaj Kumar and Balwinder Raj, "Performance analysis of majority gate SET based 1-bit full adder," *International Journal of Computer and Communication Engineering*, vol. 2, no. 4, 2013, ISSN: 2010–3743.
61. Anil Kumar Bhardwaj, Sumeet Gupta and B Raj, "Investigation of parameters for Schottky barrier (SB) height for Schottky barrier based carbon nanotube field effect transistor device," *Journal of Nanoelectronics and Optoelectronics*, vol. 15, pp. 783–791, July 2020.

62. Priya Bansal and B. Raj, "Memristor: a versatile nonlinear model for dopant drift and boundary issues," *JCTN,* vol. 14, no. 5, pp. 2319–2325, May 2017.
63. Neeraj Jain and B. Raj, "An analog and digital design perspective comprehensive approach on fin-FET (fin-field effect transistor) technology – a review," *Reviews in Advanced Sciences and Engineering (RASE),* vol. 5, pp. 1–14, 2016.
64. Balwinder Raj, A. K. Saxena and S. Dasgupta, "Analytical modeling for the estimation of leakage current and subthreshold swing factor of nanoscale double gate FinFET device," *Microelectronics International,* vol. 26, pp. 53–63, 2009.
65. Shailendra Singh Soniya, Girish Wadhwa and Balwinder Raj, "An analytical modeling for dual source vertical tunnel field effect transistor," *International Journal of Recent Technology and Engineering (IJRTE),* vol. 8, no. 2, pp. 603–608, July 2019.
66. Shailendra Singh and B. Raj, "Design and analysis of hetrojunction vertical T-shaped tunnel field effect transistor," *Journal of Electronics Material,* vol. 48, no. 10, pp. 6253–6260, Oct. 2019.
67. Candy Goyal, Jagpal Singh Ubhi and B Raj, "A low leakage CNTFET based inexact full adder for low power image processing applications," *International Journal of Circuit Theory and Applications*, vol. 47, no. 9, pp. 1446–1458, Sep. 2019.
68. B. Raj, A. K. Saxena and S. Dasgupta, "A compact drain current and threshold voltage quantum mechanical analytical modeling for FinFETs," *Journal of Nanoelectronics and Optoelectronics*, vol. 3, no. 2, pp. 163–170, 2008.
69. Girish Wadhwa and B. Raj, "An analytical modeling of charge plasma based tunnel field effect transistor with impacts of gate underlap region," *Superlattices and Microstructures,* vol. 142, pp. 106512, June 2020.
70. Shailendra Singh and B. Raj,"Modeling and simulation analysis of SiGe hetrojunction double gatevertical t-shaped tunnel FET," *Superlattices and Microstructures*, vol. 142, p. 106496, June 2020.
71. Neeraj Jain and Balwinder Raj, "Device and circuit co-design perspective comprehensive approach on FinFET technology – a review," *Journal of Electron Devices*, vol. 23, no. 1, pp. 1890–1901, 2016.
72. Sunil Kumar and B. Raj, "Analysis of I_{ON} and ambipolar current for dual-material gate-drain overlapped DG-TFET," *Journal of Nanoelectronics and Optoelectronics*, vol. 11, pp. 323–333, June 2016.
73. Naveed Anjum, Tarun Bali and B. Raj, "Design and simulation of handwritten multiscript character recognition," *International Journal of Advanced Research in Computer and Communication Engineering,* vol. 2, no. 7, pp. 2544–2549, July 2013.
74. Karmjit Singh and B. Raj, "Performance and analysis of temperature dependent multi-walled carbon nanotubes as global interconnects at different technology nodes," *Journal of Computational Electronics,* vol. 14 no. 2, pp. 469–476, June 2015.
75. Sunil Kumar and B. Raj, "Compact channel potential analytical modeling of DG-TFET based on Evanescent–mode approach," *Journal of Computational Electronics*, vol. 14, no. 2, pp. 820–827, July 2015.

76. Karmjit Singh and B. Raj, "Temperature dependent modeling and performance evaluation of multi-walled CNT and single-walled CNT as global interconnects," *Journal of Electronic Materials*, vol. 44, no. 12, pp. 4825–4835, Dec. 2015.
77. V. K. Sharma, M. Pattanaik and B. Raj, "INDEP approach for leakage reduction in nanoscale CMOS circuits," *International Journal of Electronics*, vol. 102, no. 2, pp. 200–215, 2014.
78. Karmjit Singh and B. Raj, "Influence of temperature on MWCNT bundle, SWCNT bundle and copper interconnects for nanoscaled technology nodes," *Journal of Materials Science: Materials in Electronics,*, vol. 26, no. 8, pp. 6134–6142, 2015.
79. Naveed Anjum, Tarun Bali and B Raj, "Design and simulation of handwritten Gurumukhi and Devanagri numerical recognition," *International Journal of Computer Applications*, vol. 73, no. 12, pp. 16–21, 2013.
80. S. Khandelwal, V. Gupta, B. Raj and R. D. Gupta, "Process variability aware low leakage reliable nano scale DG-FinFET SRAM cell design technique," *Journal of Nanoelectronics and Optoelectronics*, vol. 10, no. 6, pp. 810–817, Dec. 2015.
81. V. K. Sharma, M. Pattanaik and B Raj, "ONOFIC approach: low power high speed nanoscale VLSI circuits design," *International Journal of Electronics*, vol. 101, no. 1, pp. 61–73, 2014.
82. S. Khandelwal, Balwinder Raj and R. D. Gupta, "FinFET based 6T SRAM cell design: analysis of performance metric, process variation and temperature effect," *Journal of Computational and Theoretical Nanoscience*, vol. 12, pp. 2500–2506, 2015.
83. Sumit Singh, Shekhar Yadav, Jagdeep Rahul, Anurag Srivastava and B. Raj, "Impact of HfO_2 in graded channel dual insulator double gate MOSFET," *Journal of Computational and Theoretical Nanoscience,* , vol. 12, no. 6, pp. 950–953, April 2015.
84. Vijay Kumar Sharma, Manisha Pattanaik and B. Raj, "PVT variations aware low leakage INDEP approach for nanoscale CMOS Circuits," *Microelectronics Reliability*, vol. 54, pp 90–99, 2014.
85. B. Raj, A. K. Saxena and S. Dasgupta, "Quantum mechanical analytical modeling of nanoscale DG FinFET: evaluation of potential, threshold voltage and source/drain resistance," *Journal of Material Science in Semiconductor Processing*, vol. 16, no. 4, pp. 1131–1137, 2013.
86. Maisagalla Gopal, Siva Sankar D Prasad and Balwinder Raj, "8T SRAM cell design for dynamic and leakage power reduction," *International Journal of Computer Applications*, vol. 71, no. 9, pp. 43–48, June 2013.
87. Manisha Pattanaik, B. Raj, Shashikant Sharma and Anjan Kumar, "Diode based trimode multi-threshold CMOS technique for ground bounce noise reduction in static CMOS adders," *Advanced Materials Research*, vol. 548, pp 885–889, 2012.
88. Balwinder Raj, A. K. Saxena and S. Dasgupta, "Nanoscale FinFET based SRAM cell design: analysis of performance metric, process variation, underlapped FinFET and temperature effect," *IEEE Circuits and System Magazine,* vol. 11, no. 2, pp. 38–50, 2011.

89. V. K. Sharma, M. Pattanaik and Balwinder Raj, "Leakage current ONOFIC approach for deep submicron VLSI circuit design," *International Journal of Electrical, Computer, Electronics and Communication Engineering,* vol. 7, no. 4, pp. 239–244, 2013.
90. Parminder Kaur, Sandeep Singh Gill and B Raj, "Comparative analysis of OFETs materials and devices for sensor applications," *Journal of Silicon,* vol 14, pp. 4463–4471, 2022.
91. Sanjeev Kumar Sharma, Parveen Kumar, Balwant Raj and B. Raj, "$In_{1-x}Ga_x$as double metal gate-stacking cylindrical nanowire MOSFET for highly sensitive photo detector," *Journal of Silicon*, vol. 14, pp. 3535–3541, 2022.
92. B. Raj, A. K. Saxena and S. Dasgupta, "Analytical modeling of quasi planar nanoscale double gate FinFET with source/drain resistance and field dependent carrier mobility: a quantum mechanical study," *Journal of Computer (JCP)*, vol. 4, no. 9, pp. 1–8, 2009.
93. S. Bhushan, S. Khandelwal and B. Raj. "Analyzing different mode FinFET based memory cell at different power supply for leakage reduction," *Seventh International Conference on Bio-Inspired Computing: Theories and Application, (BIC-TA 2012) Advances in Intelligent Systems and Computing,* vol. 202, pp. 89–100, 2013.
94. Jeetendra Singh and B. Raj, "Temperature dependent analytical modeling and simulations of nanoscale memristor," *Engineering Science and Technology, an International Journal*, vol. 21, pp. 862–868, Oct. 2018.
95. Shradhya Singh, Shashi Bala, Balwant Raj and B. Raj, "Improved sensitivity of dielectric modulated junctionless transistor for nanoscale biosensor design," *Sensor Letter*, vol. 18, pp. 328–333, April 2020.
96. Vivek Kumar, Santosh Kumar Vishvakarma and B Raj, "Design and performance analysis of ASIC for IoT applications," *Sensor Letter ASP*, vol. 18, pp. 31–38, Jan. 2020.
97. Akanksha Jaiswal, R. K. Sarin, B. Raj and Shikha Sukhija, "A novel circular slotted microstrip-fed patch antenna with three triangle shape defected ground structure for multiband applications," *Advanced Electromagnetic (AEM)*, vol. 7, no. 3, pp. 56–63, Aug. 2018.
98. Girish Wadhwa and B. Raj, "Label free detection of biomolecules using charge-plasma-based gate underlap dielectric modulated junctionless TFET," *Journal of Electronic Materials (JEMS)*, vol. 47, no. 8, pp 4683–4693, Aug. 2018.
99. Gurmohan Singh, R. K. Sarin and B. Raj, "Design and performance analysis of a new efficient coplanar quantum-dot cellular automata adder," *Indian Journal of Pure & Applied Physics*, vol. 55, pp. 97–103, Feb. 2017.
100. Parminder Kaur, Vikas Pandey and B Raj, "Comparative study of efficient design, control and monitoring of solar power using IoT," *Sensor Letter*, vol. 18, pp. 419–426, May 2020.
101. Anil Kumar Bhardwaj, Sumeet Gupta and B Raj, "Development & analysis of compact model for double gate Schottky barrier CNTFET," *Journal of Nanoelectronics and Optoelectronics*, vol. 15, pp. 1199–1208, Aug. 2020.
102. Girish Wadhwa, Priyanka Kamboj, Jeetendra Singh and B Raj, "Design and investigation of junctionless DGTFET for biological molecule recognition," *Transactions on Electrical and Electronic Materials*, vol. 22, pp. 282–289, 2021.

103. Sachin Kumar Verma, Shailendra Singh, Girish Wadhwa and B Raj, "Detection of biomolecules using charge-plasma based gate underlap dielectric modulated dopingless TFET," *Transactions on Electrical and Electronic Materials (TEEM)*, vol. 21, pp. 528–535, June 2020.
104. Neeraj Jain, B. Raj, "Impact of underlap spacer region variation on electrostatic and analog/RF performance of symmetrical high-k SOI FinFET at 20 nm channel length," *Journal of Semiconductors*, vol. 38, no. 12, pp. 122002, Dec. 2017.
105. Shailendra Singh and B. Raj, "Analytical modeling and simulation analysis of T-shaped III-V heterojunction vertical T-FET," *Superlattices and Microstructures*, vol. 147, p. 106717, Nov. 2020.
106. R. K. Sarin Gurmohan Singh and B. Raj, "Design and analysis of area efficient QCA based reversible logic gates," *Journal of Microprocessors and Microsystems*, vol. 52, pp. 59–68, May 2017.
107. Shailendra Singh and B. Raj, "Parametric variation analysis on hetero-junction Vertical t-shape TFET for supressing ambipolar conduction," *Indian Journal of Pure and Applied Physics*, vol. 58, pp. 478–485 June 2020.
108. Shailendra Singh Soniya, Girish Wadhwa and Balwinder Raj, "Design and analysis of dual source vertical tunnel field effect transistor for high performance," *Transactions on Electrical and Electronics Materials*, vol. 21, pp. 74–82, Oct. 2019.
109. Manjit Kaur, Neena Gupta, Sanjeev Kumar, B. Raj and Arun Kumar Singh, "RF performance analysis of intercalated graphene nanoribbon based global level interconnects," *Journal of Computational Electronics*, vol. 19, pp. 1002–1013, June 2020
110. Girish Wadhwa and B. Raj, "Design and performance analysis of junctionless TFET biosensor for high sensitivity," *IEEE Nanotechnology*, vol. 18, pp. 567–574, 2019.
111. Jeetendra Singh and B. Raj, "Enhanced nonlinear memristor model encapsulating stochastic dopant drift," *Journal of Nanoelectronics and Optoelectronics*, vol. 14, pp. 958–963, 2019.
112. Anjana Bhardwaj, Pradeep Kumar, B. Raj and Sunny Anand, "Design and performance optimization of doping-less vertical nanowire TFET using gate stack technique," *Journal of Electronic Materials* vol. 41, no. 7, pp. 4005–4013, 2022.
113. Jeetendra Singh and B. Raj, "Tunnel current model of asymmetric MIM structure levying various image forces to analyze the characteristics of filamentary memristor," *Applied Physics A*, vol. 125, no. 3, pp. 203.1–203.11, Feb. 2019.
114. Candy Goyal, Jagpal Singh Ubhi and B. Raj, "Low leakage zero ground noise nanoscale full adder using source biasing technique," *Journal of Nanoelectronics and Optoelectronics*, vol. 14, pp. 360–370, March 2019.
115. Gurmohan Singh, R. K. Sarin and B. Raj, "A novel robust exclusive-OR function implementation in QCA nanotechnology with energy dissipation analysis," *Journal of Computational Electronics*, vol. 15, no. 2, pp. 455–465, June 2016.
116. Girish Wadhwa, Priyanka Kamboj and Balwinder Raj" Design optimisation of junctionless TFET biosensor for high sensitivity," *Advances in Natural Sciences: Nanoscience and Nanotechnology*, vol. 10, no. 7, p. 045001, 2019.

Chapter 5

Characterizations and applications of pomegranate synthesized biogenic silver nanoparticles

Sunil Sankathala, Devanshi Soni, Aditya Shrivastav, Yadnyesh Khapekar, Aryan Vasa, Susanta Das, and Debanjali Barman Roy

5.1 INTRODUCTION

The profound interest in AgNPs and their oxides stems from their minute size, typically ranging from 1 nm to 100 nm, which offers a high surface-to-volume ratio and remarkable physical, chemical, and biological attributes [1–3]. These unique properties have catalyzed significant scientific and technological exploration, leading to diverse applications across several domains such as nanobiomedicine, nanotechnology, food industry, animal husbandry, disinfection, environmental remediation, agriculture, and ceramics, among others [1–4].

AgNPs can be synthesized using physical, chemical, and biological routes [5]. Various techniques such as laser ablation, mechanical/ball milling, vapor deposition, pyrolysis (spray, aerosol, and laser), sol–gel, emulsion, atomic and molecular condensation, chemical and electrochemical methods are used to synthesize metallic nanoparticles and their oxides [5].

However, these physical/chemical techniques require expensive instruments and therefore are costly. They are time consuming and require the use of hazardous chemicals [6]. Furthermore, the synthesis process involves a number of steps and various controlled thermal, physical, and chemical parameters to achieve the optimum synthesis process, which are crucial in accomplishing stable nanoparticles with intended properties for their planned applications. Additionally, physically/chemically synthesized AgNPs can have an adverse impact on the society and living beings. It is in this context that the biosynthesis of AgNPs has received considerable attention from the scientific community in recent years [6].

This approach involves synthesizing biogenic nanoparticles using environmentally benign materials, such as plant extracts, microorganisms, or other biocompatible substances. This eco-friendly, non-toxic, cost-effective, and time-efficient approach has drawn significant interest compared to conventional chemical and physical processes.

The integration of plant-based agents in fabricating AgNPs stands as a pivotal realm in contemporary research. Phytonanotechnology represents a domain where plant-derived materials, specifically bioactive compounds

DOI: 10.1201/9781003612797-5

and phytochemicals, find application within the landscape of nanotechnology [7]. This avenue delves into leveraging plant-based substances to create nanomaterials or nanoparticles, exploring their diverse utility across technological and biomedical arenas.

Plant metabolites or polyphenols act as reducing and capping agents. They reduce ions into metal as well as stabilize the nanoparticles [8]. Extracts derived from various parts of medicinal plants have facilitated the creation of diverse nanoparticles and their oxides such as gold NPs (AuNPs), AgNPs, silver oxide NPs (Ag_2ONPs), zinc oxide NPs (ZnONPs), copper NPs (CuNPs), platinum NPs (PtNPs), and iron NPs (FeNPs), each exhibiting a myriad of beneficial properties covering nanobiomedicine and nanotechnology to a sustainable environment [6].

Punica granatum (pomegranate) is an edible fruit. It belongs to the family *Punicaceae* or *Lythraceae* [9]. Recent advancements in scientific research have spotlighted the pomegranate (*Punica granatum L.*) fruit. The diverse array of biomolecules in pomegranate serves as a source of nutrients and therapeutic agents for combating diseases [9]. Moreover, pomegranate-mediated green synthesis of AgNPs exhibits immense potential across a wide spectrum of domains.

This chapter extensively covers the green synthesis methodologies employed using extracts derived from various components of the pomegranate plant for generating AgNPs, their comprehensive characterization, and multifaceted applications across diverse fields.

5.2 LITERATURE REVIEW

In this section, the biosynthesis of AgNPs using various parts of pomegranate, their characterizations using various spectroscopic techniques, and their numerous applications have been discussed. Figure 5.1 illustrates a schematic diagram detailing the steps involved in extracting, synthesizing, characterizing, and applying AgNPs derived from diverse components of pomegranate fruits.

Saratale and his group [10] biosynthesized AgNPs utilizing the leaf extract of pomegranate, characterized them using various methods, compared their efficacies with acarbose (an antibiotic used for diabetes mellitus), and investigated their anticancer, scavenging, and antibacterial activities. Fresh pomegranate leaves were collected, purified using chemical techniques, dried for several days at room temperature to eliminate the moisture, crushed to powder, and kept in bottle in a dark place to avoid oxidation [11–16]. Subsequently, 5 g of this powder was mixed with distilled water, processed, and cooled under controlled conditions. The resulting mixture was filtered to produce the leaf extract, which was then combined with $AgNO_3$ solution at appropriate chemical concentrations under

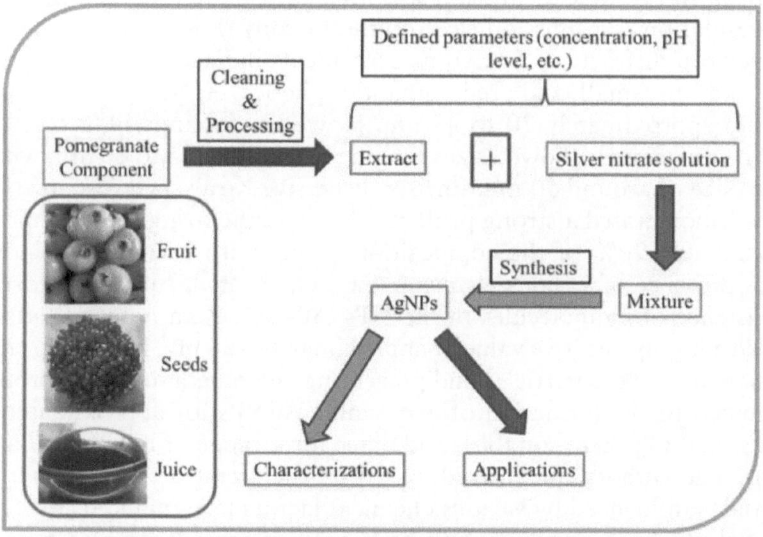

Figure 5.1 A schematic diagram illustrating the process of preparation, characterization, and diverse applications of pomegranate-assisted green synthesis of AgNPs.

ambient conditions. The time-dependent formation of NPs was monitored through an ultraviolet–visible (UV–Vis) spectrometer. The color of the mixture turned into dark reddish-brown. Moreover, a strong surface plasmon resonance (SPR) peak appeared at 415 nm in a very short time [17–23]. All these results indicated the active roles played by various biological macromolecules (e.g., phenolics, flavonoids, tannins, and alkaloids) of the leaf extract in reducing the Ag^+ into neutral Ag metals, capping them, and ultimately forming highly stable AgNPs. The resulting AgNPs were pelleted, purified, and subjected to further characterization. X-ray diffraction (XRD) patterns demonstrated the formation of face-centered cubic (*fcc*) crystal lattice structure with highly oriented (111) planes, a characteristic of AgNPs. X-ray photoelectron spectroscopy (XPS) signals at 370 eV and further analysis of it revealed the strong presence of AgNPs and binding energies of various associated valence states [24–29]. The Fourier transform infra-red (FTIR) spectrum of the AgNPs displayed distinct peaks corresponding to vibrations and stretches of various functional groups, namely flavonoids, alkenes, aromatic rings, and alcohols inherent in the leaf extract. The peaks observed at wavenumbers (cm^{-1}) around O-H (~3400), C=O (~3385), C-H (~2930), and C-C (~1630) signify the prevalent presence of these biomolecules contributing to the composition of the AgNPs. A comparison of the spectrum with that of the leaf extract revealed slight shifts in the synthesized NPs, suggesting the significant roles played by various biomolecules in reducing Ag ions to neutral Ag metals, capping, forming, and stabilizing AgNPs. Field emission scanning electron microscopy (FESEM), High-resolution transmission electron microscopy (HRTEM), and selected area electron diffraction (SAED) measurements indicated the formation of polycrystalline, small, and predominantly spherical AgNPs with sizes ranging from approximately 20 to 45 nm. However, dynamic light scattering (DLS) measurements showed sizes ranging from about 35 to 60 nm, with an average size of around 50 nm. Energy-dispersive X-ray spectroscopy (EDS) results demonstrated a strong peak at 3 keV specific to AgNPs, comprising approximately 90% of the composition, along with small peaks indicating the presence of various elements (e.g., C, O, etc.), further confirming the existence of biomolecules in the NPs [30–36]. Zeta potential analysis revealed a highly negative value of approximately –26 mV, indicating strong repulsion among the particles and preventing agglomeration. This property contributed to the formation of very stable AgNPs for approximately six months, making them suitable candidates for a range of biomedical applications. The authors emphasized, based on the literature review, that zeta potential is influenced by various chemical factors (e.g., pH and concentration of electrolyte), and NPs with high positive or negative zeta potential demonstrate strong particle repulsion. Consequently, they do not conglomerate, and the synthesized NPs remain highly stable for several months, a characteristic of green synthesized NPs [37–43]. The antidiabetic activities of these NPs were evaluated using α-amylase and α-glucosidase enzymes, which play roles in carbohydrate digestion and blood sugar regulation. The results showed that these NPs effectively created inhibition zones against these enzymes, with efficacy falling between pomegranate and acarbose, a standard antibiotic for diabetes mellitus. The dose-dependent cytotoxic properties of AgNPs on HepG2 liver cancer cells were tested using the 3-(4,5-dimethylthiazol-2-yl)-2,5-diphenyl-2*H*-tetrazolium bromide (MTT) assay. Various concentrations, ranging from low to high (5 to 200 μg/mL), were used, and the results indicated a significant dose-dependent decrease in cell viability with increasing concentration. The potential antioxidant applications of AgNPs were measured using the DPPH (2,2-diphenyl-1-picrylhydrazyl) and ABTS (2,2-azinobis(3-ethylbenzothiazoline-6-sulfon ic acid)) assays, showing an increase in antioxidant activity with higher nanoparticles' concentrations. Additionally, the results were compared with pomegranate and standard antioxidants: catechol, commonly used in various industries, for the DPPH assay, and ascorbic acid, a dietary supplement, for the ABTS assay. Both results demonstrated that the antioxidant efficacies of AgNPs lie between leaf extract and standard antioxidants [44–48]. The applications of AgNPs as antibacterial agents were investigated on *Escherichia coli* (gram-positive) and *Staphylococcus aureus* (gram-negative) bacteria, revealing different effective inhibition zones (17 and 15 mm, respectively) against these two bacteria. The difference was attributed to thickness and composition of cell, and the interaction mechanism between bacteria and AgNPs. Furthermore, the authors studied the combined effects of AgNPs and various antibiotics on both types of bacteria to understand potential synergies. They observed varying degrees of enhancement in the formation of effective zones of inhibition for both types of bacteria. The authors suggested that these biogenic NPs have promising applications in nanobiomedicine and industries [49–52].

Joshi *et al.* [11] biosynthesized AgNPs from the peel extract of pomegranate, an agrowaste containing several nutrients and biological macromolecules, in the presence of $AgNO_3$. The researchers studied the properties of these AgNPs and investigated their photocatalytic degradation properties on methylene blue dye. To prepare the peel extract, pomegranate peels were separated from the arils of fresh, thoroughly cleaned pomegranate fruits [53–56]. The peels were then cut into small pieces, boiled in pure water for some time, cooled, and finally filtered. The extract was mixed with 1 mM $AgNO_3$ solution, and the reaction was continuously monitored using UV-Vis spectroscopy. A rapid change of color of the mixture and the appearance of a broad SPR peak in the range of 432 to 442 nm clearly indicated the reduction of Ag^+ to neutral metal Ag, capping, and formation of stable AgNPs by the botanical compounds present in the peel extract. The XRD pattern confirmed the formation of crystalline cubic structure [57–64]. FTIR spectrum exhibited different vibration/stretching peaks corresponding to various groups present in the peel extract, such as O-H at 3286 and N-H at 1636 cm^{-1}, signifying the important roles played by the biomolecules present in the peel extract in this green synthesis method. The particle size ranged from approximately 58 to 142 nm. A high zeta potential of around −69 mV indicated significant particle repulsion, resulting in a stable system. Considerable degradation of methylene blue dye was found when the AgNPs interacted with it under sunlight, demonstrating the viable applications of these environmentally benign, non-toxic AgNPs as a photocatalyst [65–69].

Pavan Akkiraju *et al.* [12] biosynthesized AgNPs from the Aarakata, a variant of pomegranate, with the help of an $AgNO_3$ solution, characterized them, and studied their activities against different bacteria. The exocarp, mesocarp, and seed were separated from the fresh and cleaned Aarakata, and chemically processed (boiled, cooled, and filtered) to prepare the exocarp and mesocarp extracts and seed juice, which were kept at a low temperature for further studies. Each of these samples was then mixed with $AgNO_3$ solution in appropriate proportions at ambient conditions for some time to synthesize three different types of AgNPs [70–76]. The change of color and appearance of characteristic SPR peaks of AgNPs at 556 nm for exocarp, at 506 nm for mesocarp, and at 453 nm for seed juice in UV–Vis measurements at different times confirmed the crucial roles played by different biomolecules in the synthesis. In the FTIR spectra of three different samples, numerous distinct bands and peaks due to bending, vibrations, and stretches of functional groups as well as shifting of the peaks from one sample to another were observed, which indicated the presence of various biomolecules associated with three different AgNPs [77–83]. These bands encompassed the range of 3200 to 3600 cm^{-1}, signifying the presence of O-H in all three AgNPs. In exocarp-assisted AgNPs, additional peaks (cm^{-1}) were observed at 1714 for C = O, at 1641 for C = C, at 1392 for C-N, 1217 and at 1278 cm^{-1} for C-O. In mesocarp-assisted AgNPs, additional peaks appeared at 2083 for C≡C, at 1645 for C = C, and at 1215 cm^{-1} for C-O. Lastly, in seed juice-assisted AgNPs, additional peaks were observed at 2088 for C≡C, at 1643 for C = O, and at 1217 cm^{-1} for C-O. Principal component analysis further confirmed the association of various biomolecules with AgNPs. Antibacterial efficiency tests were conducted on *Escherichia coli*, *Pseudomonas aeruginosa*, and *Proteus vulgaris* using three different biogenic samples – exocarp-assisted, mesocarp-assisted, and seed juice-assisted AgNPs. Various inhibition zones ranging from 8 to 16 mm (for both exocarp and mesocarp-assisted AgNPs) and from 9 to 16 mm (for seed juice-assisted AgNPs) were observed at different concentrations ranging from 10 to 40 μL. The findings were contrasted with the outcomes derived from the conventional ampicillin antibiotic, demonstrating its superior performance in comparison to the three samples [84–88].

Sonia Sarkar *et al.* [13] prepared a powder from the cleaned and dried pomegranate leaves. This powder was then mixed with distilled water, boiled, cooled, and subsequently filtered to create the aqueous leaf extract. The extract was mixed with $AgNO_3$ in appropriate proportions at ambient conditions for synthesis. The formation of stable and biocompatible AgNPs was confirmed when the color of the mixture changed from golden yellowish to dark brownish and a sharp SPR peak (450 nm) in UV–Vis measurements. DLS measurements showed the particle size was around 46 nm. A high value of approximately –67 mV in zeta potential analysis demonstrated the significant repulsion among the AgNPs, which prevented the particles from coalescing and made the AgNPs highly stable. The appearance of three peaks 667, 1638, and 3339 cm^{-1}, corresponding to vibration/stretching of functional groups C-Cl, C=O, and O-H, respectively, in the FTIR spectrum exhibited the signature of various phytochemicals in the synthesized AgNPs and confirmed their contributions in the reduction of Ag^+ into neutral Ag metals, capping, and finally forming AgNPs. Scanning electron microscopy (SEM) images showed the particle size varied from 40 to 70 nm. The EDS spectrum revealed a strong peak at 3.2 keV, a characteristic of AgNPs, confirming the presence of the Ag element. XRD pattern substantiated the formation of crystalline AgNPs. The anticancer activities of these nanoparticles were investigated on the HeLa human cervical cancer cells using MTT assay, lactate dehydrogenase (LDH) cell cytotoxicity, and DNA fragmentation assay. The viability of cells was found to decrease with increasing concentrations of AgNPs, indicating that these AgNPs can potentially be used for anticancer treatment [89–93].

Devanesan and colleagues [14] began by isolating the peel from freshly cleaned pomegranates, followed by washing, drying, and pulverizing the peel into a fine powder. This powder was subsequently mixed with deionized distilled water under ambient conditions for approximately one day and then filtered to obtain the peel extract. The resulting extract underwent a chemical interaction with $AgNO_3$ for several hours. The initially colorless mixture transformed into a brown hue, indicating the successful synthesis of biogenic AgNPs. Further chemical processing was applied to refine and purify the AgNPs, which were stored at low temperatures for subsequent analyses. UV–Vis measurements taken at regular intervals (24, 48, and 72 hours) revealed an SPR peak at 378 nm with increasing intensity over time. The increasing intensity of this peak indicated a continuous reduction of Ag^+ by the biomolecules present in the peel extract over time [94–98]. Consequently, the escalation in reaction time corresponded to the synthesis of a greater quantity of silver nanoparticles. The XRD graph exhibited the formation of *fcc* crystal structure, with the dominant presence of AgNPs, specifically the (111) plane. Transmission electron microscopy (TEM) analysis revealed the production of spherical AgNPs with sizes ranging from 20 to 40 nm, averaging approximately 27 nm, in this environmentally friendly nanosynthesis process. SEM imaging reinforced these findings, indicating a uniform distribution of particles, suggesting that these AgNPs could be used for nanodrug delivery as the particle size is less than 100 nm. Energy-dispersive X-ray spectroscopy (EDX) measurements indicated a substantial Ag element content, approximately 70%, with the remaining 30% comprised of other elements such as oxygen, copper, and potassium in the AgNPs [99–103].

Antibacterial efficacies were evaluated by observing distinct effective inhibition zones when these biocompatible AgNPs, at concentrations ranging from 25 to 100 μg/mL, were tested against both gram-positive and gram-negative strains, including *Escherichia coli*, *Pseudomonas aeruginosa*, *Proteus vulgaris*, *Salmonella typhi*, *Staphylococcus aureus*, *Staphylococcus epidermidis*, and *Klebsiella pneumoniae*. Variable zones of inhibition, reaching up to 14 mm, were observed against most bacteria, correlating with the AgNPs' concentrations and bacterial types, except for *Salmonella typhi*. This suggests the potential effectiveness of low-concentration AgNPs against diverse bacterial strains. Additionally, the researchers investigated the cytotoxic properties of green-synthesized AgNPs on colon cancer cells, noting a time-dependent impact on cell viability. Their findings led to the conclusion that these non-toxic, cost-effective, and environmentally friendly AgNPs hold significant potential as candidates for nanodrug carriers and for the treatment of bacteria-related diseases, including cancer [104–107].

Hui and his group [15] recorded the phytosynthesis of an $Ag/Ag^+/Ag^{3+}$ nanocomposite/nanoparticle ($Ag/Ag_2O/AgO$ nanocomposite) via peel extract, acting as the reducing agent for Ag^+. They employed diverse spectroscopic techniques to characterize the $Ag/Ag^+/Ag^{3+}$ nanoparticles and evaluated their antibacterial efficacy against both gram-positive and gram-negative bacteria. They extracted peel from the fruit, subjected it to rigorous cleaning, and combined it with purified water under controlled thermal and chemical conditions. Subsequently, the resulting mixture underwent filtration to obtain the peel extract essential for synthesizing the $Ag/Ag^+/Ag^{3+}$ nanocomposite [108–113]. An appropriate volume of peel extract was combined with $AgNO_3$ at specific concentrations and under controlled pH conditions. The transition from a light-yellow hue to a dark brown shade indicated the successful formation of $Ag/Ag^+/Ag^{3+}$ nanocomposite. Following this, the nanocomposite underwent further chemical processing and was subsequently dried for several hours, resulting in the powdered form of the $Ag/Ag^+/Ag^{3+}$ nanocomposite. The synthesis process involved the reduction of Ag^+ to neutral Ag^0 and the presence of Ag^+ in the nanocomposite, the oxidation of Ag^+ to Ag^{3+} as well as presence of Ag^{3+} in the nanocomposite, contributing to the formation of this nanocomposite. UV–Vis measurements revealed a distinct and intense SPR peak at 450 nm, indicating the active involvement of diverse biological compounds present in the peel extract in crafting a remarkably stable and biocompatible $Ag/Ag^+/Ag^{3+}$ nanocomposite. High-resolution TEM imaging showcased the predominant formation of spherical nanocomposites ranging in size from 15 to 35 nm, exhibiting exceptional crystallinity. The XRD pattern further confirmed the formation of $Ag/Ag^+/Ag^{3+}$ nanocomposite ($Ag/Ag_2O/AgO$ nanocomposite) with average sizes of 19.5, 34.3, and 33.1 nm [114–120].

FTIR measurements conducted on the $Ag/Ag^+/Ag^{3+}$ nanocomposite revealed specific vibrations and stretches attributed to the bonds within the biomolecules present in the extract, crucial for composite formation. Comparative analysis of the FTIR spectra between the $Ag/Ag^+/Ag^{3+}$ nanocomposite and the peel extract showcased shifts toward red (or blue) wavelengths and variations in peak intensities within the $Ag/Ag^+/Ag^{3+}$ nanocomposite, indicating alterations relative to the peel extract. The $Ag/Ag^+/Ag^{3+}$ nanocomposite exhibited significant antibacterial effectiveness against both *Staphylococcus aureus* and *Escherichia coli*. This notable efficacy can be attributed to the higher charges and smaller radii of the Ag^+ and Ag^{3+} ions within the $Ag/Ag^+/Ag^{3+}$ nanocomposite, rendering them more efficient in penetrating and damaging bacterial cells [121–126].

Swilam *et al.* [16] isolated polyphenols and flavonoids from pomegranate leaf extract, identifying ethyl acetate as the primary fraction containing these compounds. Subsequent profiling of ethyl acetate revealed 23 identified compounds, with ellagic acid being the most predominant contributor.

After synthesizing AgNPs using this extract, UV–Vis measurements exhibited a robust SPR peak at 425 nm. FTIR peaks observed at 1635 cm^{-1} (C=C bond) and 3300 cm^{-1} (OH bond) highlighted the active involvement of various peel extract derivatives in the reduction and formation of stable biogenic silver nanoparticles [127–132].

Light scattering analysis revealed an average AgNP diameter of approximately 26 nm. XRD data displayed four intense silver characteristic peaks (1 1 1), (2 0 0), (2 2 0), and (3 1 1) with respective interplanar spacings of 2.378, 2.0561, 1.451, and 1.236 Å, indicating the formation of an *fcc* crystal lattice, with an average size of ~32 nm. SEM imaging showcased the formation of polygonal-shaped particles, with an average diameter ranging from 20 to 40 nm, with the high-resolution image showing the presence of biomolecules on the surface of nanoparticles [133].

The antibacterial efficacy of these biogenic AgNPs was assessed against gram-positive (*Bacillus subtilis, Staphylococcus aureus,* and *Sarcina lutea*) and gram-negative bacteria (*Salmonella paratyphi, Escherichia coli,* and *Pseudomonas aeruginosa*) as well as the fungus *Candida albicans* using different concentrations of nanoparticles. The observed zones of inhibition varied based on the concentrations of AgNPs used and the specific type of microorganism [134–141].

Shalini Chauhan and a team of researchers [17] initiated the process by mixing pomegranate leaves with distilled water. The mixture was heated for a brief period and then subjected to filtration, resulting in the creation of the leaf extract. This leaf extract was combined with $AgNO_3$ in appropriate proportions under ambient conditions for a specified duration, facilitating the biosynthesis process. The transition in color from clear to yellowish, accompanied by the subsequent emergence of an SPR peak at 430 nm in the UV–Vis spectrum, provided conclusive evidence of the pivotal roles played by the biomolecules within the leaf extract in the formation of AgNPs. Analysis through XRD spectra confirmed the presence of crystalline AgNPs. TEM imaging unveiled the production of spherical particles with sizes ranging from 10 nm to 15 nm. Subsequently, the team delved into exploring the antibacterial and antifungal properties of these AgNPs. Tests were conducted on *Escherichia coli*, *Pseudomonas aeruginosa*, and *Aspergillus flavus*, revealing the formation of inhibition zones. These findings suggest that these eco-friendly and cost-effective biogenic AgNPs hold potential for application in the treatment of bacterial and fungal infections and associated diseases.

Al-Othman Monira R *et al.* [18] began the process by preparing a powder derived from pomegranate peel through a series of chemical procedures involving cleaning, drying, and crushing. This resulting powder was then combined with distilled water, subjected to boiling, and finally filtered to yield the peel extract. Subsequently, the extract underwent a purification process in preparation for AgNPs synthesis. The purified peel extract was meticulously mixed with an $AgNO_3$ solution in carefully measured proportions and allowed to chemically interact at a moderate temperature for an extended duration. The transformation of the color from yellow to a rich dark brown, accompanied by the subsequent emergence of an SPR peak at 437 nm in UV–Vis measurements, provided compelling evidence supporting the key role played by the various phytochemicals present in the peel extract in facilitating the biosynthesis of AgNPs. Furthermore, the FTIR spectrum analysis revealed four distinct peaks corresponding to the vibration and stretching of different functional groups (including amide, carboxyl, and hydroxyl) at 3281, 1635, 431, and 421 cm^{-1}. This reaffirmed the involvement of various biomolecules associated with the peel extract in the reduction of Ag^+ into neutral Ag metal and, ultimately, in the formation of AgNPs. The XRD spectrum clearly revealed the formation of AgNPs exhibiting an *fcc* crystal structure. The most prominent peak in the spectrum corresponded to the (111) plane of silver. SEM and TEM images provided visual evidence of the synthesis of spherical nanoparticles, with particle sizes spanning the range from 5 to 50 nm. The researchers conducted a dose-dependent evaluation of the antiaflatoxin properties of AgNPs on aflatoxin B1 production in *Aspergillus flavus* fungus. They observed a decrease in the rate of aflatoxin B1 production with increasing AgNPs concentration.

Gnana Jobitha *et al.* [19] cleaned and dried the fresh pomegranate fruit (arils or seeds), crushed the seeds, and mixed them with purified water. The resulting mixture was filtered twice to prepare the fruit extract for the synthesis of AgNPs. This extract was then combined with an $AgNO_3$ (metal precursor) solution in specific proportions. The color change from pink to brown announced the formation of AgNPs in this eco-friendly and cost-effective biosynthesis method. The mixture underwent centrifugation at a moderate temperature multiple times and was subsequently dried to obtain a pure sample for further characterization and analysis.

Time-dependent UV–Vis spectroscopic measurements revealed a maximum SPR peak between 450 nm and 470 nm, indicating the formation of AgNPs through the biocompounds present in the fruit extract. Magnified SEM images displayed spherical nanoparticles with diameters ranging from 30 nm to 40 nm. EDAX analysis exhibited multiple peaks, with the dominant peak at 3 keV attributed to the AgNPs, while minor peaks were associated with different chemical elements from the fruit extract. XRD patterns confirmed the crystalline nature of the AgNPs. FTIR spectra displayed peaks/bands at approximately 601, 651, 750, 810, 1336, 1193, 1456, 1672, 2929, 3780, and 3697 cm^{-1}, corresponding to vibration, stretching, and bending of various groups such as C-Cl, C=O, C-F, C-H, C-F, and C-OH present in the alkenes, alkyl halides, amines, carboxylic acids, carbonyls, alcohols, aromatics, ethers, and esters. These results indicate the vital role of phytochemicals from the fruit extract in reducing, capping, and forming AgNPs. The antibacterial efficacy of the biocompatible AgNPs was tested on both gram-positive (*Bacillus subtilis*) and gram-negative (*Klebsiella planticolae*) bacteria using various concentrations at room temperature for several hours. The AgNPs demonstrated higher effectiveness compared to common antibiotics.

Ibrahim *et al.* [20] prepared the pomegranate fruit juice extract by combining fruit juice and purified water using a microwave at 65°C for a brief duration (1 minute). Subsequently, the juice extract was preserved at subzero temperatures for the synthesis of AgNPs. An $AgNO_3$ solution was prepared by mixing $AgNO_3$ with purified water at room temperature and stirring the mixture for a period. The juice extract and $AgNO_3$ solution were then combined in specific proportions using a microwave for 5 minutes, undergoing successive heating processes up to 65°C. The change in color indicated the formation of AgNPs.

UV–Vis measurements displayed a distinct SPR peak at approximately 417 nm, a characteristic signature of AgNPs. FTIR measurements showed various peaks at 3427, 2931, 1343, and 1797 cm^{-1} corresponding to vibration, stretching, and bending of various bonds such as O-H, C-H, and C=O bonds. The peaks of AgNPs were compared with the peaks of fruit juice. FTIR measurements of fruit extract showed the peaks similar to the reported literature values of biocompounds. Discrepancies in peaks between AgNPs and fruit juice were observed. SEM images illustrated the fabrication of spherical silver nanoparticles with an average size of approximately 1.3 μm, while TEM images confirmed spherical particles with diameters less than 30 nm and an average size of ~22 nm. EDS analysis revealed a prominent characteristic peak of AgNPs at 3 keV, alongside low-intensity signals corresponding to various elements such as C and O. XRD patterns indicated that the AgNPs possessed an *fcc* structure.

The antibacterial properties of both the fruit juice extract and nanoparticles were evaluated against *Staphylococcus aureus* and *Escherichia coli* bacteria using different concentrations. The AgNPs exhibited a larger inhibition zone against *Escherichia coli* in comparison to *Staphylococcus aureus*, while the juice extract showed no inhibition zones against these bacteria. These results strongly suggest that AgNPs synthesized from the juice extract possess superior antibacterial properties compared to the juice extract alone. Therefore, these pomegranate juice-extracted AgNPs hold potential for antibacterial treatments.

Shanmugavadivu *et al.* [21] utilized pomegranate fruit peel extract to synthesize AgNPs and examined their antibacterial properties. They procured fresh peels from the fruit, thoroughly washed and dried them. Subsequently, they blended the dried peels with distilled water in appropriate proportions, boiled the mixture for a certain duration, and filtered it to generate the extract. This extract was then combined with a silver nitrate solution under specific thermal and chemical conditions, allowing it to undergo a chemical reaction for several hours. The colorless solution transformed into a brown hue, signifying the formation of AgNPs facilitated by the botanical compounds present in the solution.

The time-dependent UV–Vis spectra showed an intense peak at approximately 370 nm corresponding to the characteristics SPR peak of AgNPs.

FTIR measurements showed various peaks (cm^{-1}) (such as at 1373, 1635, 2924, and 3371) associated with the vibration/stretching of different biomolecules present in the solution. These findings suggested that these biomolecules played a role in the bioreduction of Ag^{+} ions into neutral silver particles (Ag^{0}) and in the capping and formation of stable AgNPs. SEM images depicted the production of nanoparticles exhibiting a wide size distribution, with average sizes ranging from 5 to 50 nm.

The antibacterial efficacy of these biologically synthesized AgNPs proved effective in tests conducted on *Staphylococcus aureus*, *Pseudomonas aeruginosa,* and *Escherichia coli* species.

Nisha *et al.* [22] employed both pomegranate seed and peel extracts as bioreduction agents to synthesize AgNPs from $AgNO_3$ solution, investigating their potential as antibacterial agents. The seeds and peels were extracted from the fruit, meticulously cleaned, pulverized, mixed separately with purified water in appropriate proportions, and filtered to produce seed and peel extracts, respectively, for AgNP synthesis. These extracts were then separately combined with $AgNO_3$ solution under controlled chemical conditions and proper concentrations. The change in color from pink to brown for the seed/AgNO3 solution and yellow to tea brown for the peel/$AgNO_3$ solution indicated the formation of AgNPs. Various molecules within the extracts played pivotal roles in reducing Ag^{+} ions to neutral silver metals (Ag^{0}), followed by capping and the formation of stable biogenic AgNPs. Subsequently, the AgNPs underwent further processing at high temperatures to obtain purified AgNPs in powder form.

UV–Vis spectra exhibited distinct SPR peaks at 472 and 448 nm for seed- and peel-mediated AgNPs, respectively. FTIR measurements revealed peaks corresponding to stretching, vibration, and bending bands of various biochemical compounds (e.g., alkynes, alkyl halides, amines, aromatics, etc.) present in the extracts. In the case of seed-mediated AgNPs, peaks appeared at 3449 (O-H), 2073 (alkynes), 1400 (C-C), 1638 (N-H), 1053 (C-N), and 675 cm^{-1} (C-H), whereas for peel-mediated AgNPs, peaks were observed at 3364 (O-H), 2931 (C-H), 1639 (N-H), 1090 (C-H), and 694 cm^{-1} (C-H). XRD patterns indicated the formation of cubic structured AgNPs in both cases, with sizes of 10 to 30 nm for seed-mediated AgNPs and 15 to 35 nm for peel-mediated AgNPs. SEM images depicted the formation of flower- and spherical-shaped AgNPs, with diameters ranging approximately from 10 nm to 30 nm in both cases. EDAX measurements showed a strong SPR signal at approximately 3 keV, the characteristic peak of AgNPs.

The antibacterial efficacy of both seed- and peel-mediated AgNPs was assessed against *Pseudomonas*, *Bacillus cereus*, *Staphylococcus albus*, and *Proteus* pathogens. The AgNPs exhibited varying zones of inhibition, ranging from 0 mm to 14 mm (17 mm), depending on the type of bacteria and seed (peel) extract used. These findings suggest that these cost-effective and environmentally friendly biogenic AgNPs hold promise as agents for treating bacteria-related diseases.

Khan *et al.* [23] showcased the successful utilization of pomegranate peel extract in the synthesis of AgNPs with remarkable characteristics and promising applications in the realms of antibacterial and anticancer therapies. They conducted an experiment involving the preparation of pomegranate peel extract and its subsequent utilization in the synthesis of AgNPs. The dried pomegranate peel powder was mixed with a controlled combination of methanol and water, filtered, and then further purified. The resultant peel extract was stored at low temperatures before being mixed with water at suitable concentrations. This peel extract solution was later combined with $AgNO_3$ at room temperature and allowed to undergo a processing period. The transition of the mixture's color from colorless to dark brown indicated the active involvement of various biomolecules from the peel extract in the creation of stable biogenic AgNPs. Subsequently, the sample underwent characterization processes and was evaluated for its antibacterial and anticancer applications.

UV--Vis spectrophotometry revealed the SPR peak at approximately at 450 nm. TEM images illustrated the formation of spherical nanoparticles, showing a diameter range of 15 to 30 nm. DLS analysis indicated an average particle size of approximately 20 nm. Additionally, zeta potential measurements displayed a surface charge of around –31 mV.

Antimicrobial and antibacterial activities of these AgNPs were evaluated against *Listeria monocytogenes* and MDA-MB-231 breast cancer cells. Various concentrations of AgNPs were employed using diverse techniques, demonstrating remarkable effectiveness. These findings highlighted the potential of peel-extracted AgNPs as cost-effective, non-toxic, and eco-friendly agents for the treatment of bacterial infections and cancer.

Tao *et al.* [24] conducted a synthesis of AgNPs using polyphenols extracted from pomegranate peel. The nanoparticles were thoroughly characterized using UV–Vis spectroscopy, TEM, and particle size distribution analysis. Their catalytic properties were evaluated for the degradation of Acid Red 6 with the support of $NaBH_4$. The study suggested that biogenic AgNPs hold promise for mitigating environmental pollution caused by toxic dyes.

Initially, they combined the polyphenols extracted from pomegranate peel with an $AgNO_3$ solution in the proper ratio and observed a distinct change in the mixture's color, indicating the formation of AgNPs.

UV–Vis spectroscopy confirmed the presence of a characteristic SPR peak for AgNPs at 433 nm, further substantiating their formation. Particle size distribution analysis highlighted the presence of two distinct sets of AgNPs: one ranging from 1 nm to approximately 10 nm and the other from 10 nm to around 100 nm. TEM imaging displayed the formation of various nanoparticle shapes, including spherical and ellipsoidal structures.

Catalytic properties of the AgNPs were tested for the degradation of Acid Red 6 dye with $NaBH_4$. UV–Vis measurements revealed a significant decrease in dye intensity within a short duration when AgNPs were present. In contrast, measurements conducted solely with $NaBH_4$ (without AgNPs) showed no significant reduction in dye intensity even after an extended period. These findings clearly demonstrated that AgNPs, aided by $NaBH_4$, efficiently degrade the dye, presenting a potential environmentally friendly solution.

Thilagavathi *et al.* [25] prepared pomegranate peel extract by cutting freshly washed pomegranate peel into small pieces. These pieces were mixed with purified water in specific proportions and under controlled conditions, followed by filtration to obtain the extract. The obtained peel extract was gradually combined with a chemically prepared $AgNO_3$ solution. Over time, a gradual color change from pale pink to chocolate brown signaled the successful formation of biogenic AgNPs.

The XRD pattern revealed a cubic crystal structure with an average size of approximately 35 nm. FTIR measurements highlighted the presence of diverse biocompounds within the peel extract, evidenced by peaks at various wavenumbers corresponding to the bending or stretching of different bonds (such as 2340: C-C; 1632: C=O; 667 cm^{-1}: C-H, etc.) associated with these biomolecules. These findings unequivocally demonstrated that the biocompounds present in the peel extract served as effective reducing agents for transforming Ag^+ ions into neutral Ag^0 metal, leading to the capping and subsequent formation of stable, environmentally friendly AgNPs.

The phytoconstituents found predominantly in the Omani peel extract are flavonoids and polyphenols [26]. Quercetin and gallic acid stand out as the primary compounds within the flavonoid and polyphenol content of the pomegranate peel extract. They are known for various medicinal properties [26].

Saad *et al.* [26] conducted a study to synthesize silver nanoparticles (AgNPs) using an eco-friendly approach employing Omani pomegranate peel extract (P) and two polyphenolic natural compounds, quercetin (Q) and gallic acid (GA), as reducing agents. The researchers characterized these AgNPs, assessed their biological activities, and performed comparative analyses.

Initially, fresh and clean pomegranate fruit peels were dried, powdered, and processed into extracts using ethanol under proper thermal and chemical conditions. AgNPs were synthesized separately by mixing $AgNO_3$ solutions with each of the peel extract, quercetin, and gallic acid under specific chemical conditions. This process resulted in three distinct types of AgNPs: peel extract-mediated (P-AgNPs), quercetin-assisted (Q-AgNPs), and gallic acid-mediated (GA-AgNPs). Color changes from pale yellow to deep brown for P-AgNPs and Q-AgNPs, and colorless to brown for GA-AgNPs confirmed the successful formation of AgNPs in all cases.

UV–Vis measurements revealed distinct SPR peaks at 425 nm, 421 nm, and 413 nm for Q-AgNPs, P-AgNPs, and GA-AgNPs, indicating significant differences in their characteristics. The results further suggested that various phytochemicals present in the peel extract were more effective than the two polyphenolic natural products (quercetin and gallic acid) in reducing Ag^+ to neutral Ag^0 metal, capping, and forming stable AgNPs. The XRD spectra for each of the three samples exhibited analogous patterns, signifying the generation of AgNPs characterized by a cubic crystal structure. SEM and TEM visualizations unveiled the presence of diverse shapes and aggregated AgNPs across all samples. Notably, spherical nanoparticles dominated the shapes in all three scenarios. EDX spectra demonstrated a prominent peak representing AgNPs, accompanied by minor peaks attributed to the attachment of various biomolecules on the AgNP surfaces. Particle size analysis indicated an average size of 43.6 nm for Q-AgNPs, 31.6 nm for P-AgNPs, and 21.7 nm for GA-AgNPs, highlighting the difference in sizes among the three variants. Zeta potential measurements confirmed the stability of all samples, with P-AgNPs showing intermediate surface potential (–21.5 mV) between Q-AgNPs (–18.01 mV) and GA-AgNPs (–27.9 mV).

Biological properties including antioxidant, antibacterial, and cytotoxic activities of the three sample variants were rigorously assessed and compared. This comparative analysis was carried out meticulously, following specific protocols and varying concentrations of the samples to ensure accurate evaluation. Four bacteria strains (two gram-positive and two gram-negative), namely, *Staphylococcus aureus*, *Bacillus subtilis*, *Escherichia coli*, and *Proteus vulgaris,* were used for antibacterial properties of Q-AgNPs, P-AgNPs, and GA-AgNPs. Antibacterial assays against four bacterial strains demonstrated varied inhibition zones (from 0 mm to approximately 18 mm), contingent upon the sample concentration and bacteria type. The dose (concentration)-dependent antioxidant properties of all three samples were tested on 2,2-diphenyl-1-picrylhydrazyl (DPPH) with P-AgNPs exhibiting the highest potency, followed by Q-AgNPs and GA-AgNPs. Cytotoxicity evaluation using brine shrimp eggs revealed P-AgNPs as the most cytotoxic, followed by GA-AgNPs and Q-AgNPs.

This study underscores the promising antioxidant, antibacterial, and cytotoxic properties of P-AgNPs, Q-AgNPs, and GA-AgNPs. Additionally, it highlights the distinct physical, chemical, and biological differences among these synthesized AgNPs despite the presence of quercetin and gallic acid in the pomegranate peel extract.

Nisha *et al.* [27] utilized the pomegranate leaves to synthesize AgNPs. The process involved cleaning and pulverizing pomegranate leaves, then mixing them with distilled water in appropriate ratios and filtering to obtain a leaf extract suitable for synthesis. This extract was combined with an $AgNO_3$ solution at controlled concentrations. A notable change in color indicated the rapid formation of biogenic and stable AgNPs. This transformation suggested the involvement of various biomolecules from the leaf extract in the reduction, capping, and formation of AgNPs. Subsequently, the particles underwent chemical processing to achieve purified AgNPs.

UV–Vis measurements displayed a distinctive and robust SPR peak of AgNPs at 437 nm. FTIR spectrum analysis of the leaf extract exhibited multiple bands corresponding to various phytochemicals present in the extract, which played a significant role in nanoparticle formation. XRD measurements confirmed the presence of AgNPs with a cubic crystal structure, ranging in size from approximately 10 nm to 15 nm. SEM imaging indicated the formation of AgNPs with a nanorod shape, with particle diameters varying between 10 nm and 30 nm. The EDAX spectrum exhibited a pronounced peak at 3 keV, characteristic of AgNPs, attributed to the SPR.

The antibacterial efficacy of these biocompatible AgNPs was evaluated using the disc diffusion technique against *Pseudomonas*, *Bacillus cereus*, *Staphylococcus albus*, and *Proteus* bacterial strains. A zone of inhibition measuring up to 15 mm was observed for the first three bacterial strains, while the *Proteus* strain did not display any zone of inhibition.

The contaminated drinking water harbors numerous harmful bacteria that can lead to various diseases in the human body. Additionally, hazardous wastewater significantly impacts the environment. To address these detrimental effects on both human health and the environment, a range of techniques has been employed [28]. These methods encompass precipitation, adsorption, ion exchange, reverse osmosis, activated carbon, ozonation, UV photolysis, membrane (micro, ultra, nano) filtration, flotation, oxidation, biosorption, and more techniques [29]. The extensive exploration of nanoparticle applications for water purification stems from their reduced sizes, extensive surface areas, and distinctive physical, chemical, and biological attributes [29].

Hassoon *et al.* [30] reported the potential use of AgNPs in purifying polluted drinking water, contributing to the creation of an eco-friendly environment. They collected water samples from three distinct sources in Basra, Iraq – river water, tap water, and reverse osmosis (RO) water filters. The pomegranate peel was separated from the fruit, dried, and crushed into powder. An appropriate amount of this powder was mixed with pure water and processed for approximately 2 days to create a solution, which was then filtered to obtain the peel extract. A specific quantity of this extract was mixed with an $AgNO_3$ solution to synthesize AgNPs at room temperature. The change in color from gray to brown within 3 hours indicated the successful formation of biogenic, non-toxic, eco-friendly, and stable AgNPs. The synthesized samples were subjected to characterization using UV–Vis spectroscopy, AFM, and TEM. AFM measurements revealed a surface roughness of approximately 56 nm and a particle size of 42 nm, TEM images further confirmed the size of the AgNPs.

The antibacterial properties of these AgNPs were evaluated against both gram-positive and gram-negative bacteria, including *Staphylococcus epidermidis*, *Staphylococcus aureus*, *Escherichia coli*, *Klebsiella species*, and *Enterobacter species*. Diverse zones of inhibition were observed. The findings suggested the potential use of biogenic AgNPs for purifying water contaminated by various microorganisms responsible for causing diseases due to the consumption of polluted water.

The presence of foodborne pathogens like bacteria, viruses, parasites, and fungi poses risks to human health, the food packaging industry, and the environment [31]. Bintis *et al.* [31] conducted an in-depth analysis of these pathogens, including the epidemics/pandemics they've caused. Nanoparticles have proven effective in managing these foodborne pathogens due to their size, substantial surface-to-volume ratio, and distinctive properties compared to their bulk counterparts. Ajose *et al.* [32] conducted a review focusing on the utilization of plant-synthesized bionanoparticles for managing antibiotic-resistant pathogens in animal husbandry (such as in milk and meat production).

Farouk *et al.* [33] analyzed the properties of pomegranate peel extract, synthesized AgNPs from this extract, characterized the resulting AgNPs using diverse analytical methods, and assessed their antimicrobial effectiveness against a range of foodborne pathogens. The scientists combined a specific amount of pomegranate peel with distilled water and processed the mixture to obtain the pomegranate peel extract at room temperature. Subsequently, they combined an $AgNO_3$ solution with the peel extract using precise concentrations and let the mixture react under controlled conditions in darkness. Over time, the mixture's color transitioned from a pale yellow to a dark orange hue, signifying the successful synthesis of stable and biocompatible AgNPs through this eco-friendly synthesis process. The UV–Vis spectrum showed the characteristic SPR peak of AgNPs at 440 nm, demonstrating the roles of various biomolecules of peel extract in the reduction of Ag^+ ions, capping, and formation of stable AgNPs. DLS measurements exhibited the wide variation of diameters of the AgNPs ranging from approximately 75 to 4085 nm with an average value of approximately 624 nm. Zeta potential measurement showed a potential of approximately -20 mV. The researchers suggested that the wide variation in diameter and low value of zeta potential could be due to the synthesis process and the structure and concentration of various phytochemicals present in the peel extract. FTIR analysis revealed peaks corresponding to vibration and stretching of compounds of botanical compounds, thus further confirming the presence of various bioactive molecules (flavonoids, alkaloids, etc.) in the peel extract. SEM images showed the formation of semi-spherical AgNPs. The researchers additionally analyzed the composition of biomolecules within the peel extract, identifying phenols as the predominant contributors to its constituents.

The antimicrobial efficacy of these AgNPs was assessed against eight distinct foodborne pathogens using varying concentrations of AgNPs. The experiment involved six bacterial strains (*Bacillus subtilis*, *Enterococcus faecalis*, *Escherichia coli*, *Klebsiella pneumoniae*, *Salmonella typhi*, and *Shigella sonnei*), along with two fungal species (*Aspergillus flavus* and *Rhizopus oryzae*). Diverse zones of inhibition were observed, varying with the concentration of the AgNPs and the specific type of pathogen.

Khorrami *et al.* [34] synthesized AgNPs from black peel pomegranate (also known as black leather pomegranate), characterized the green AgNPs, and studied their antioxidant and anticancer activities. The black peel was extracted from fresh and cleaned black pomegranates, cut into small pieces, and mixed with distilled water in appropriate proportions. This mixture was chemically processed at 37°C for 24 hours, followed by filtration to obtain the extract in powder form, which was then dried under laboratory conditions. The powder was mixed with distilled water under controlled pH conditions. Subsequently, the solution was combined with an $AgNO_3$ solution in appropriate proportions and left to interact at room temperature. The change in color of the mixture from light brown to dark brown confirmed the formation of AgNPs. The synthesized AgNPs underwent further processing for characterization and subsequent applications.

Time-dependent UV–Vis measurements revealed a red shift of the SPR peak from 414 to 423 nm, confirming the gradual formation of AgNPs. The FTIR spectrum of the peel-assisted AgNPs closely resembled that of the peel extract, validating the involvement of various botanical compounds (such as polyphenols and flavonoids) in reducing silver ions (Ag^+) to neutral metallic silver (Ag^0), acting as capping agents, and facilitating the formation of stable and biocompatible AgNPs. Several peaks observed in the FTIR data corresponded to vibrations/stretches of different functional groups associated with the phytochemicals present in the extract (3403 cm^{-1}: O-H; 1724 cm^{-1}: C-O; 1611 cm^{-1}: N-H; and 1445 cm^{-1}: C-C bonds).

XRD results revealed the formation of AgNPs displaying an *fcc* crystal structure, exhibiting sizes of approximately 10 nm with planes (1 1 1), (2 0 0), (2 2 0), and (3 1 1). DLS measurements indicated a size of about 32 nm, a polydispersity index of 0.32, and a zeta potential of approximately −40 mV. The high zeta potential value confirmed the stable formation of AgNPs. FESEM images illustrated the formation of spherical AgNPs, with an average size of approximately 16 nm.

The antioxidant and anticancer properties of these biogenic AgNPs were evaluated using DPPH and two breast cancer cell lines. The results clearly indicated the potential of these AgNPs for use as antioxidant and anticancer agents.

Abdelmigid and colleagues [35] pursued two distinct pathways to synthesize AgNPs: one involving a chemical approach using trisodium citrate, and the other employing eco-friendly methods using pomegranate fruit peel and coffee ground waste. The synthesized nanoparticles underwent thorough characterization using a variety of spectroscopic techniques. Subsequently, their antimicrobial properties were evaluated and compared across a spectrum of pathogens.

The synthesis process commenced by extracting pomegranate peel from freshly cleaned fruits, which was then dried and processed into a powdered form. Similarly, coffee waste obtained from ground beans was dried and converted into a powdered state. These individual powdered materials derived from the peel and coffee waste were separately mixed with distilled water to generate extracts: pomegranate peel extract (PPE) and coffee ground waste extract (CE). These extracts were carefully maintained at lower temperatures to facilitate the synthesis process.

Following this preparation, the PPE and CE extracts were separately combined with $AgNO_3$ solution under controlled thermal and chemical conditions using specific ratios. A notable change in color, from pale yellow to dark brown for PPE and from pale brown to brown for CE, indicated the successful generation of biocompatible and stable silver nanoparticles: PPE-assisted silver nanoparticles (AgNPs_PPE) and coffee waste-assisted silver nanoparticles (AgNPs_CE). These nanoparticles underwent further purification. Simultaneously, precise amounts of $AgNO_3$ solution and trisodium citrate were mixed and processed. The gradual transition of the colorless solution to brown indicated the formation of chemically synthesized silver nanoparticles (AgNPs_Chem), which were subsequently purified.

The UV–Vis spectra revealed distinctive features: a sharp SPR peak at 350 nm for AgNPs_PPE, dual peaks at 320 and 440 nm for AgNPs_CE, and a singular peak at 430 nm for AgNPs_Chem. These varied peak positions across the spectra distinctly indicated the formation of silver nanoparticles (AgNPs) with different morphologies in each case, namely AgNPs_PPE, AgNPs_CE, and AgNPs_Chem. Zeta potential measurements conducted on AgNPs_Chem, AgNPs_CE, and AgNPs_PPE indicated distinct values: AgNPs_Chem displayed the highest value at −46.7 mV, followed by AgNPs_CE at -12.6 mV, and AgNPs_PPE at around -8 mV. These findings revealed that AgNPs_PPE exhibited the least stability among the three, resulting in a larger particle size of approximately 592 nm. This was due to reduced repulsion forces compared to the other two nanoparticle types, which had smaller sizes of about 274 nm for AgNPs_CE and approximately 63 nm for AgNPs_Chem. The researchers contrasted their results with previously reported literature values and identified notable discrepancies. They proposed that these differences could be attributed to variations in the concentration of the $AgNO_3$ solution and pH levels. The SEM images and XRD data from all three samples confirmed the presence of *fcc* crystal structures. In the XRD data, the four predominant peaks corresponded to Ag planes (1 1 1), (2 0 0), (2 2 0), and (3 1 1). Notably, in the XRD analysis of AgNPs_PPE, supplementary minor peaks were detected, indicating the potential presence of a diverse range of phytochemicals within the pomegranate peel extract. The FTIR measurements of AgNPs_PPE and AgNPs_CE revealed distinct peaks corresponding to the vibrations and stretching of various bonds related to quinine, ether, ester, proteins, or phenols found in the peel and/or coffee waste extracts, in the context of the green-synthesized AgNPs. However, in the FTIR data of AgNPs_Chem, the vibrations and stretching were associated with different molecules from the trisodium citrate, which acted as the reducing agent.

Antibacterial efficacy of these nanoparticles was conducted on both gram-positive and gram-negative pathogens, namely *Enterobacter aerogenes*, *Klebsiella pneumoniae*, *Pseudomonas aeruginosa*, and *Methicillin-Resistant Staphylococcus aureus* (MRSA) using different concentrations of AgNPs (2, 4, and 8 mg/ml) by the well diffusion method. Different zones of inhibition were observed depending on the concentration of the AgNPs and the types of bacteria used. The zone size increased with increasing concentration.

Abad *et al.* [36] combined fresh and thoroughly cleaned pomegranate peel with deionized water in specific proportions and processed the mixture at moderate temperature for a certain duration. The resulting solution was filtered and then introduced to an $AgNO_3$ solution under controlled chemical and thermal conditions to synthesize Ag_2O NPs. The change in color indicated the successful formation of Ag_2O NPs, suggesting the active role of various biomolecules present in the peel extract in reducing Ag^+ ions to neutral Ag metal and subsequently capping and stabilizing the biogenic Ag_2O NPs.

XRD pattern confirmed the formation of *fcc* crystal structure, estimating a size of approximately 37 nm. SEM images revealed the creation of spherical nanoparticles with an average size of about 64 nm. UV–Vis spectroscopic analysis displayed a prominent SPR peak at around 460 nm, while band gap energy (E_g) measurements indicated an E_g value close to 4 eV. Furthermore, the FTIR spectrum exhibited distinct peaks corresponding to various molecular vibrations and stretching of bonds related to molecules in the peel extract, such as peaks at 3363 (O-H), 2359 (C-N), and 763 cm^{-1} (Ag=O).

The antibacterial and antifungal efficacy of the biosynthesized Ag_2O nanoparticles was assessed against *Staphylococcus aureus, S. epidermidis, Escherichia coli* bacteria, and *Candida albicans* fungus under different conditions using low concentrations of Ag_2O NPs for several hours. The results demonstrated diverse inhibition zones. The authors concluded that Ag_2O NPs hold potential for treating bacterial and fungal infections.

Fu and his team [37] processed the pomegranate peel extract under controlled conditions to employ it as a reducing agent for synthesizing AgNPs. They utilized various analytical techniques to characterize the particles and extensively studied their cytotoxicity and anticancer properties. The XRD spectrum confirmed the formation of nanocrystals with an average size of approximately 45 nm, while FE-SEM revealed spherical particles with an average size of about 63 nm. Cytotoxicity and anticancer efficacy were evaluated through MTT assays conducted over several hours across different cell lines and AgNP concentrations. A clear dose-dependent relationship was observed between AgNPs and cancer cell viability, indicating the potential therapeutic applications of AgNPs in cancer treatment.

Rozykulyyeva *et al.* [38] studied the antibacterial properties of AgNPs synthesized from the pomegranate peel extract on oral pathogens. They procured the pomegranate fruit from the local market, removed the peel carefully from the fruit, and properly cleaned the peel in distilled water. They then dried the peel for several days without contaminating it. The dried peel was further cleaned and powdered, and chemically processed to prepare the peel extract. It was then kept at a low temperature for further investigations. The peel extract and silver nitrate solution were mixed in proportions under controlled thermal and chemical conditions. The color of the mixture changed after several hours, indicating the formation of biogenic AgNPs. Subsequently, the sample was kept at a low temperature for characterizations and investigations into its properties. UV–Vis spectrum displayed an SPR peak at 418 nm. The intensity of this peak increased over time, indicating a corresponding increase in the reaction rate. Prostate-specific antigen and DLS measurements revealed the formation of particles of high quality, with sizes distributed in the range of 40–1000 nm. The peak size ranged from 40 to 350 nm, with an average size of approximately 600 nm. Additionally, the polydispersity index (PDI) value was found to be very small (around 0.4), clearly indicating the formation of stable biogenic homogeneous AgNPs. Four pathogens, two gram-negative, namely *P. aeruginosa* and *A. actinomycetemcomitans*, and two gram-positive, namely *E. faecalis* and *S. aureus*, were used to understand the antibacterial properties of these AgNPs. Different concentrations of AgNPs were utilized to measure the minimum inhibitory concentration (MIC) zones against these bacteria. The MIC zone was found to increase with the rising concentration of AgNPs. Moreover, different bacteria exhibited varying MICs. Furthermore, the ANOVA method was employed to study the differences between the two groups of bacteria.

Shanmugavadivu *et al.* [39] prepared the pomegranate peel extract following proper cleaning, drying, grinding, and filtration in a controlled environment. The peel extract was then mixed with an appropriate amount of $AgNO_3$ solution. The color of the mixture changed after some time, indicating the successful formation of biocompatible and stable AgNPs. The UV–Vis spectrum recorded a sharp SPR peak at 371 nm, confirming once again the formation of AgNPs and the roles played by various biochemicals in the synthesis process. FTIR measurements further revealed the vibrations and stretches of various groups associated with the polyphenols and other biocompounds present in the pomegranate peel extract. SEM images revealed that the average size of the particles ranged from 5 to 50 nm. The antibacterial efficacy of AgNPs was assessed on *S. aureus*, *P. aeruginosa*, and *E. coli*. Different concentrations of AgNPs were used for this experiment. Various inhibition zones were observed for different types of bacteria and various concentrations of AgNPs, indicating the potential of AgNPs as promising antibacterial candidates.

Fernandes and a group of researchers [40] synthesized AgNPs from pomegranate peel extract, characterized them, tested their antibacterial and cytotoxicity properties, and compared them with chemically synthesized AgNPs. Peels from fresh pomegranate fruits were separated, cleaned, and processed to form the peel extract. The concentrations of gallic acid, ellagic acid, pH value, and dry matter in the extract were evaluated. In controlled physical and chemical conditions, carboxymethylcellulose, propylene glycol, $AgNO_3$ solution, pomegranate peel extract, and pure water were mixed in appropriate amounts to synthesize the AgNPs. To facilitate the synthesis of chemical (conventional) AgNPs, sodium citrate was added to a boiling $AgNO_3$ solution. The XRD pattern revealed the formation of cubic crystals, with a peak (2 2 2) attributed to the Si substrate as particles were deposited onto a Si wafer. TEM images showed that the shapes and sizes of AgNPs differed between green and chemical synthesis methods, with larger sizes observed in AgNPs synthesized using the green method compared to conventionally synthesized AgNPs. DLS measurements also revealed that pomegranate peel synthesized AgNPs exhibited larger average sizes (around 90 nm) compared to conventional AgNPs (average size approximately 20 nm), with some variation observed. Zeta potential analysis indicated values of around −46 mV and −68 mV for pomegranate-assisted AgNPs and chemical AgNPs, respectively, suggesting the formation of highly stable nanoparticles in both cases. The antibacterial efficacy of peel-extracted AgNPs was tested against *Candida albicans* and *Staphylococcus aureus*, and cytotoxicity was evaluated on mammalian L929 cell lines using different concentrations of AgNPs, with comparisons made to chemical AgNPs. Various zones of inhibition were observed in both cases.

Early blight is a dangerous fungal disease of tomato and potato, which causes a huge number of potato and tomato losses worldwide. This disease is triggered by *Alternaria solani, Alternaria tomatophila, and Alternaria protenta*. They belong to the *Alternaria* species [41–42].

Mostafa *et al.* [43] utilized pomegranate and orange peel extracts to synthesize biogenic AgNPs, characterized them, and investigated their antifungal properties against *Alternaria solani* pathogens, a major cause of early blight disease in tomatoes. They compared the properties and activities of AgNPs extracted from pomegranate peel and orange peel. Peels from pomegranates and oranges were separated, cleaned, and processed to prepare the extracts. These peel extracts were then separately mixed with $AgNO_3$ solution. A change of color occurred in both cases, confirming the formation of AgNPs and demonstrating the roles of biocompounds in facilitating the eco-friendly and non-toxic synthesis of AgNPs with the assistance of pomegranate peel extract and orange peel extract. UV–Vis spectra of samples showed sharp resonance peaks at 437 nm (pomegranate) and 450 nm (orange), further confirming the formation of AgNPs.

FTIR spectra exhibited signals at –3273, 2223, 2047, and 1972 cm^{-1} in the case of pomegranate peel-mediated, and at –3260, 1634, 1376, and 1243 cm^{-1} in the case of orange peel-mediated AgNPs, indicating the existence of various chemical compounds in both peel extracts and their roles in the synthesis, capping, and formation of stable biocompatible AgNPs. TEM images showed the formation of spherical nanoparticles without agglomeration with average particle sizes ranging from 10 to approximately 17 nm for pomegranate peel-assisted AgNPs and approximately 10 to 17 nm in the case of orange peel-assisted AgNPs. The FTIR spectra revealed characteristic signals at approximately 3273, 2223, 2047, and 1972 cm^{-1} for pomegranate peel-mediated AgNPs and at around 3260, 1634, 1376, and 1243 cm^{-1} for orange peel-mediated AgNPs. These signals suggest the presence of diverse chemical compounds within the peel extracts, underscoring their roles in mediating the synthesis, capping, and formation of stable, biocompatible AgNPs. TEM imaging further corroborated these findings, demonstrating the production of spherical nanoparticles devoid of agglomeration. The observed average particle sizes ranged from 10 to approximately 17 nm for pomegranate peel-assisted AgNPs and approximately 10 to 17 nm for orange peel-assisted AgNPs. Zeta potential and particle size analyses revealed average particle sizes of approximately 8 nm and 14 nm, with corresponding zeta potential values of approximately –17 mV and –20 mV for AgNPs synthesized with the assistance of pomegranate and orange peel extracts, respectively. The phenolic contents in both extracts were also measured using high-performance liquid chromatography (HPLC) and were found to be higher in pomegranate peel extract in comparison to orange. Antifungal activities of both samples (pomegranate peel-assisted and orange peel-assisted AgNPs) were tested using different concentrations on *A. solani* fungi. Eight isolates of *A. solani* were initially selected, grouped into four groups: Group 1 containing one single isolate (A1), Group 2 containing three isolates (A2, A3, A8), Group 3 containing two isolates (A4, A6), and Group 4 containing two isolates (A5, A7). The observation indicated that isolates A1, A3, and A5 induced greater plant infection and damage compared to isolates A2, A4, A7, and A8. Subsequently, isolate A1 was selected for conducting antifungal measurements on both samples. Various zones of inhibition were observed for both peel extracts (pomegranate and orange) and peel-assisted AgNPs, across different concentrations. The peel extract-assisted AgNPs exhibited greater efficacy compared to both the peel extracts and $AgNO_3$. Furthermore, the antifungal efficacy of pomegranate-assisted AgNPs surpassed that of orange peel-assisted AgNPs.

Pomegranate seed oil, extracted from pomegranate seeds, contains various types of acids and has numerous health benefits [44]. Sadrolhosseini *et al.* [45] immersed a silver plate in pomegranate seed oil and subjected it to laser ablation using a Q-switched Nd:YAG laser for durations of 5, 10, 15, and 20 minutes to prepare silver nanofluids of various concentrations.

UV-Vis spectra showed a blue shift of the resonance peak with increasing ablation time (411 nm: 5 minutes; 407 nm: 10 minutes; 403 nm: 15 minutes; and 399 nm: 20 minutes), indicating a decrease in particle over time. FE-TEM images revealed dispersed AgNPs void of agglomeration. Images further showed a decrease in particle size with increasing ablation time, specifically, 21 nm (5 minutes), 14 nm (10 minutes), 10 nm (15 minutes), and 2 nm (20 minutes), respectively. FTIR spectrum, taken between the interval 280 and 4000 cm^{-1}, showed various peaks corresponding to stretches and vibrations of various bio compounds, including C-H at approximately 3474 cm^{-1} and 3008 cm^{-1}, CH_2 at 2923 cm^{-1} (asymmetric) and 2855 cm^{-1} (symmetric), C=C at 1743 cm^{-1}, and many more. Photo-acoustic measurements of thermal effusivity showed that AgNPs enhanced the thermal properties of pomegranate seed oil. The researchers concluded that laser ablation could be used to successfully synthesize AgNPs in pomegranate seed oil.

Al-Askar *et al.* [46] mixed pomegranate peel extract with $AgNO_3$ solution and allowed the reaction occur for some time under controlled conditions. The color of the mixture gradually changed, indicating the formation of AgNPs.

SEM images showed that spherical AgNPs of sizes ranging from 61 to 97 nm formed. SEM images illustrated the presence of spherical AgNPs within the size range of 61 to 97 nm. TEM images provided further insight, confirming the formation of nanoparticle shapes with sizes ranging from approximately 17 to 57 nm and exhibiting a mean size of about 33 nm, albeit with some deviation. Notably, TEM images also suggested the potential encapsulation of AgNPs by nonmetallic botanical compounds. Consequently, the researchers proposed that the larger particle sizes observed in the SEM images might be attributed to this encapsulation phenomenon, indicative of a capping process occurring during the synthesis of AgNPs. UV–Vis spectrum displayed a resonance peak at approximately 485 nm, indicating the characteristic surface plasmon resonance of silver nanoparticles. The XRD spectrum revealed the formation of *fcc* crystal structure, with distinct peaks observed at specific angles (2θ) corresponding to crystallographic planes, namely approximately 32° (1 2 2), 46° (2 3 1), 54° (1 4 2), and 77° (3 1 1).

Additionally, the EDX spectrum, which provides elemental analysis, exhibited a prominent peak corresponding to Ag at energy levels between 3 and 3.5 keV, indicating the predominant presence (more than 64%) of silver in the sample. Furthermore, minor peaks corresponding to carbon and oxygen elements were also observed, constituting approximately 24% and 10% of the sample composition, respectively. Collectively, these peaks characterize the structural and elemental properties of the synthesized AgNPs.

The zeta potential measurement indicated a high value of approximately –18 mV, confirming the formation of stable silver nanoparticles (AgNPs). The FTIR spectrum exhibited several peaks corresponding to various functional groups, including 3293 cm^{-1} (O-H group of alcohols and polyphenols), 2910 cm^{-1} (C-H or alkane), 2353 cm^{-1} (C=O, i.e., aldehydes and ketones), and 1628 cm^{-1} (C=O of carbonyl and N-H of amine). Additional peaks were attributed to vibrations or stretches of C-O, N-H, and C-O, among others. This spectrum clearly indicated the active roles played by various functional groups in the formation of biogenic AgNPs. The antiviral property of the AgNPs was evaluated against the tobacco mosaic virus strain, with the virus isolated from infected tomato plants for testing purposes. The results suggested that these biocompatible AgNPs could effectively inhibit the growth of this virus.

Alharbi *et al.* [47] cleaned, dried, and pulverized the pomegranate waste. Subsequently, they mixed a suitable quantity of powdered pomegranate waste with distilled water at a specific temperature for a period of time to create the pomegranate waste extract, which was then stored at a low temperature for subsequent analysis. They mixed 5 ml of extract with 50 ml of silver nitrate (metal precursor), heated it for around 20 minutes at a temperature of 50°C at a constant pH value of 6.8. The mixture changed its color from colorless to brown, indicating the formation of AgNPs in this bottom-up technique. They measured the total phenolic content (gallic acid equivalent) and total flavonoid content (quercetin equivalent) of the extract and AgNPs using the Folin and $AlCl_3$ methods. Phenolic compounds of both extract and AgNPs were quantified using HPLC. The UV–Vis spectrum showed a peak at 430 nm, providing further evidence of AgNPs formation and the diverse roles played by phytochemicals in the extract. TEM images revealed particle sizes ranging from 15 to 55 nm. Zeta potential measurement yielded a value of approximately –26 mV, indicating the formation of stable AgNPs. FTIR analysis of both pomegranate waste extract and AgNPs revealed 8 and 11 bands at different wavenumbers. These bands corresponded to various functional groups occurring at approximately 3419 cm^{-1} (OH and NH_2), 2,983 cm^{-1} (CH), 2094 cm^{-1} (alkyne), 1651 cm^{-1} and 1373 cm^{-1} (alkanes, alkenes, and aromatic), 1150 cm^{-1} (ester), and 618 cm^{-1} (halide).

They further assessed the antioxidant properties of the extract and AgNPs against DPPH radicals. It was observed that AgNPs scavenged almost 94% of the radicals. Additionally, the researchers evaluated the antibacterial and antifungal properties of AgNPs against various pathogens (e.g., *Listeria monocytogenes*, *Staphylococcus aureus*, *Campylobacter jejuni*, and *Salmonella typhi*) using the disc assay method. Inhibition zones ranging from 16 to 45 mm were observed. Furthermore, they tested the antiviral efficacy of AgNPs against SARS-CoV-2 and investigated their anticancer properties on breast cancer and human colon cancer cell lines. The results demonstrated a significant reduction in the growth of cancer cell lines in the presence of AgNPs. Additionally, different concentrations of AgNPs were added to the diet of rats to understand their effect on rat growth rate. Moreover, coated AgNPs were examined for their preservative properties and found to be effective for up to 60 days. These results suggest that AgNPs synthesized from pomegranate waste extract can serve as promising candidates for antioxidant, anticancer, antiviral, and antimicrobial activities, as well as for food preservation purposes.

Jasim *et al.* [48] prepared the pomegranate peel extract using standard practices and mixed it with an $AgNO_3$ solution under controlled conditions to synthesize the AgNPs. The change in color of the mixture demonstrated that the peel extract acted as a reducing agent for $AgNO_3$, facilitating the synthesis of AgNPs. UV–Vis spectrum exhibited a peak at 450 nm. FTIR analysis showed the appearance of different bands/peaks at various wavenumbers, indicating the existence of various biocompounds and hence their active roles in the synthesis. SEM images showed the formation of spherically shaped nanoparticles with an average size of approximately 50 nm. Zeta potential measurements indicated a value of approximately –29 mV, suggesting the formation of stable AgNPs. The antifungal activity of the AgNPs against *Aspergillus flavus* was assessed using various concentrations, revealing a dose-dependent relationship.

Various types of aflatoxins (B1, B2, G1, etc.) are produced by *Aspergillus flavus*, which cause major health issues [49]. Monira R *et al.* [49] synthesized green AgNPs from pomegranate peel extract, analyzed the nanoparticles, and measured their antifungal efficacy against the toxic aflatoxin produced by *Aspergillus flavus* fungus. They collected pomegranate fruits, cleaned and separated the peel, dried it at a moderate temperature for several hours, and then powdered it. They mixed the powder with water in appropriate proportions, boiled the mixture for some time, and further processed it to prepare the pomegranate peel extract. 10 ml of peel extract was then mixed with 100 ml of $AgNO_3$ solution at high temperature for around 20 minutes. The color of the mixture changed from yellow to dark brown, signifying the formation of biogenic AgNPs. UV–Vis results showed a peak at 437 nm, which further confirmed the formation of AgNPs. The UV–Vis results showed a peak at 437 nm, further confirming the formation of AgNPs. The XRD pattern indicated strong peaks at angles (2θ) {38.25° (1 1 1), 44.31° (2 0 0), 68.31° (2 2 0), and 77.35° (3 1 1)}, confirming the crystalline nature of the synthesized AgNPs. FTIR signals approximately at 3281, 1636, 431, and 421 cm^{-1} corresponded to vibrations and stretching of various bonds in the functional groups of phytochemicals present in the peel extract. SEM and TEM images revealed the formation of well-dispersed spherically shaped nanoparticles with average particle size ranging from 5 to 50 nm. The antifungal efficacy of AgNPs was tested against aflatoxin B_1 using various concentrations of nanoparticles. Different zones of inhibition were observed, with the maximum value observed at 150 ppm of AgNPs.

Helminths, parasitic worms, cause various types of infections in human bodies and are considered one of the most prevailing diseases in many parts of the world. Helminth diseases can be caused by ingestion of contaminated food or water or through contact with contaminated soil [50].

Thagfan *et al.* [51] used pomegranate peel extract to biologically synthesize AgNPs, studied their characteristics, and investigated their anthelmintic properties. They separated the peel from the fresh pomegranate fruit, cleaned it, and dried it. The dried peel was then chopped and mixed with methanol for several hours. The solution was further processed to prepare the peel extract. The extract was then combined with $AgNO_3$ solution for the synthesis of AgNPs. TEM images showed the formation of spherical nanoparticles with sizes ranging from 10 nm to 30 nm. The FTIR spectrum exhibited numerous peaks at specific wavenumbers (e.g., 3293.63 cm^{-1}, 1716.87 cm^{-1}, 1614.74 cm^{-1}, 1224.69 cm^{-1}, 1340.12 cm^{-1}, 919.19 cm^{-1}, 867.06 cm^{-1}, etc.), corresponding to the stretching and vibration frequencies of various bonds (e.g., O-H, C-H, C = C, etc.) associated with the phenolic groups present in the peel extract. The cytotoxicity of AgNPs was measured using the MTT assay on liver, colon, and breast cancer cell lines at different concentrations of AgNPs. The anthelmintic property of AgNPs was tested on the *Eisenia fetida* parasite using different doses (5, 2.5, and 1.25 mg/mL) of AgNPs. The results showed a dose-dependent efficacy of AgNPs. Consequently, the researchers concluded that pomegranate peel-mediated green AgNPs can be utilized as a potent anthelmintic candidate.

5.3 CONCLUSION

Summarizing our extensive exploration, we investigated eco-friendly and non-toxic green synthesis methods to produce biogenic AgNPs using diverse parts of the pomegranate fruit. The botanical components within the extract were identified as reducing agents for Ag^+ ions, contributing to the capping process and the formation of stable and biocompatible AgNPs. The color change confirmed the formation of AgNPs, while UV–Vis measurements revealed sharp SPR peaks. FTIR analysis indicated the presence of various compounds associated with phytochemicals, affirming their role in the synthesis process. We discussed other characterization techniques such as XRD, TEM, SEM, DLS, and zeta potential analysis to determine particle size, morphology, crystal structure, Ag plane orientation, and zeta potential of the nanoparticles.

Additionally, we assessed the potential applications of these AgNPs in nanobiomedicine (antibacterial and anticancer applications), water disinfection, antioxidants, and cytotoxicity. Our observations emphasized the significant impact of parameters including pomegranate parts (peels, leaves, and seeds), extraction process, concentrations of pomegranate extract and $AgNO_3$ solution, pH levels, temperature, and other factors on the properties of the synthesized AgNPs. These factors significantly influenced various characteristics of the nanoparticles, including the position and shifting of SPR peaks observed in UV-Vis and FTIR spectra, as well as the sizes, shapes, zeta potential, stability, and aggregation of the nanoparticles, consequently shaping their potential applications.

While our findings demonstrated promising applications, further detailed and systematic investigations are necessary for the synthesis and systematic characterization of these AgNPs to fully realize their potential applications across diverse fields.

REFERENCES

1. P. Prasher and M. Sharma, *Silver Nanoparticles: Synthesis, Functionalization and Applications*, Bentham Books, 2022.
2. A. Facibeni, *Silver Nanoparticles: Synthesis, Properties, and Applications*, CRC Press, 2023.
3. E. Conley (Ed.), *Applications of Silver Nanoparticles*, Nova Science Publisher, 2023.
4. Anam Pushpa Singh, T. K. Srivastava and R. R. Verma, "Nanoparticles applications in agriculture", Scientific Publisher, 2022, ISBN: 9789391418229.
5. Nadeem Baig, Irshad Kammakakam and Wail Falath, "Nanomaterials: a review of synthesis methods, properties, recent progress, and challenges", *Material Advances*, vol. 2, p. 1821, 2021.
6. S. Kanchi and S. Ahmed (Ed.), *Green Metal Nanoparticles: Synthesis, Characterization and Their Applications*, Scrivener Publishing/Wiley, 2018.
7. Maulin P. Shah and Arpita Roy, *Phytonanotechnology*, Springer, 2023, https://doi.org/10.1007/978-981-19-4811–4.
8. A. Rana, M. Samtiya, T., V. Mishra and R. E. Aluko, "Health benefits of polyphenols: a concise review", *Journal of Food Biochemistry*, vol. 46, no. 10, p. e14264, 2022.
9. Panagiotis Kandylis and Evangelos Kokkinomagoulos, "Food applications and potential health benefits of pomegranate and its derivatives", *Foods*, vol. 9, p. 122, 2020.
10. Rijuta G. Saratale, Han Seung Shin, Gopalakrishnan Kumar, Giovanni Benelli, Dong-Su Kim and Ganesh D. Saratale, "Exploiting antidiabetic activity of silver nanoparticles synthesized using Punica granatum leaves and anticancer potential against human liver cancer cells (HepG2)", *Artificial Cells, Nanomedicine, and Biotechnology"*, vol. 46, no. 1, pp. 211–222, 2018.
11. Sanket J. Joshi, S. J. Geetha, Saif Al-Mamari and Ahlam Al-Azkawi, "Green synthesis of silver nanoparticles using pomegranate peel extracts and its application in photocatalytic degradation of methylene blue", *Jundishapur Journal of Natural Pharmaceutical Products*, vol. 13, no. 3, p. e67846, 2018.
12. C. Pavan Akkiraju, S. Pooja Tathe and Srilakshmi Mamillapalli, "Green synthesis of silver nanoparticles from *Punica granatum* L. and its antimicrobial activity", *Advances in Applied Science Research*, vol. 8, no. 1, pp. 42–49, 2017.
13. Sonia Sarkar and Venkatesan Kotteeswaran, "Green synthesis of silver nanoparticles from aqueous leaf extract of Pomegranate (*Punica granatum*) and their anticancer activity on human cervical cancer cells" *Advances in Natural Sciences: Nanoscience and Nanotechnology*, vol. 9, p. 25014, 2018.

14. Sandhanasamy Devanesan, Mohamad S AlSalhi, Radhakrishnan Vishnu Balaji, Amirtham Jacob A Ranjitsingh, Anis Ahamed, Akram A Alfuraydi, Fulwah Y AlQahtani, Fadilah S Aleanizy and Ahmed H Othman, "Antimicrobial and cytotoxicity effects of synthesized silver nanoparticles from punica granatum peel extract", *Nanoscale Research Letters*, vol. 13, p. 315, 2018.
15. Yang Hui, Ren Yan-yu, Wang Tao and Wang Chuang, "Preparation and antibacterial activities of Ag/Ag+/Ag3+ nanoparticle composites made by pomegranate (*Punica granatum*) rind extract", *Results in Physics,* vol. 6, pp. 299–304, 2016.
16. Noha Swilam and Khaled A. Nematallah, "Polyphenols profile of pomegranate leaves and their role in green synthesis of silver nanoparticles", *Scientific Reports,* vol. 10, p. 14851, 2020.
17. Shalini Chauhan, Mukesh Kumar Upadhyay, Sushma Rishi and Priyanka Lokwani, "Phytofabrication of silver nanoparticles through leaf extract of pomegranate fruit", *International Journal of Pharmaceutical Sciences Review and Research*, vol. 12, no. 1, pp. 61–63, 2012.
18. R. Al-Othman Monira, R. M. Abd El-Aziz Abeer, A. Mahmoud Mohamed and A. Hatamleh Ashraf, "Green biosynthesis of silver nanoparticles using pomegranate peel and inhibitory effects of the nanoparticles on aflatoxin production", *Pakistan Journal of Botany*, vol. 49, pp. 751–756, 2017.
19. G. Gnana Jobitha, S. Rajeshkumar, G. Annadurai and C. Kannan, "Preparation and characterization of fruit-mediated silver nanoparticles using pomegranate extract and assessment of its antimicrobial activities", *J. Environ. Nanotechnol.*, vol. 2, no. 1, pp. 4–10, 2013.
20. M. H. Ibrahim, A. A. Ibrahiem and T. R. Dalloul, "Biosynthesis of silver nanoparticles using pomegranate juice extract and its antibacterial activity", *International Journal of Applied Sciences and Biotechnology*, vol. 4, no. 3, pp. 254–258, 2016.
21. M. Shanmugavadivu, Selvam Kuppusamy and R. Ranjithkumar, "Synthesis of pomegranate peel extract mediated silver nanoparticles and its antibacterial activity", *American Journal of Advanced Drug Delivery*, vol. 2, no. 2, pp. 174–182, 2015.
22. M. Haniff Nisha, R. Tamileaswari and Sr. S. Jesurani, "Analysis of anti-bacterial activity of silver nanoparticles from pomegranate (*Punica granatum*) seed and peel extracts", *International Journal of Engineering Research & Technology*, vol. 4, no. 4, pp. 1044–1048, 2015.
23. Azmat Ali Khan, Amer M. Alanazi, Nawaf Alsaif, Tanveer A. Wani and Mashooq A. Bhat, "Pomegranate peel induced biogenic synthesis of silver nanoparticles and their multifaceted potential against intracellular pathogen and cancer", *Saudi Journal of Biological Sciences*, vol. 28, pp. 4191–4200, 2021.
24. R. Tao, P. Yao, J. Zhang, L. Xu, H. -Y. Wu, M. -M. Ding and X-S. Zhu, "Biogenic synthesis of silver nanoparticles using pomegranate peel polyphenols and its catalytic action in reduction of acid red 6", *International Conference on Civil, Transportation and Environment (ICCTE 2016)*, Atlantis Press, pp. 1128–1131, 2016.

25. T. Thilagavathi, R. Renuka and R. Sathiya Priya, "Bio-synthesis of silver nanoparticles using *Punica granatum* (pomegranate) peel extract: a novel approach towards waste utilization", *International Journal of Advanced Science and Engineering.*, vol. 3, no. 1, pp. 234–236, 2016.
26. Pasent Gharib Saad, Rohan Daniel Castelino, Vimal Ravi, Issa Sulaiman Al-Amri and Shah Alam Khan, "Green synthesis of silver nanoparticles using Omani pomegranate peel extract and two polyphenolic natural products: characterization and comparison of their antioxidant, antibacterial, and cytotoxic activities", *Beni-Suef University Journal of Basic and Applied Sciences*, vol. 10, no. 29, 2021.
27. M. H. Nisha, R. Tamileswari, Sr. S. Jesurani, S. Kanagesan, M. Hashim, S. Catherine and P. Alexander, "Green synthesis of silver nanoparticles from pomegranate (*Punica granatum*) leaves and analysis of anti-bacterial activity", *International Journal of Advanced Technology in Engineering and Science*, vol. 3, no. 6, pp. 1–10, 2015.
28. Li Lin, Haoran Yang and Xiaocang Xu, "Effects of water pollution on human health and disease heterogeneity: a review", *Frontiers in Environmental Science*, vol. 10, p. 880246, 2022.
29. M. T. Amin, A. A. Alazba and U. Manzoor, "A review of removal of pollutants from water/wastewater using different types of nanomaterials", *Advances in Materials Science and Engineering*, vol. 2014, Article ID 825910, p. 24, 2014.
30. H. Hassoon, M. Ahmed and Y. Qaddoori, "Efficiency of silver nanoparticles against bacterial contaminants isolated from water in Basra", *The Eurasia Proceedings of Science, Technology & Mathematics*, vol. 7, pp. 164–174, 2019.
31. T. Bintsis, "Foodborne pathogens", *AIMS Microbiology*, vol. 3, no. 3, pp. 529–563, 2017.
32. Daniel Jesuwenu Ajose, Tesleem Olatunde Abolarinwa, Bukola Opeyemi Oluwarinde, Peter Kotsoana Montso, Omolola Esther Fayemi, Adeyemi Oladapo Aremu and Collins Njie Ateba, "Application of plant-derived nanoparticles (PDNP) in food-producing animals as a bio-control agent against antimicrobial-resistant pathogens", *Biomedicines*, vol. 10, p. 2426, 2022.
33. S. M. Farouk, S. H. Abu-Hussain, B. T. Abd-Elhalim, R. M. Mohamed, N. A. Arabe, A. A. T. Hussain, M. E. Mostafa, B. Hemdan, S. M. El-Sayed, A. Bakry, N. M. Ebeed, M. Salah, H. Elhariry and A. Galal, "Biosyntheis and characterization of silver nanoparticles from Punica granatum (pomegranate) peel waste and its application to inhibit foodborne pathogens", *Scientific Reports*, vol. 13, p. 19469, 2023.
34. Sadegh Khorramia, Atefeh Zarepour and Ali Zarrabi, "Green synthesis of silver nanoparticles at low temperature in a fast pace with unique DPPH radical scavenging and selective cytotoxicity against MCF-7 and BT-20 tumor cell lines", *Biotechnology Reports*, vol. 24, p. e00393, 2019.
35. Hala M. Abdelmigid, Maissa M. Morsi, Nahed Ahmed Hussien, Amal Ahmed Alyamani and Noha Moslah Al Sufyani, "Comparative analysis of nanosilver particles synthesized by different approaches and their antimicrobial efficacy", *Journal of Nanomaterials*, vol. 2021, Article ID 2204776, p. 12 pages, 2021.

36. W. K. Abad, A. N. Abd and N. F. Habubi. "Synthesis of Ag2O nanoparticles via fresh pomegranate peel extract for bioapplications", *Nano Biomedicine and Engineering*, vol.15, pp. 363–368, 2023, http://doi.org/10.26599/NBE.2023.9290032.
37. Y. Fu, J. Li and M. Almasi, "Pomegranate peel extract-mediated green synthesis of silver nanoparticles: evaluation of cytotoxicity, antioxidant, and anti-esophageal cancer effects", *Chemistry Select*, vol. 8, p. e202204841, 2023.
38. Lale Rozykulyyeva, Suryani Dyah Astuti, Andi Hamim Zaidan, Anak Agung Surya Pradhana and Putri Suryaning Puspita, "Antibacterial activities of green synthesized silver nanoparticles from *Punica granatum* peel extract", *AIP Conf. Proc.*, vol. 2314, p. 60012, 2020).
39. M. Shanmugavadivu, Selvam Kuppusamy and Rajamani Ranjithkumar, "Synthesis of pomegranate peel extract mediated silver nanoparticles and its antibacterial activity", *American Journal of Advanced Drug Delivery*, vol. 2, pp. 174–182, 2014.
40. R. A. Fernandes, A. A. Berretta, E. C. Torres, A. F. M. Buszinski, G. L. Fernandes, C. C. Mendes-Gouvêa, F. N. de Souza-Neto, L. F. Gorup, E. R. de Camargo and D. B. Barbosa, "Antimicrobial potential and cytotoxicity of silver nanoparticles phytosynthesized by pomegranate peel extract", *Antibiotics*, vol. 7, p. 51, 2018.
41. Žarko Ivanović, Jovana Blagojević, Gordana Jovanović, Borko Ivanović and Danica Žeželj, "New insight in the occurrence of early blight disease on potato reveals high distribution of *Alternaria solani* and *Alternaria protenta* in Serbia", *Frontiers in Microbiology*, vol. 13, Article 856898, 2022.
42. Pragya Adhikari, Yeonyee Oh and Dilip R. Panthee, "Current status of early blight resistance in tomato: an update", *International Journal of Molecular Science,* vol. 18, p. 2019, 2017.
43. Yasser S. Mostafa, Saad A. Alamri, Sulaiman A. Alrumman, Mohamed Hashem and Zakaria A. Baka, "Green synthesis of silver nanoparticles using pomegranate and orange peel extracts and their antifungal activity against *Alternaria solani*, the causal agent of early blight disease of tomato", *Plants*, vol. 10, p. 2363, 2021.
44. Shadi Basiri, "Evaluation of antioxidant and antiradical properties of Pomegranate (*Punica granatum* L.) seed and defatted seed extracts", *J Food Sci Technol.*, vol. 52, no. 2, pp. 1117–1123, 2015.
45. A. R. Sadrolhosseini, S. A. Rashid, A.S.M. Noor, A. Kharazmi and L. A. Mehdipour, "Fabrication of silver nanoparticles in pomegranate seed oil with thermal properties by laser ablation technique", *Digest Journal of Nanomaterials and Biostructures*, vol. 10, no. 3, pp. 1009–1018, 2015.
46. Abdulaziz A. Al-Askar, Dalia G. Aseel, Hamada El-Gendi, Sherien Sobhy, Marwa A. Samy, Esraa Hamdy, Sarah El-Messeiry, Said I. Behiry, Toufic Elbeaino and Ahmed Abdelkhalek, "Antiviral activity of biosynthesized silver nanoparticles from pomegranate (*Punica granatum* L.) peel extract against tobacco mosaic virus", *Plants*, vol. 12, p. 2103, 2023.
47. Asmaa Ali Alharbi, Amira M. Alghamdi, Soha Talal Al-Goul, Aminah Allohibi, Roua S. Baty, Safa H. Qahl and Eman A. Beyari, "Valorizing pomegranate wastes by producing functional silver nanoparticles with antioxidant,

anticancer, antiviral, and antimicrobial activities and its potential in food preservation", *Saudi Journal of Biological Sciences*, vol. 31, no. 1, p. 103880, 2024.
48. Jasim Younis Jasim and Shahbaa Khalil Ebrahim Al-Taee, "Evaluation the antifungal activity of green biosynthesis silver nanoparticles using pomegranate peel extract", *IOP Conference Series: Earth and Environmental Science*, vol. 1158, p. 102001, 2023.
49. A.-O. Monira, A. E-A Abeer and A. A. Hatamieh, "Green biosynthesis of silver nanoparticles using pomegranate peel and inhibitory effects of the nanoparticles on aflatoxin production", *Pakistan Journal of Botany*, vol. 15, pp. 751–756, 2017.
50. Peter J. Hotez, Paul J. Brindley, Jeffrey M. Bethony, Charles H. King, Edward J. Pearce and Julie Jacobson, "Helminth infections: the great neglected tropical diseases", *The Journal of Clinical Investigation*, vol. 118, no. 4, pp. 1311–1321, 2008.
51. Felwa Abdullah Thagfan, Rewaida Abdel-Gaber, Amal Marzouk Marey, Mohammed Ibrahim Alquraishi, Murad Ali Mubaraki and Mohamed Abdelmonem Dkhil, "Role of Punica granatum-derived nanosilver in helminth protection", *Food Science and Technology*, vol. 43, p. e14123, 2023.
52. Shailendra Singh and B. Raj, "A 2-D analytical surface potential and drain current modeling of double-gate vertical t-shaped Tunnel FET", *Journal of Computational Electronics,* vol. 19, pp.1154–1163, April 2020.
53. Jeetendra Singh and B. Raj, "An accurate and generic window function for non-linear memristor model", *Journal of Computational Electronics*, vol. 18 no. 2, pp. 640–647 June 2019.
54. Manjit Kaur, Neena Gupta, Sanjeev Kumar, B. Raj and Arun Kumar Singh, "Comparative RF and crosstalk analysis of carbon based nano interconnects", *IET Circuits, Devices & Systems*, vol.15, no. 6, pp. 493–503, Feb. 2021.
55. Nehru Kandasamy, Firdous Ahmad, D. Ajitha, B. Raj and Nagarjuna Telagam, "Quantum dot cellular automata based scan flip flop and boundary scan register", *IETE Journal of Research*, vol. 66, pp. 535–548, 2020.
56. Jeetendra Singh and B. Raj, "Design and investigation of 7T2M NVSARM with enhanced stability and temperature impact on store/restore energy", *IEEE Transactions on Very Large Scale Integration Systems,* vol. 27, no. 6, pp. 1322–1328, June 2019.
57. Anil Kumar Bhardwaj, Sumeet Gupta, B. Raj and Amandeep Singh, "Impact of double gate geometry on the performance of carbon nanotube field effect transistor structures for low power digital design", *Computational and Theoretical Nanoscience*, vol. 16, pp. 1813–1820, 2019.
58. Neeraj Jain and B. Raj, "Thermal stability analysis and performance exploration of asymmetrical dual-k underlap spacer (ADKUS) SOI FinFET for security and privacy applications", *Indian Journal of Pure & Applied Physics (IJPAP)*, vol. 57, pp. 352–360, May 2019.
59. Neeraj Jain and Balwinder Raj, "Dual-k spacer region variation at the drain side of asymmetric SOI FinFET structure: performance analysis towards the analog/RF design applications", *Journal of Nanoelectronics and Optoelectronics*, vol. 14, pp. 349–359, March 2019.

60. Jeetendra Singh, Sanjeev Sharma, B. Raj and Mamta Khosla, "Analysis of barrier layer thickness on performance of In1-xGaxAs based gate stack cylindrical gate nanowire MOSFET", *Journal of Nanoelectronics and Optoelectronics*, vol. 13, pp. 1473–1477, Oct. 2018.
61. Neeraj Jain and B. Raj, "Analysis and performance exploration of high-k SOI FinFETs over the conventional low-k SOI FinFET toward analog/RF design", *Journal of Semiconductors (JoS),* vol. 39, no. 12, pp. 124002-1–7, Dec. 2018.
62. Candy Goyal, Jagpal Singh Ubhi and B. Raj, "A reliable leakage reduction technique for approximate full adder with reduced ground bounce noise", *Journal of Mathematical Problems in Engineering*, vol. 2018, Article ID 3501041, p. 16, 15 Oct. 2018.
63. Jeetendra Singh Anuradha, B. Raj and Mamta Khosla, "Design and performance analysis of nano-scale memristor-based nonvolatile SRAM", *Journal of Sensor Letter*, vol. 16, pp. 798–805, Oct. 2018.
64. Girish Wadhwa and B. Raj, "Parametric variation analysis of charge-plasma-based dielectric modulated JLTFET for biosensor application", *IEEE Sensor Journal*, vol. 18, no. 15, pp. 6070–6077, 1 Aug. 2018.
65. Jeetendra Singh and B. Raj, "Comparative analysis of memristor models for memories design", *Journal of Semiconductors*, vol. 39, no. 7, p. 074006-1–12, July 2018.
66. Divya Yadav, Shailesh Singh Chouhan, Santosh Kumar Vishvakarma and B. Raj, "Application specific microcontroller design for IoT based WSN", *Sensor Letter*, vol. 16, pp. 374–385, May 2018.
67. Gurmohan Singh, R. K. Sarin and B. Raj, "Fault-tolerant design and analysis of quantum-dot cellular automata based circuits", *IEEE/IET Circuits, Devices & Systems*, vol. 12, pp. 638–664, 2018.
68. Jeetendra Singh and B. Raj, "Modeling of mean barrier height levying various image forces of metal insulator metal structure to enhance the performance of conductive filament based memristor model", *IEEE Nanotechnology*, vol. 17, no. 2, pp. 268–267, March 2018.
69. Aakash Jain, Sanjeev Sharma and B. Raj, "Analysis of triple metal surrounding gate (TM-SG) III–V nanowire MOSFET for photosensing application", *Opto-Electronics Journal*, vol. 26, no. 2, pp. 141–148, May 2018.
70. Aakash Jain, Sanjeev Sharma and B. Raj, "Design and analysis of high sensitivity photosensor using cylindrical surrounding gate MOSFET for low power sensor applications", *Engineering Science and Technology, an International Journal*, vol. 19, no. 4, pp. 1864–1870, Dec. 2016.
71. G. Saiphani Kumar, Amandeep Singh and B. Raj, "Design and analysis of gate all around CNTFET based SRAM cell design", *Journal of Computational Electronics*, vol. 17, no. 1, pp 138–145, March 2018.
72. Gurinder pal Singh, B. S. Sohi and Balwinder Raj, "Material properties analysis of graphene base transistor (GBT) for VLSI analog circuits", *Indian Journal of Pure & Applied Physics (IJPAP)*, vol. 55, pp. 896–902, Dec. 2017.
73. Sunil Kumar and B. Raj, "Estimation of stability and performance metric for inward access transistor based 6T SRAM cell design using n-type/p-type DMDG-GDOV TFET", *IEEE VLSI Circuits and Systems Letter,* vol.3, no. 2, pp. 25–39, June 2017

74. Shashikant Sharma, Anjan Kumar, Manisha Pattanaik and B. Raj, "Forward body biased multimode multi-threshold CMOS technique for ground bounce noise reduction in static CMOS adders", *International Journal of Information and Electronics Engineering*, vol. 3, no. 3, pp. 567–572, 2013.
75. Hamendra Singh, Pankaj Kumar and Balwinder Raj, "Performance analysis of majority gate SET based 1-bit full adder", *International Journal of Computer and Communication Engineering (IJCCE),* vol. 2, no. 4, pp. 1–8, 2013.
76. Anil Kumar Bhardwaj, Sumeet Gupta and B. Raj, "Investigation of parameters for schottky barrier (SB) height for Schottky barrier based carbon nanotube field effect transistor device", *Journal of Nanoelectronics and Optoelectronics*, vol. 15, pp. 783–791, July 2020.
77. Priya Bansal and B. Raj, "Memristor: a versatile nonlinear model for dopant drift and boundary issues," *Journal of Computational and Theoretical Nanoscience*, vol. 14, no. 5, pp. 2319–2325, May 2017.
78. Neeraj Jain and B. Raj, "An analog and digital design perspective comprehensive approach on Fin-FET (Fin-Field Effect transistor) technology - a review", *Reviews in Advanced Sciences and Engineering,* vol. 5, pp. 1–14, 2016.
79. Balwinder Raj, A. K. Saxena and S. Dasgupta, "Analytical modeling for the estimation of leakage current and subthreshold swing factor of nanoscale double gate FinFET device", *Microelectronics International*, vol. 26, pp. 53–63, 2009.
80. Shailendra Singh Soniya, Girish Wadhwa and Balwinder Raj, "An analytical modeling for dual source vertical tunnel field effect transistor", *International Journal of Recent Technology and Engineering*, vol. 8, no. 2, pp. 603–608, July 2019.
81. Shailendra Singh and B. Raj, "Design and analysis of hetrojunction vertical T-shaped tunnel field effect transistor", *Journal of Electronics Material,* vol. 48, no. 10, pp 6253–6260, Oct. 2019.
82. Candy Goyal, Jagpal Singh Ubhi and B. Raj, "A low leakage CNTFET based inexact full adder for low power image processing applications", *International Journal of Circuit Theory and Applications*, vol.47, no.9, pp. 1446–1458, Sep. 2019.
83. B. Raj, A. K. Saxena and S. Dasgupta, "A compact drain current and threshold voltage quantum mechanical analytical modeling for FinFETs", *Journal of Nanoelectronics and Optoelectronics,* vol. 3, no. 2, pp. 163–170, 2008.
84. Girish Wadhwa and B. Raj, "An analytical modeling of charge plasma based tunnel field effect transistor with impacts of gate underlap region", *Superlattices and Microstructures,* vol. 142, p.106512, June 2020.
85. Shailendra Singh and B. Raj,"Modeling and simulation analysis of SiGe hetrojunction double gatevertical t-shaped tunnel FET", *Superlattices and Microstructures*, vol. 142, p. 106496, June 2020.
86. Neeraj Jain and Balwinder Raj, "Device and circuit co-design perspective comprehensive approach on FinFET technology – a review", *Journal of Electron Devices*, vol. 23, no. 1, pp 1890–1901, 2016.
87. Sunil Kumar and B. Raj, "Analysis of I_{ON} and Ambipolar current for dual-material gate-drain overlapped DG-TFET", *Journal of Nanoelectronics and Optoelectronics*, vol. 11, pp. 323–333, June 2016.

88. Naveed Anjum, Tarun Bali and B. Raj, "Design and simulation of handwritten multiscript character recognition", *International Journal of Advanced Research in Computer and Communication Engineering*, vol. 2, no. 7, pp. 2544–2549, July 2013.
89. Karmjit Singh and B. Raj, "Performance and analysis of temperature dependent multi-walled carbon nanotubes as global interconnects at different technology nodes", *Journal of Computational Electronics*, vol. 14 no. 2, pp. 469–476, June 2015.
90. Sunil Kumar and B. Raj, "Compact channel potential analytical modeling of DG-TFET based on evanescent–mode approach", *Journal of Computational Electronics,* vol. 14, no. 2, pp.820–827, July 2015.
91. Karmjit Singh and B. Raj, "Temperature dependent modeling and performance evaluation of multi-walled CNT and single-walled CNT as global interconnects", *Journal of Electronic Materials*, vol. 44, no. 12, pp 4825–4835, Dec. 2015.
92. V. K. Sharma, M. Pattanaik and B. Raj, "INDEP approach for leakage reduction in nanoscale CMOS circuits", *International Journal of Electronics*, vol.102, no.2, pp. 200–215, 2014.
93. Karmjit Singh and B. Raj, "Influence of temperature on MWCNT bundle, SWCNT bundle and copper interconnects for nanoscaled technology nodes", *Journal of Materials Science: Materials in Electronics*, vol. 26, no. 8, pp. 6134–6142, 2015.
94. Naveed Anjum, Tarun Bali and B. Raj, "Design and simulation of handwritten Gurumukhi and Devanagri numerical recognition", *International Journal of Computer Applications, Published by Foundation of Computer Science*, vol. 73, no.12, pp. 16–21, 2013.
95. S. Khandelwal, V. Gupta, B. Raj and R. D. Gupta, "Process variability aware low leakage reliable nano scale DG-FinFET SRAM cell design technique", *Journal of Nanoelectronics and Optoelectronics*, vol. 10, no. 6, pp. 810–817, Dec. 2015.
96. V. K. Sharma, M. Pattanaik and B. Raj, "ONOFIC approach: low power high speed nanoscale VLSI circuits design", *International Journal of Electronics*, vol. 101, no. 1, pp. 61–73, 2014.
97. S. Khandelwal, Balwinder Raj and R. D. Gupta, "FinFET based 6T SRAM cell design: analysis of performance metric, process variation and temperature effect", *Journal of Computational and Theoretical Nanoscience*, vol. 12, pp. 2500–2506, 2015.
98. Sumit Singh, Yadav, Shekhar, Rahul, Jagdeep, Srivastava, Anurag and B. Raj, "Impact of HfO_2 in graded channel dual insulator double gate MOSFET", *Journal of Computational and Theoretical Nanoscience*, vol. 12, no. 6, pp. 950–953, April 2015.
99. Vijay Kumar Sharma, Manisha Pattanaik and B. Raj, "PVT variations aware low leakage INDEP approach for nanoscale CMOS circuits", *Microelectronics Reliability*, vol. 54, pp. 90–99, 2014.
100. B. Raj, A. K. Saxena and S. Dasgupta, "Quantum mechanical analytical modeling of nanoscale DG FinFET: evaluation of potential, threshold voltage and source/drain resistance", *Journal of Material Science in Semiconductor Processing*, vol. 16, no. 4, pp. 1131–1137, 2013.

101. Maisagalla Gopal, Siva Sankar D Prasad and Balwinder Raj, "8T SRAM cell design for dynamic and leakage power reduction", *International Journal of Computer Applications*, vol. 71, no. 9, pp. 43–48, June 2013.
102. Manisha Pattanaik, B. Raj, Shashikant Sharma and Anjan Kumar, "Diode based trimode multi-threshold CMOS technique for ground bounce noise reduction in static CMOS adders", *Advanced Materials Research*, vol. 548, pp. 885–889, 2012.
103. Balwinder Raj, A. K. Saxena and S. Dasgupta, "Nanoscale FinFET based SRAM cell design: analysis of performance metric, process variation, underlapped FinFET and temperature effect", *IEEE Circuits and System Magazine*, vol. 11, no. 2, pp. 38–50, 2011.
104. V. K. Sharma, M. Pattanaik and Balwinder Raj, "Leakage current ONOFIC approach for deep submicron VLSI circuit design", *International Journal of Electrical, Computer, Electronics and Communication Engineering*, vol. 7, no. 4, pp. 239–244, 2013.
105. Parminder Kaur, Sandeep Singh Gill and B. Raj, "Comparative analysis of OFETs materials and devices for sensor applications", *Journal of Silicon*, vol. 14, pp. 4463–4471, 2022.
106. Sanjeev Kumar Sharma, Parveen Kumar, Balwant Raj and B. Raj· "$In_{1-x}Ga_xAs$ double metal gate-stacking cylindrical nanowire MOSFET for highly sensitive photo detector", *Journal of Silicon*, vol.14, pp. 3535–3541, 2022.
107. B. Raj, A. K. Saxena and S. Dasgupta, "Analytical modeling of quasi planar nanoscale double gate FinFET with source/drain resistance and field dependent carrier mobility: a quantum mechanical study", *Journal of Computer (JCP)*, vol. 4, no. 9, pp.1–8, 2009.
108. S. Bhushan, S. Khandelwal, and B. Raj, "Analyzing different mode FinFET based Memory cell at different power supply for leakage reduction", *Seventh International Conference on Bio-Inspired Computing: Theories and Application, (BIC-TA 2012) Advances in Intelligent Systems and Computing*, vol. 202, pp. 89–100, 2013.
109. Jeetendra Singh and B. Raj, "Temperature dependent analytical modeling and simulations of nanoscale memristor", *Engineering Science and Technology, an International Journal*, vol. 21, pp. 862–868, Oct. 2018.
110. Shradhya Singh, Shashi Bala, Balwant Raj and B. Raj, "Improved sensitivity of dielectric modulated junctionless transistor for nanoscale biosensor design", *Sensor Letter*, vol.18, pp. 328–333, April 2020.
111. Vivek Kumar, Santosh Kumar Vishvakarma and B. Raj, "Design and performance analysis of ASIC for IoT applications", *Sensor Letter*, vol. 18, pp. 31–38, Jan. 2020.
112. Akanksha Jaiswal, R. K. Sarin, B. Raj and Shikha Sukhija, "A novel circular slotted microstrip-fed patch antenna with three triangle shape defected ground structure for multiband applications", *Advanced Electromagnetic*, vol. 7, no. 3, pp. 56–63, Aug. 2018.
113. Girish Wadhwa and B. Raj, "Label free detection of biomolecules using charge-plasma-based gate underlap dielectric modulated junctionless TFET", *Journal of Electronic Materials (JEMS)*, vol. 47, no. 8, pp 4683–4693, Aug. 2018.

114. Gurmohan Singh, R. K. Sarin and B. Raj, "Design and performance analysis of a new efficient coplanar quantum-dot cellular automata adder", *Indian Journal of Pure & Applied Physics*, vol. 55, pp. 97–103, Feb. 2017.
115. Parminder Kaur, Vikas Pandey and B. Raj, "Comparative study of efficient design, control and monitoring of solar power using IoT", *Sensor Letter*, vol. 18, pp. 419–426, May 2020.
116. Anil Kumar Bhardwaj, Sumeet Gupta and B. Raj, "Development & analysis of compact model for double gate Schottky barrier CNTFET", *Journal of Nanoelectronics and Optoelectronics*, vol. 15, pp. 1199–1208, Aug. 2020.
117. Girish Wadhwa, Priyanka Kamboj, Jeetendra Singh and B. Raj, "Design and investigation of junctionless DGTFET for biological molecule recognition", *Transactions on Electrical and Electronic Materials,* vol.22, pp. 282–289, 2021.
118. Sachin Kumar Verma, Shailendra Singh, Girish Wadhwa and B. Raj, "Detection of biomolecules using charge-plasma based gate underlap dielectric modulated dopingless TFET", *Transactions on Electrical and Electronic Materials*, vol. 21, pp. 528–535, June 2020.
119. Neeraj Jain and B. Raj, "Impact of underlap spacer region variation on electrostatic and analog/RF performance of symmetrical high-k SOI FinFET at 20 nm channel length", *Journal of Semiconductors*, vol. 38, no. 12, pp. 122002, Dec. 2017.
120. Shailendra Singh and B. Raj, "Analytical modeling and simulation analysis of T-shaped III-V heterojunction vertical T-FET", *Superlattices and Microstructures*, vol. 147, pp. 106717, Nov. 2020.
121. Gurmohan Singh, R. K. Sarin and B. Raj, "Design and analysis of area efficient QCA based reversible logic gates", *Journal of Microprocessors and Microsystems,* vol. 52, pp. 59–68, May 2017.
122. Shailendra Singh and B. Raj, "Parametric variation analysis on hetero-junction Vertical t-shape TFET for Supressing ambipolar conduction", *Indian Journal of Pure and Applied Physics*, vol. 58, pp. 478–485, June 2020.
123. Shailendra Singh Soniya, Girish Wadhwa and Balwinder Raj, "Design and analysis of dual source vertical tunnel field effect transistor for high performance", *Transactions on Electrical and Electronics Materials,* vol. 21, pp. 74–82, Oct. 2019.
124. Manjit Kaur, Neena Gupta, Sanjeev Kumar, B. Raj and Arun Kumar Singh, "RF performance analysis of intercalated graphene nanoribbon based global level interconnects", *Journal of Computational Electronics,* vol. 19, pp. 1002–1013, June 2020
125. Girish Wadhwa and B. Raj, "Design and performance analysis of junctionless TFET biosensor for high sensitivity", *IEEE Nanotechnology*, vol.18, pp. 567–574, 2019.
126. Jeetendra Singh and B. Raj, "Enhanced nonlinear memristor model encapsulating stochastic dopant drift, *Journal of Nanoelectronics and Optoelectronics*, vol. 14, pp. 958–963, 2019.
127. Anjana Bhardwaj, Pradeep Kumar, B. Raj and Sunny Anand, "Design and performance optimization of doping-less vertical nanowire TFET using gate stack technique", *Journal of Electronic Materials*, vol. 41, no. 7, pp. 4005–4013, 2022.

128. Jeetendra Singh and B. Raj, "Tunnel current model of asymmetric MIM structure levying various image forces to analyze the characteristics of filamentary memristor", *Applied Physics A*, vol. 125, no.3, pp. 203.1–203.11, Feb. 2019.
129. Candy Goyal, Jagpal Singh Ubhi and B. Raj, "Low leakage zero ground noise nanoscale full adder using source biasing technique", *Journal of Nanoelectronics and Optoelectronics*, vol. 14, pp. 360–370, March 2019.
130. Gurmohan Singh, R. K. Sarin and B. Raj, "A novel robust exclusive-OR function implementation in QCA nanotechnology with energy dissipation analysis", *Journal of Computational Electronics*, vol. 15, no. 2, pp. 455–465, June 2016.
131. Girish Wadhwa, Priyanka Kamboj and Balwinder Raj, "Design optimisation of junctionless TFET biosensor for high sensitivity", *Advances in Natural Sciences: Nanoscience and Nanotechnology*, vol. 10, pp. 045001, 2019.
132. Priya Bansal and B. Raj, "Memristor modeling and analysis for linear dopant drift kinetics", *Journal of Nanoengineering and Nanomanufacturing*, vol. 6, pp. 1–7, 2016.
133. Ashima, D Vaithiyanathan and B. Raj, "Performance analysis of charge plasma induced graded channel Si nanotube", *Journal of Engineering Research*, EMSME Special no. pp. 146–154, Aug. 2021.
134. Abhishek Singh Tomar, Vijay Kumar Magraiya and B. Raj, "Scaling of access and data transistor for high performance DRAM cell design", *Quantum Matter*, vol. 2, pp. 412–416, Oct. 2013.
135. Neeraj Jain and B. Raj, "Parasitic capacitance and resistance model development and optimization of raised source/drain SOI FinFET structure for analog circuit applications", *Journal of Nanoelectronics and Optoelectronics*, vol. 13, pp. 531–539, April 2018.
136. Shradhya Singh, S. K. Vishvakarma and B. Raj, "Analytical modeling of split-gate junction-less transistor for a biosensor application", *Sensing and Bio-Sensing*, vol. 18, pp. 31–36, April 2018.
137. Maisagalla Gopal and Balwinder Raj, "Low power 8T SRAM cell design for high stability video applications", *ITSI Transaction on Electrical and Electronics Engineering*, vol. 1, no. 5, pp. 91–97, 2013.
138. Balwinder Raj, Jatin Mitra, Deepak Kumar Bihani, Rangharajan V, A.K. Saxena and S. Dasgupta, "Analysis of noise margin, power and process variation for 32 nm FinFET based 6T SRAM cell", *Journal of Computer*, vol. 5, no. 6, pp. 1–8, 2010.
139. Divya Sharma, Rajesh Mehra and B. Raj, "Comparative analysis of photovoltaic technologies for high efficiency solar cell design", *Superlattices and Microstructures*, vol. 153, pp. 106861, May 2021.
140. Pawandeep Kaur, Avtar Singh Buttar and B. Raj, "A comprehensive analysis of nanoscale transistor based biosensor: a review", *Indian Journal of Pure and Applied Physics*, vol. 59, pp. 304–318, April 2021.
141. Divya Yadav, Balwant Raj and B. Raj, "Design and simulation of low power microcontroller for IoT applications", *Journal of Sensor Letters*, vol. 18, pp. 401–409, May 2020.

Chapter 6

Comparative analysis of 6T SRAM and ROM cell using CMOS technology at 200 nm, 45 nm, and 7 nm

Suman Lata Tripathi and Balwinder Raj

6.1 INTRODUCTION

predictive technology modelRandom access memory (RAM) is volatile memory, which means that the data is lost when the power is turned off. There are two types of RAM: SRAM (static RAM) and DRAM (dynamic RAM) [1, 2]. The use of different transistor designs like double-gate metal–ōxide–semiconductor field-effect transistor (MOSFET), fin field-effect transistor (FINFET), or nanowires supports performance optimization at lower technology nodes [3–6]. Memory based on FinFETs shows low power consumption and faster memory access time. Here, 6T SRAM (static random access memory) was designed and simulated using different model files of 200 nm, 45 nm, and 7 nm. The design and simulation work was carried out using a predictive technology model model file [7] on Tanner electronic design automation.

6.2 SCHEMATIC DIAGRAM

Figure 6.1 shows the schematic diagram of 6T SRAM (6 Transistor static random access memory). The core of the cell is a bistable complementary metal–oxide–semiconductor (CMOS) inverter or two identical cross-coupled inverter circuits. This feedback loop stabilizes the inverters to their respective states. The access transistors, two NMOS, are connected to the word and bit lines, WL, BL, and BLB. These are used to read and write from or to the cell. Moreover, the outputs are Q and Qbar [8–13].

6.3 ANALYSIS

6.3.1 200 nm

Figure 6.2 shows the transient analysis of 6T SRAM using 200 nm file. If the word line WL is high, the output Q will copy the bit line, BL

DOI: 10.1201/9781003612797-6

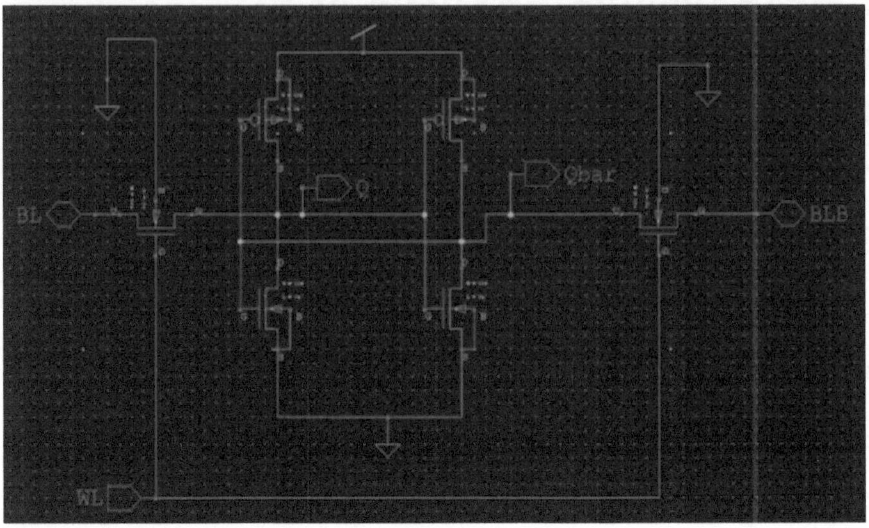

Figure 6.1 Schematic diagram of 6T static RAM.

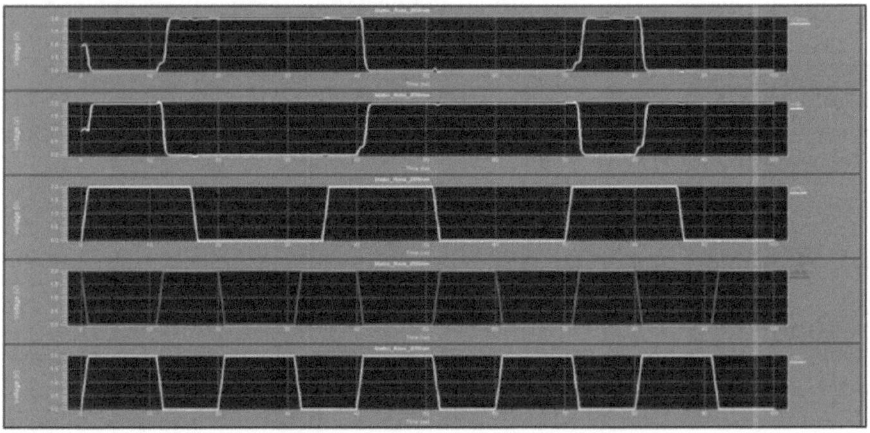

Figure 6.2 Transient analysis of 6T static RAM using 200 nm file.

(lowermost line). Moreover, if the WL is low, Q will copy the previous output. Observe that Q and Qbar, and B and BLB complements each other [14–19].

Figure 6.3 shows the DC characteristics of 6T SRAM using 200 nm file. BL is the lowermost line, BLB is the V(BLB), WL is the V(WL), Q is the V(Q), and Qbar is the uppermost line. Observe that when vBL is increasing, there is an increase in the output. This increase is being complemented in Qbar [20–27].

Figure 6.4 shows the power consumption of 6T SRAM using 200 nm file. The average power consumed is 10.4873 μW. The maximum power is 12 nW while the minimum power is 67.6875 nW.

Figure 6.5 shows the delay of 6T SRAM using 200 nm file. The image on the upper shows the graph obtaining the response time of gate from low to high transition with t_{pLH} = 850.45 ps [28–33]. The image on the lower shows high to low transition with t_{pHL} = 527.86 ps. To determine the delay, the following is computed using the dx of two graphs.

$$\text{Delay} = \frac{tp_{LH} + tp_{HL}}{2}$$

$$\text{Delay} = \frac{850.45\ \text{ps} + 527.86\text{ps}}{2}$$

$$\text{Delay} = 689.155\ \text{ps}$$

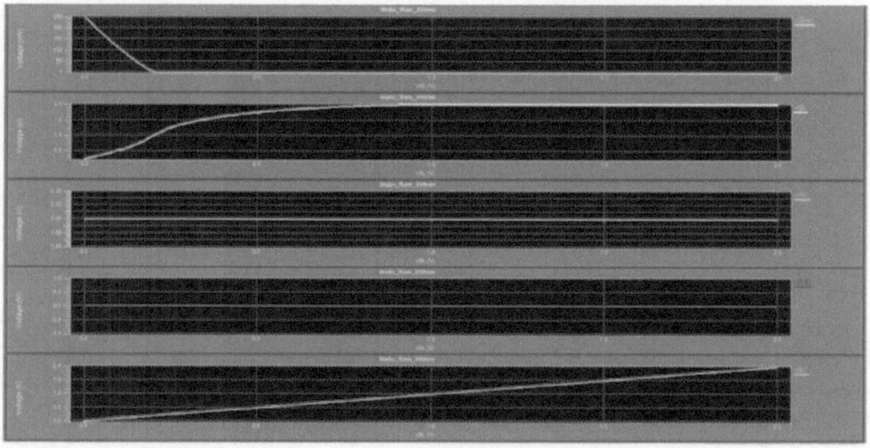

Figure 6.3 DC characteristics of 6T static RAM using 200 nm file.

```
Power Results
Vdd from time 1e-009 to 1e-005
Average power consumed -> 1.048737e-007 watts
Max power 3.063349e-004 at time 1.2e-008
Min power 1.847170e-012 at time 6.76875e-008
```

Figure 6.4 Power consumption of 6T static RAM using 200 nm file.

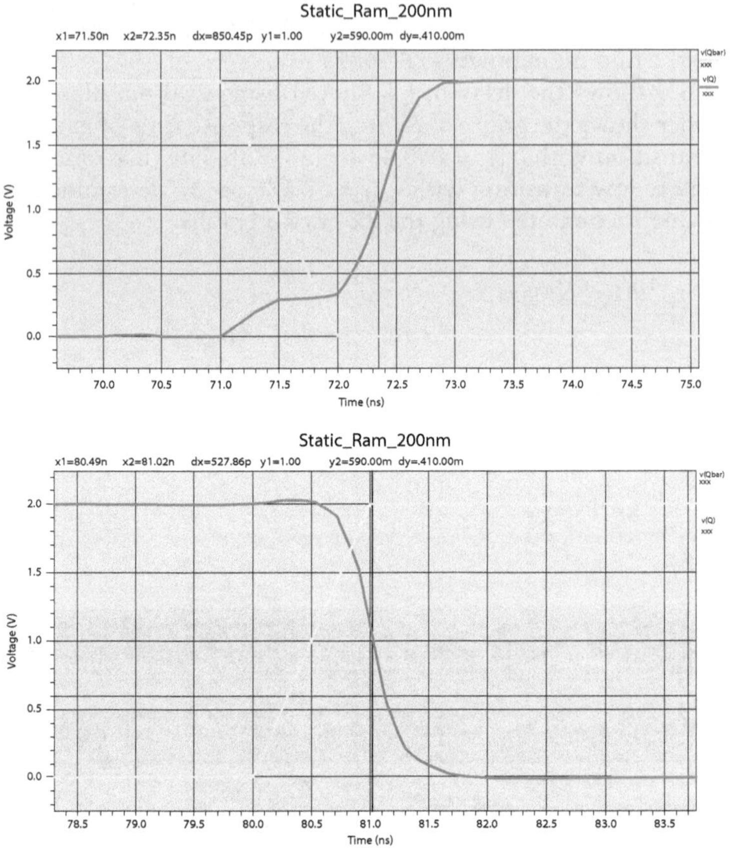

Figure 6.5 **Propagation delay of 6T static RAM using 200 nm file.**

Figure 6.6 shows the noise margin of 6T SRAM using the 200 nm file. V_{OH} is equal to 1.85 V, V_{OL} is equal to 140.32 mV, V_{IL} is equal to 71.63 mV, and V_{IH} is equal to 596.28 V [34–41]. The noise margins N_{MH} and N_{ML} are computed as follows:

$$N_{MH} = V_{OH} - V_{IH}$$

$$N_{MH} = 1.85V - 596.28V$$

$$N_{MH} = 1.2537\ V \quad N_{ML} = V_{IL} - V_{OL}$$

$$N_{ML} = 71.63\ mV - 140.32\ mV$$

$$N_{ML} = 68.69\ mV$$

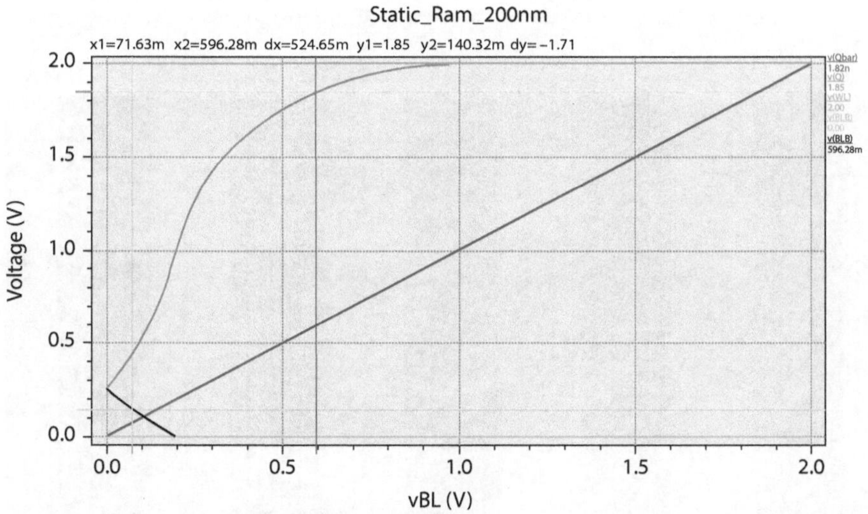

Figure 6.6 Noise margin of 6T static RAM using 200 nm file.

6.3.2 45 nm SRAM

Figure 6.7 shows the transient analysis of 6T SRAM using the 45 nm file. If the word line WL V(Q) is high, the output Q V(Q) will copy the bit line, BL (lowermost line). Moreover, if WL is low, Q will copy the previous output [42–48]. Observe that Q and Qbar, and B and BLB complements each other.

Figure 6.8 shows the DC characteristics of 6T SRAM using 45 nm file. BL is the lowermost line, BLB is the V(BLB), WL is the V(WL), Q is the V(Q), and Qbar is the uppermost line [49–55]. Observe that when vBL is increasing, there is an increase in the output. This increase is being complemented in Qbar.

Figure 6.9 shows the power consumption of 6T SRAM using 45 nm file. The average power consumed is 1.3361 nW. The maximum power is 1 nW, while the minimum power is 3.6875 nW.

Figure 6.10 shows the delay of 6T SRAM using 45 nm file. The image on the upper shows the graph obtaining the response time of gate from low to high transition with t_{pLH} = 21.64 ps. The image on the lower shows high to low transition with t_{pHL} = 1.02 ns. To determine the delay, the following is computed using the dx of two graphs.

$$\text{Delay} = \frac{tp_{LH} + tp_{HL}}{2}$$

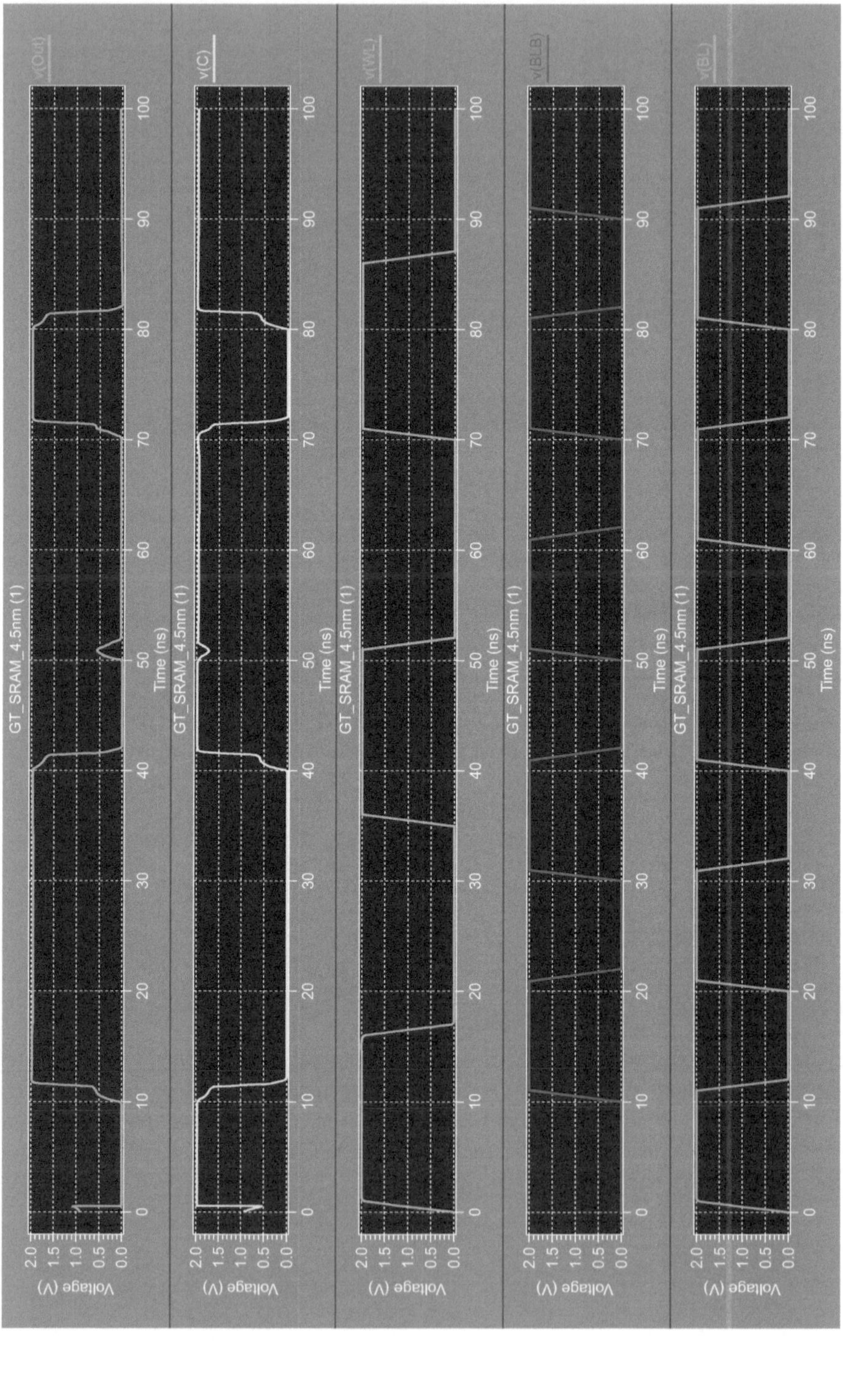

Figure 6.7 Transient analysis of 6T static RAM using 45 nm file.

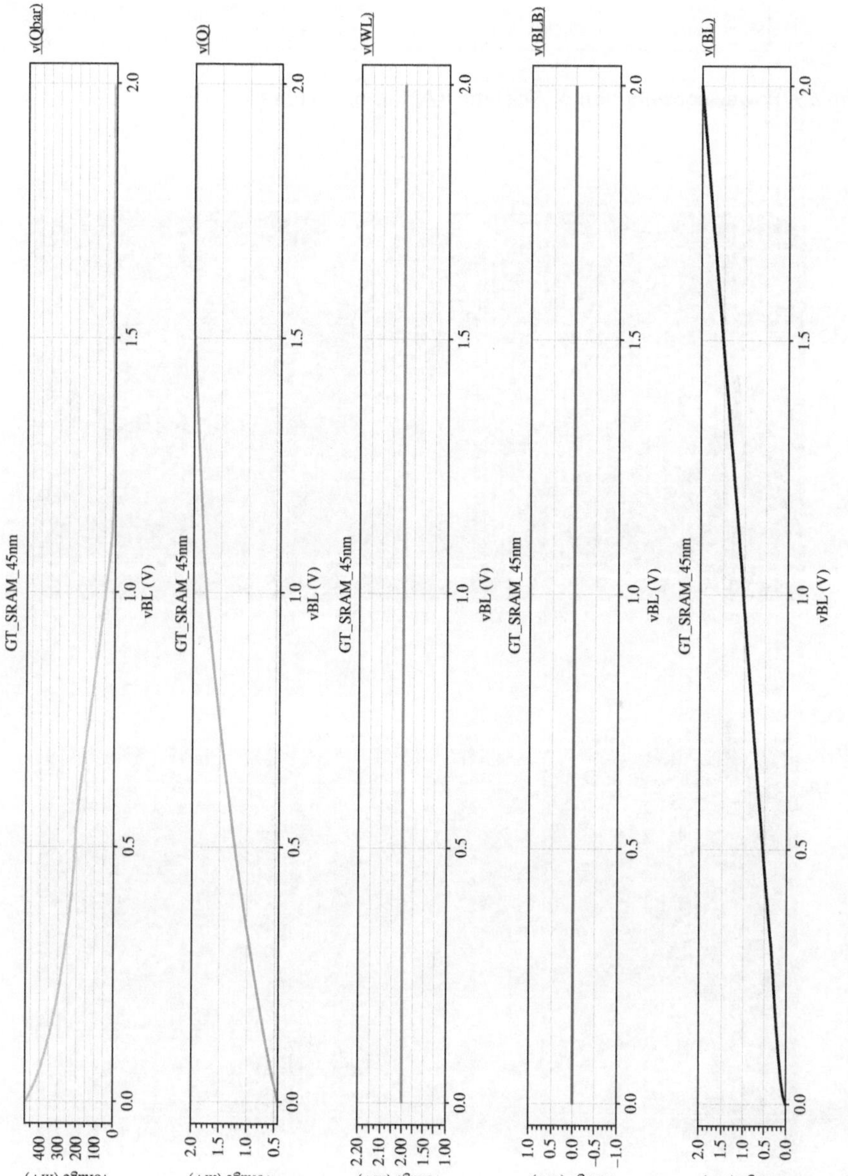

Figure 6.8 DC characteristics of 6T static RAM using 45 nm file.

```
Power Results
Vdd from time 1e-009 to 1e-005
Average power consumed -> 1.336141e-009 watts
Max power 5.484160e-006 at time 1e-009
Min power 1.398802e-009 at time 3.6875e-009
```

Figure 6.9 Power consumption of 6T static RAM using 45 nm file.

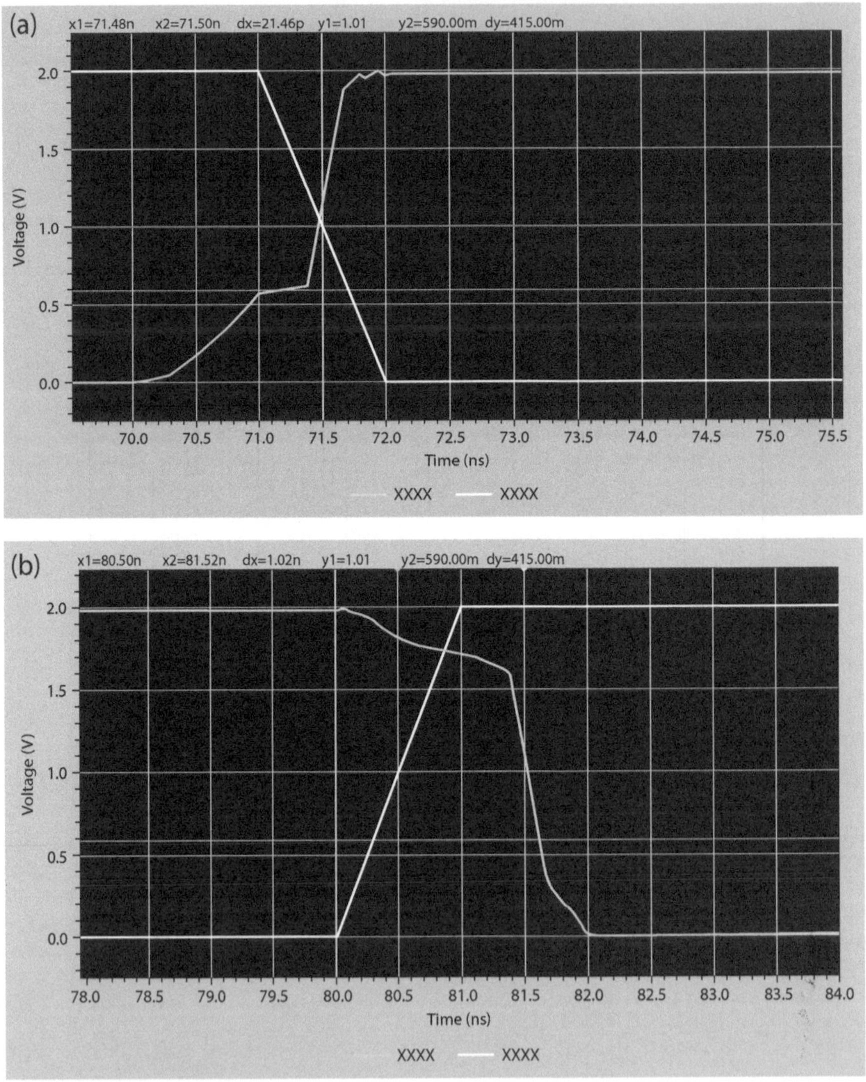

Figure 6.10 Propagation delay of 6T static RAM using 45 nm file.

$$\text{Delay} = \frac{21.64\ \text{ps} + 1.02\ \text{ns}}{2}$$

$$\text{Delay} = 520.82\ \text{ps}$$

Figure 6.11 shows the noise margin of 6T SRAM using 45 nm file. V_{OH} is equal to 1.70 V, V_{OL} is equal to 295.48 mV, V_{IL} is equal to 220.47 mV, and V_{IH} is equal to 949.77 mV. The noise margins N_{MH} and N_{ML} are computed as follows:

$$\text{N}_{\text{MH}} = \text{V}_{\text{OH}} - \text{V}_{\text{IH}}$$

$$\text{N}_{\text{MH}} = 1.70\ \text{V} - 949.77\ \text{mV}$$

$$\text{N}_{\text{MH}} = 0.7502\,\text{V} \quad \text{N}_{\text{ML}} = \text{V}_{\text{IL}} - \text{V}_{\text{OL}}$$

$$\text{N}_{\text{ML}} = 220.47\ \text{mV} - 294.48\ \text{mV}$$

$$\text{N}_{\text{ML}} = 74.01\ \text{mV}$$

6.3.3 *7 nm SRAM*

Figure 6.12 shows the transient analysis of 6T SRAM using 7 nm file. If the word line WL V(WL) is high, the output Q V(Q) will copy the bit line BL (lowermost line). Moreover, if the WL is low, Q will copy the previous

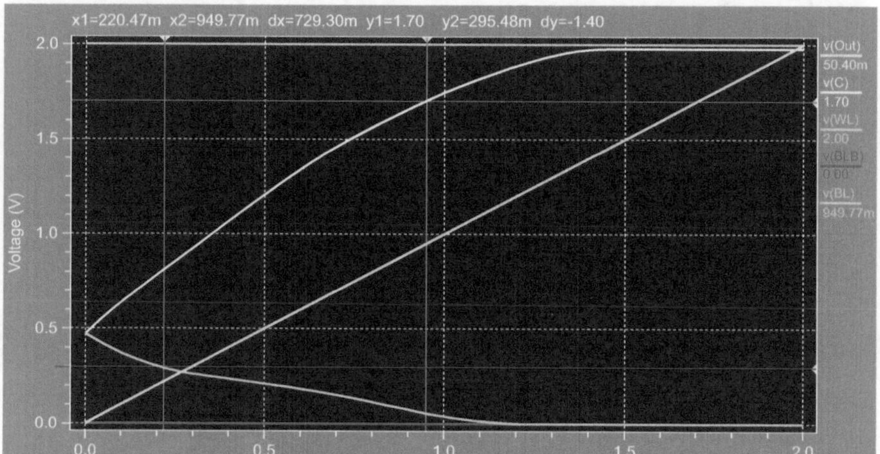

Figure 6.11 Noise margin of 6T static RAM using 45 nm file.

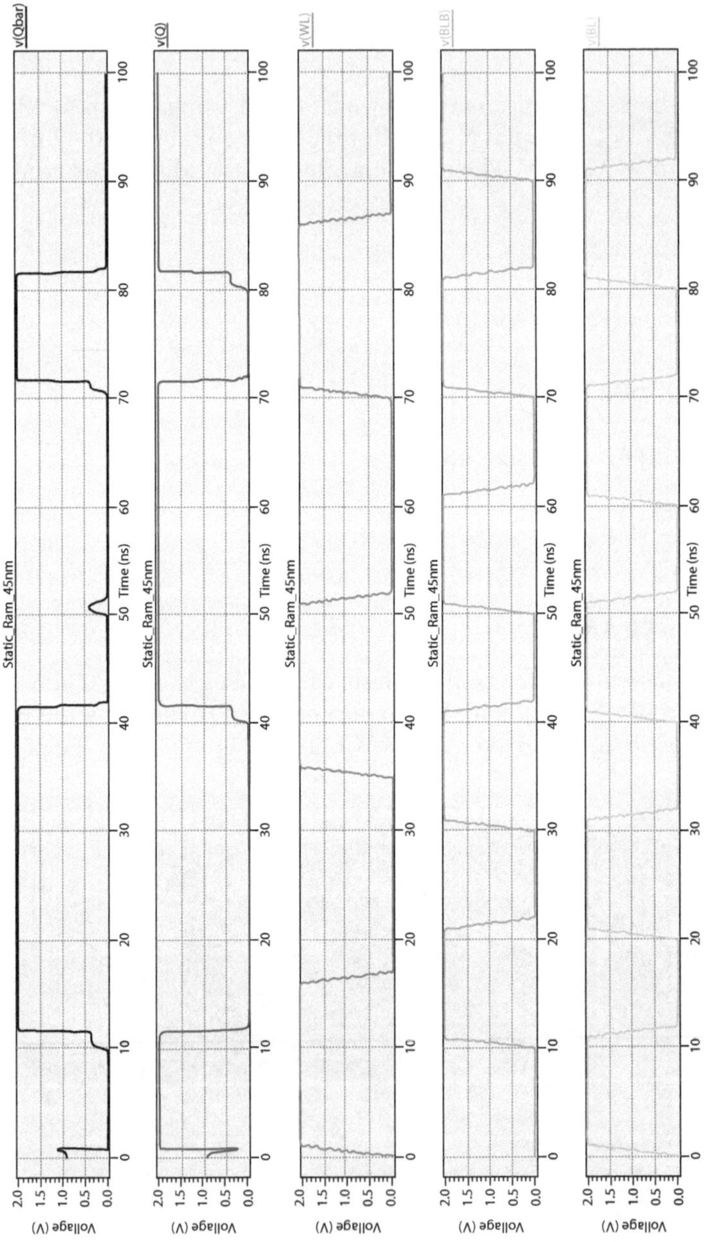

Figure 6.12 Transient analysis of 6T static RAM using 7 nm file.

output. Observe that Q and Qbar, and B and BLB complements each other [56–62].

Figure 6.13 shows the DC characteristics of 6T SRAM using 7 nm file. BL is the lowermost line, BLB is the V(BLB), WL is the V(WL), Q is the V(Q), and Qbar is the uppermost line [63–68]. Observe that when vBL is increasing, there is an increase in the output. This increase is being complemented in Qbar.

Figure 6.14 shows the power consumption of 6T SRAM using 7 nm file. The average power consumed is 487.129 pW. The maximum power is 0.7149 nW while the minimum power is 0.3606 nW.

Figure 6.15 shows the delay of 6T SRAM using 7 nm file. The image on the upper shows the graph obtaining the response time of gate from low to high transition with t_{pLH} = 7.36 ps [69–74]. The image on the lower shows high to low transition with t_{pHL} = 1.03 ns. To determine the delay, the following is computed using the dx of two graphs.

$$\text{Delay} = \frac{tp_{LH} + tp_{HL}}{2}$$

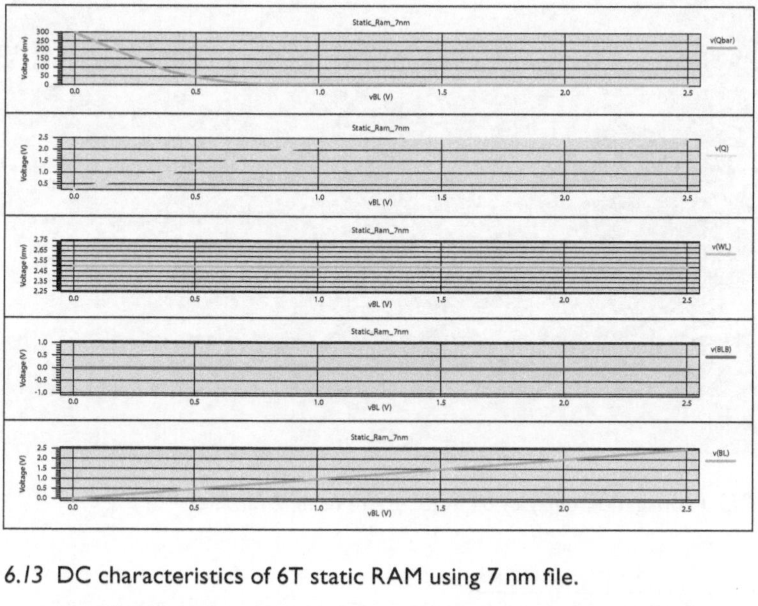

Figure 6.13 DC characteristics of 6T static RAM using 7 nm file.

```
Power Results
Vdd from time 1e-009 to 1e-005
Average power consumed -> 4.875179e-010 watts
Max power 9.984344e-006 at time 7.15486e-008
Min power 6.586115e-012 at time 3.60625e-008
```

Figure 6.14 Power consumption of 6T static RAM using 7 nm file.

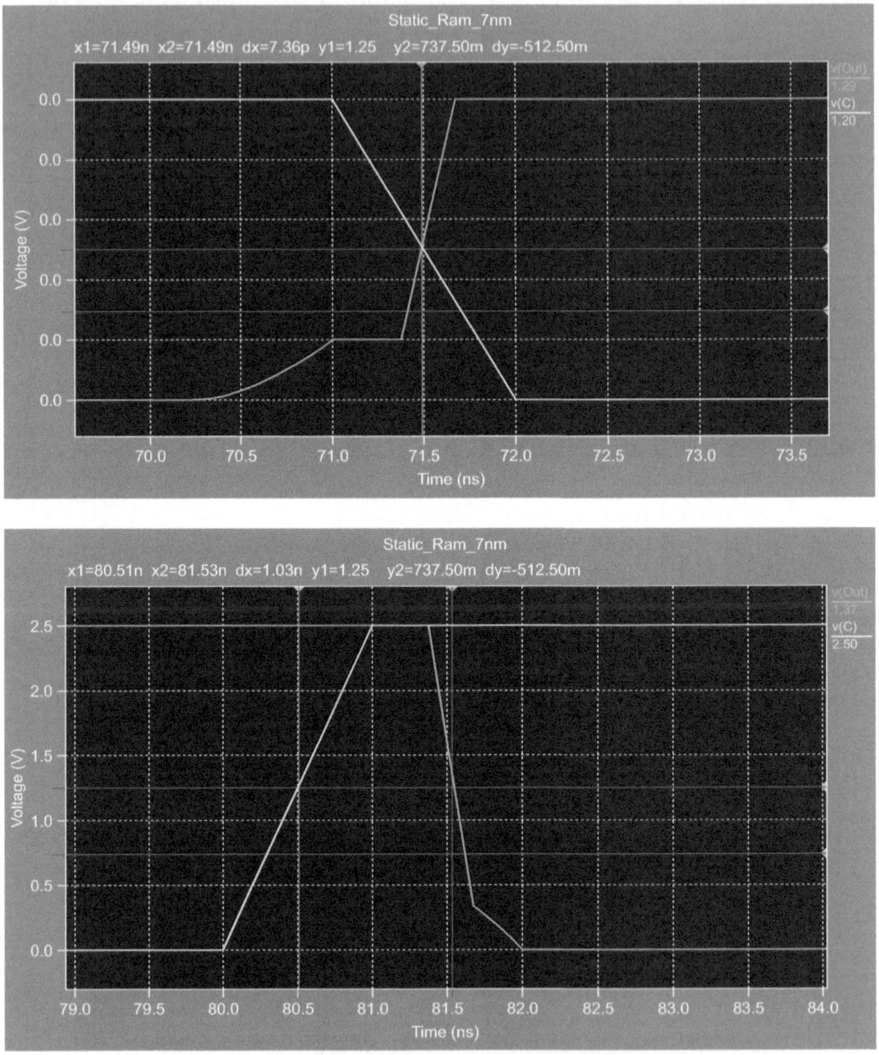

Figure 6.15 Propagation delay of 6T static RAM using 7 nm file.

$$\text{Delay} = \frac{7.36\ \text{ps} + 1.03\ \text{ns}}{2}$$

$$\text{Delay} = 518.68\ \text{ps}$$

Figure 6.16 shows the noise margin of 6T SRAM using 7 nm file. V_{OH} is equal to 940.41 mV, V_{OL} is equal to 31.33 mV, V_{IL} is equal to 199.10 mV,

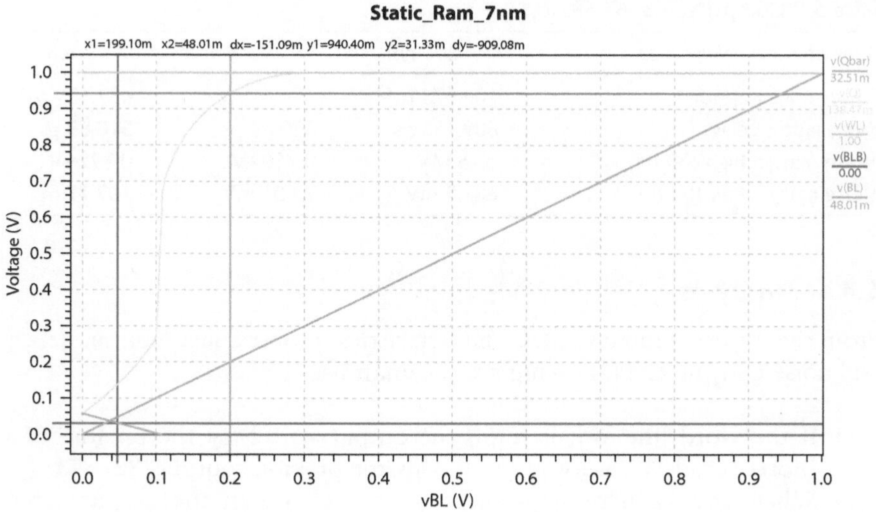

Figure 6.16 Noise margin of 6T static RAM using 7 nm file.

and V_{IH} is equal to 48.01 mV. The noise margins N_{MH} and N_{ML} are computed as follows:

$$N_{MH} = V_{OH} - V_{IH}$$

$$N_{MH} = 940.41\text{ mV} - 48.01\text{ V}$$

$$N_{MH} = 892.4\text{ mV}\quad N_{ML} = V_{IL} - V_{OL}$$

$$N_{ML} = 199.10\text{ mV} - 31.33\text{ mV}$$

$$N_{ML} = 167.77\text{ mV}$$

6.4 COMPARISON

Table 16.1 shows the comparison for the 6T SRAM in terms of power consumption, delay, and noise margin with different model files. It can be observed that the 6T SRAM with 200 nm file has the greatest power consumption with 10.4873 μW, while 6T SRAM with 7 nm has the least power consumption with 487.129 pW [75–81]. Moreover, the highest propagation delay is for 200 nm with 689.155 ps, while the lowest delay is for 7 nm with 518.68 ps. Regarding noise margin, the highest NMH is for 200 nm and the highest NML is for 7 nm.

Table 6.1 Comparison of 6T SRAM

Parameter/technology node	*200 nm*	*45 nm*	*7 nm*
Power consumption	10.4873 μW	1.3361 nW	487.129 pW
Propagation delay	689.155 ps	520.82 ps	518.68 ps
Noise margin high (NMH)	1.2537 V	0.7502 V	0.8924 V
Noise margin low (NML)	68.69 mV	74.01 mV	167.77 mV

6.4.1 Analysis

From the transient analysis, DC characteristics, power consumption, delay, and noise margin, the following are the analysis:

i. If the word line WL is high, the output will copy the bit line BL. Moreover, if WL is low, Q will copy the previous output [82–87].
ii. When vBL is increasing, there is an increase in the output. This increase is being complemented in Qbar.
iii. With the decrease in transistor size, there will also be a decrease in power consumption
iv. As transistor size decreases, propagation delay increases; thus, small-sized transistors have greater switching speed [88–94].
v. The higher the transistor size, the higher the noise margin-high (NMH).
vi. Design half-byte ROM (4 × 4 ROM) using CMOS technology for 200 nm, 45 nm, and 7 nm and compare the results in terms of power consumption, delay, and noise margin [95–102].

6.5 COMPARISON

Table 16.2 shows the comparison for the 4 × 4 ROM in terms of power consumption, delay, and noise margin with different model files. It can be observed that the 4 × 4 ROM with 200 nm file has the greatest power consumption with 23.6684 μW, while 4 × 4 ROM with 7 nm has the least power consumption with 0.6428 μW. Moreover, the highest propagation delay is at 200 nm with 899.325 ps, while the lowest delay is at 7 nm with 115.03 ps. With noise margin [103–106], the highest NMH is at 200 nm with 0.90 V, and the highest NML is at 7 nm with 472.1 mV.

6.5.1 Analysis

From the transient analysis, DC characteristics, power consumption, delay, and noise margin, the following are the analysis:

Table 6.2 Comparison of 4 × 4 ROM

	200 nm	*45 nm*	*7 nm*
Power consumption	23.6684 μW	12.3880 μW	0.6428 μW
Propagation delay	899.325 ps	362.07 ps	115.03 ps
Noise margin high (NMH)	0.90 V	0.50 V	0.4706 mV
Noise margin low (NML)	0.91311 V	608.81 mV	472.1 mV

i. If the inputs are both low, the output is read as 0001. If input A0 is low and A1 is high, the output is read as 0010. If input A0 is high and A1 is low, the output is read as 0100. If all inputs are high, the output is read as 1000.
ii. When vA0 is increasing, there is an increase in the output, y2, and a decrease in the output, y0.
iii. With the decrease in transistor size, there will also be a decrease in power consumption.
iv. As transistor size decreases, propagation delay increases; thus, small-sized transistors have greater switching speed.
v. The higher the transistor size, the higher the noise margin-high (NMH), and the lower the transistor size, the lower the noise margin-low (NML).

6.6 FUTURE SCOPE

SRAM and DRAM blocks are available in various topologies at different technology nodes. Apart from conventional CMOS, new architectures like DGMOSFET, FinFET [8–11], and nanowires [11–12] can be explored for performance improvement and increased stability. The designing of bulk memories is the primary focus of researchers for Internet of Things (IoT)-based applications [13–14]. Research shows different leakage reduction techniques to improve subthreshold performance of transistors at lower technology node [15–16]. So, the circuit made with steep subthreshold performance leads to further decreases in power consumption.

REFERENCES

1. P. Fantini, G. Servalli and Tessariol, "Semiconductor memory technologies", M. Rudan, R. Brunetti and S. Reggiani, (Eds.), *Springer Handbook of Semiconductor Devices*. Springer Handbooks. Springer, Cham, 2023.
2. K. Itoh, "An introduction to memory chip design", *VLSI Memory Chip Design. Springer Series in Advanced Microelectronics*, vol. 5, Springer, Berlin, Heidelberg, 2001.

3. Namrata Mendiratta and Suman Lata Tripathi "A review on performance comparison of advanced MOSFET structures below 45 nm technology node", *Journal of Semiconductor*, vol. 41, no. 6, pp. 1–10, 2020.
4. Shekhar Verma, Vishal Narula and Suman Lata Tripathi, "Performance analysis of multi-channel-multi-gate-based junctionless field effect transistor", *IETE Journal of Research*, vol. 70, pp. 1–11, 2023. Doi:10.1080/03772063.2023.2218318.
5. T. Santosh Kumar and S. L. Tripathi, "Low power and suppressed noise 6T, 7T SRAM cell using 18 nm FinFET", *Wireless Pers Commun,* vol.130, pp. 103–112, 2023.
6. S. Verma and S. L. Tripathi, "Design and analysis of heterojunction inverted-T P-FinFET on 14 nm technology node for use in low-power digital circuits", *Silicon,* vol. 15, pp. 3725–3736, 2023. https://doi.org/10.1007/s12633-023-02294-w.
7. PTM models, http://ptm.asu.edu/.
8. K. B. Madhavi and S. L. Tripathi, "Electrical characterization of highly stable 10nm triple-gate FinFET for different contacts and oxide region materials", *Silicon* vol. 14, pp. 12281–12291, 2022.
9. T. S. Kumar and S. L. Tripathi, "Process evaluation in FinFET based 7T SRAM cell", *Analog Integr Circ Sig Process*, vol. 109, pp. 545–551, 2021.
10. G. Singh Patel, S. Lata Tripathi and S. Awasthi, "Performance enhanced unsymmetrical FinFET and its applications", *2018 IEEE Electron Devices Kolkata Conference (EDKCON)*, 2018, pp. 222–227.
11. Mohinder Bassi, Suman Lata Tripathi and Shekhar Verma, "Analysis and design of high-K material nanowire transistor for improved performance", *IEEE IEMCON*, Vancouver, BC Canada, pp. 613–618, Oct. 2019.
12. S. L. Tripathi, S. K. Sinha and P. Gupta, "Design of triple material junctionless CG MOSFET", *2018 International Conference on Intelligent Circuits and Systems (ICICS),* 2018, pp. 42–45.
13. Suman Lata Tripathi, "Low power high performance tunnel FET: analysis for IOT applications", *Handbook of Research on the Internet of Things Applications in Robotics and Automation*, IGI Global Publisher.
14. S. Bhattacharya, S. L. Tripathi and G. H. Nayana, "Design transmission gates using double-gate junctionless TFETs", *Silicon,* vol. 16, pp. 3359–3372, 2024. https://doi.org/10.1007/s12633-024-02927–8.
15. S. L. Tripathi, P. Pathak, A. Kumar, *et al.* "Improved drain current with suppressed short channel effect of p+Pocket double-gate MOSFET in sub-14 nm technology node", *Silicon,* vol. 14, pp. 10881–10891, 2022.
16. T. S. Kumar and S. L. Tripathi, "Leakage reduction in 18 nm FinFET based 7T SRAM cell using self controllable voltage level technique", *Wireless Pers Commun*, vol. 116, pp. 1837–1847, 2020.
17. Anjana Bhardwaj, Pradeep Kumar, B. Raj and Sunny Anand, "Design and performance optimization of doping-less vertical nanowire TFET using gate stack technique", *Journal of Electronic Materials*, vol. 41, no. 7, pp. 4005–4013, 2022.

18. Jeetendra Singh and B. Raj, "Tunnel current model of asymmetric MIM structure levying various image forces to analyze the characteristics of filamentary memristor", *Applied Physics A*, vol. 125, no. 3, pp. 203.1–203.11 Feb. 2019.
19. Candy Goyal, Jagpal Singh Ubhi and B. Raj, "Low leakage zero ground noise nanoscale full adder using source biasing technique", *Journal of Nanoelectronics and Optoelectronics*, vol. 14, pp. 360–370, March 2019.
20. Gurmohan Singh, R. K. Sarin and B. Raj, "A novel robust exclusive-OR function implementation in QCA nanotechnology with energy dissipation analysis", *Journal of Computational Electronics*, vol. 15, no. 2, pp. 455–465, June 2016.
21. Girish Wadhwa, Priyanka Kamboj and Balwinder Raj, "Design optimisation of junctionless TFET biosensor for high sensitivity", *Advances in Natural Sciences: Nanoscience and Nanotechnology*, vol. 10, no. 7, pp. 45001, 2019.
22. Priya Bansal and B. Raj, "Memristor modeling and analysis for linear dopant drift kinetics", *Journal of Nanoengineering and Nanomanufacturing*, vol. 6, pp. 1–7, 2016.
23. Ashima, D Vaithiyanathan and B. Raj, "Performance analysis of charge plasma induced graded channel Si nanotube", *Journal of Engineering Research*, EMSME Special Issue. pp. 146–154, Aug. 2021.
24. Abhishek Singh Tomar, Vijay Kumar Magraiya and B. Raj, "Scaling of access and data transistor for high performance DRAM cell design", *Quantum Matter*, vol. 2, pp. 412–416, Oct. 2013.
25. Neeraj Jain and B. Raj, "Parasitic capacitance and resistance model development and optimization of raised source/drain SOI FinFET structure for analog circuit applications", *Journal of Nanoelectronics and Optoelectronics*, vol. 13, pp. 531–539, April 2018.
26. Shradhya Singh, S. K. Vishvakarma and B. Raj, "Analytical modeling of split-gate junction-less transistor for a biosensor application", *Sensing and Bio-Sensing*, vol. 18, pp. 31–36, April 2018.
27. Maisagalla Gopal and Balwinder Raj, "Low power 8T SRAM cell design for high stability video applications", *ITSI Transaction on Electrical and Electronics Engineering*, vol. 1, no. 5, pp. 91–97, 2013.
28. Balwinder Raj, Jatin Mitra, Deepak Kumar Bihani, Rangharajan V, A.K. Saxena and S. Dasgupta, "Analysis of noise margin, power and process variation for 32 nm FinFET based 6T SRAM cell", *Journal of Computer*, vol. 5, no. 6, pp. 1–8, 2010.
29. Divya Sharma, Rajesh Mehra and B. Raj, "Comparative analysis of photovoltaic technologies for high efficiency solar cell design", *Superlattices and Microstructures*, vol. 153, p. 106861, May 2021.
30. Pawandeep Kaur, Avtar Singh Buttar and B. Raj, "A comprehensive analysis of nanoscale transistor based biosensor: a review", *Indian Journal of Pure and Applied Physics*, vol. 59, pp. 304–318, April 2021.
31. Divya Yadav, Balwant Raj and B. Raj, "Design and simulation of low power microcontroller for IoT applications", *Journal of Sensor Letters*, vol. 18, pp. 401–409, May 2020.
32. Shailendra Singh and B. Raj, "A 2-D analytical surface potential and drain current modeling of double-gate vertical t-shaped tunnel FET", *Journal of Computational Electronics*, vol. 19, pp. 1154–1163, April 2020.

33. Jeetendra Singh and B. Raj, "An accurate and generic window function for non-linear Memristor model", *Journal of Computational Electronics*, vol. 18, no. 2, pp. 640–647, June 2019.
34. Manjit Kaur, Neena Gupta, Sanjeev Kumar, B. Raj and Arun Kumar Singh, "Comparative RF and crosstalk analysis of carbon based nano interconnects", *IET Circuits, Devices & Systems*, vol. 15, no. 6, pp. 493–503, Feb. 2021.
35. Nehru Kandasamy, Firdous Ahmad, D. Ajitha, B. Raj and Nagarjuna Telagam "Quantum dot cellular automata based scan flip flop and boundary scan register", *IETE Journal of Research*, vol. 66, pp. 535–548, 2020.
36. Jeetendra Singh and B. Raj, "Design and investigation of 7T2M NVSARM with Enhanced stability and temperature impact on store/restore energy", *IEEE Transactions on Very Large Scale Integration Systems*, vol. 27, no. 6, pp. 1322–1328, June 2019.
37. Anil Kumar Bhardwaj, Sumeet Gupta, B. Raj and Amandeep Singh, "Impact of double gate geometry on the performance of carbon nanotube field effect transistor structures for low power digital design", *Computational and Theoretical Nanoscience*, vol. 16, pp. 1813–1820, 2019.
38. Neeraj Jain and B. Raj, "Thermal stability analysis and performance exploration of asymmetrical dual-k underlap spacer (ADKUS) SOI FinFET for security and privacy applications", *Indian Journal of Pure & Applied Physics*, vol. 57, pp. 352–360, May 2019.
39. Neeraj Jain and Balwinder Raj, "Dual-k spacer region variation at the drain side of asymmetric SOI FinFET structure: performance analysis towards the analog/RF design applications", *Journal of Nanoelectronics and Optoelectronics*, vol. 14, pp. 349–359, March 2019.
40. Jeetendra Singh, Sanjeev Sharma, B. Raj and Mamta Khosla, "Analysis of barrier layer thickness on performance of In1-xGaxAs based gate stack cylindrical gate nanowire MOSFET", *Journal of Nanoelectronics and Optoelectronics*, vol. 13, pp. 1473–1477, Oct. 2018.
41. Neeraj Jain and B. Raj, "Analysis and performance exploration of high-k SOI FinFETs over the conventional low-k SOI FinFET toward analog/RF design", *Journal of Semiconductors*, vol. 39, no. 12, pp. 124002–1–7, Dec. 2018.
42. Candy Goyal, Jagpal Singh Ubhi and B. Raj, "A reliable leakage reduction technique for approximate full adder with reduced ground bounce noise", *Journal of Mathematical Problems in Engineering*, vol. 2018, Article ID 3501041, p. 16, 15 Oct. 2018.
43. Anuradha, Jeetendra Singh, B. Raj and Mamta Khosla, "Design and performance analysis of nano-scale memristor-based nonvolatile SRAM", *Journal of Sensor Letter*, vol. 16, pp. 798–805, Oct. 2018.
44. Girish Wadhwa, B. Raj, "Parametric variation analysis of charge-plasma-based dielectric modulated JLTFET for biosensor application", *IEEE Sensor Journal*, vol. 18, no. 15, pp. 6070–6077, 1 Aug. 2018.
45. Jeetendra Singh and B. Raj, "Comparative analysis of Memristor models for memories design", *Journal of Semiconductors*, vol. 39, no. 7, pp. 074006-1–12, July 2018.
46. Divya Yadav, Shailesh Singh Chouhan, Santosh Kumar Vishvakarma and B. Raj, "Application specific microcontroller design for IoT based WSN", *Sensor Letter*, vol. 16, pp. 374–385, May 2018.

47. Gurmohan Singh, R. K. Sarin and B. Raj, "Fault-tolerant design and analysis of quantum-dot cellular automata based circuits", *IEEE/IET Circuits, Devices & Systems*, vol. 12, pp. 638–64, 2018.
48. Jeetendra Singh and B. Raj, "Modeling of mean barrier height levying various image forces of metal insulator metal structure to enhance the performance of conductive filament based Memristor model", *IEEE Nanotechnology,* vol. 17, no. 2, pp. 268–267, March 2018.
49. Aakash Jain, Sanjeev Sharma and B. Raj, "Analysis of triple metal surrounding gate (TM-SG) III–V nanowire MOSFET for photosensing application", *Opto-Electronics Journal*, vol. 26, no. 2, pp. 141–148, May 2018.
50. Aakash Jain, Sanjeev Sharma and B. Raj, "Design and analysis of high sensitivity photosensor using cylindrical surrounding gate MOSFET for low power sensor applications", *Engineering Science and Technology, an International Journal*, vol. 19, no. 4, pp. 1864–1870, Dec. 2016.
51. G. Saiphani Kumar, Amandeep Singh and B. Raj, "Design and analysis of gate all around CNTFET based SRAM cell design", *Journal of Computational Electronics*, vol. 17, no. 1, pp 138–145, March 2018.
52. Gurinder pal Singh, B. S. Sohi and Balwinder Raj, "Material properties analysis of graphene base transistor (GBT) for VLSI analog circuits", *Indian Journal of Pure & Applied Physics*, vol. 55, pp. 896–902, Dec. 2017.
53. Sunil Kumar and B. Raj, "Estimation of stability and performance metric for inward access transistor based 6T SRAM cell design using n-type/p-type DMDG-GDOV TFET", *IEEE VLSI Circuits and Systems Letter,* vol. 3, no. 2, pp. 25–39, June 2017.
54. Shashikant Sharma, Anjan Kumar, Manisha Pattanaik and B. Raj, "Forward body biased multimode multi-threshold CMOS technique for ground bounce noise reduction in static CMOS adders", *International Journal of Information and Electronics Engineering*, vol. 3, no. 3, pp. 567–572, 2013.
55. Hamendra Singh, Pankaj Kumar and Balwinder Raj, "Performance analysis of majority gate SET based 1-bit full adder", *International Journal of Computer and Communication Engineering*, vol. 2, no. 4, pp. 1–8, 2013.
56. Anil Kumar Bhardwaj, Sumeet Gupta and B. Raj, "Investigation of parameters for Schottky barrier (SB) height for Schottky Barrier based carbon nanotube field effect transistor device", *Journal of Nanoelectronics and Optoelectronics*, vol. 15, pp. 783–791, July 2020.
57. Priya Bansal and B. Raj, "Memristor: a versatile nonlinear model for dopant drift and boundary issues", *Journal of Computational and Theoretical Nanoscience*, vol. 14, no. 5, pp. 2319–2325, May 2017.
58. Neeraj Jain and B. Raj, "An analog and digital design perspective comprehensive approach on Fin-FET (Fin-Field Effect transistor) technology: a review", *Reviews in Advanced Sciences and Engineering*, vol. 5, pp. 1–14, 2016.
59. Balwinder Raj, A. K. Saxena and S. Dasgupta, "Analytical modeling for the estimation of leakage current and subthreshold swing factor of nanoscale double gate FinFET device", *Microelectronics International,* vol. 26, pp. 53–63, 2009.

60. Soniya, Shailendra Singh, Girish Wadhwa and Balwinder Raj, "An analytical modeling for dual source vertical tunnel field effect transistor", *International Journal of Recent Technology and Engineering*, vol. 8, no. 2, pp. 603–608, July 2019.
61. Shailendra Singh and B. Raj, "Design and analysis of hetrojunction vertical T-shaped tunnel field effect transistor", *Journal of Electronics Material,* vol. 48, no. 10, pp 6253–6260, Oct. 2019.
62. Candy Goyal, Jagpal Singh Ubhi and B. Raj, "A low leakage CNTFET based inexact full adder for low power image processing applications", *International Journal of Circuit Theory and Applications*, vol. 47, no.9, pp. 1446–1458, Sep. 2019.
63. B. Raj, A. K. Saxena and S. Dasgupta, "A compact drain current and threshold voltage quantum mechanical analytical modeling for FinFETs", *Journal of Nanoelectronics and Optoelectronics*, vol. 3, no. 2, pp. 163–170, 2008.
64. Girish Wadhwa and B. Raj, "An analytical modeling of charge plasma based tunnel field effect transistor with impacts of gate underlap region", *Superlattices and Microstructures*, vol. 142, pp. 106512, June 2020.
65. Shailendra Singh and B. Raj, "Modeling and simulation analysis of SiGe hetrojunction double GateVertical t-shaped tunnel FET", *Superlattices and Microstructures*, vol. 142, pp. 106496, June 2020.
66. Neeraj Jain and Balwinder Raj, "Device and circuit co-design perspective comprehensive approach on FinFET technology - a review", *Journal of Electron Devices*, vol. 23, no. 1, pp. 1890–1901, 2016.
67. Sunil Kumar and B. Raj, "Analysis of I_{ON} and ambipolar current for dual-material gate-drain overlapped DG-TFET", *Journal of Nanoelectronics and Optoelectronics*, vol. 11, pp. 323–333, June 2016.
68. Naveed Anjum, Tarun Bali and B Raj, "Design and simulation of handwritten multiscript character recognition", *International Journal of Advanced Research in Computer and Communication Engineering*, vol. 2, no. 7, pp. 2544–2549, July 2013.
69. Karmjit Singh and B. Raj, "Performance and analysis of temperature dependent multi-walled carbon nanotubes as global interconnects at different technology nodes", *Journal of Computational Electronics*, vol. 14, no. 2, pp. 469–476, June 2015.
70. Sunil Kumar and B. Raj, "Compact channel potential analytical modeling of DG-TFET based on evanescent–mode approach", *Journal of Computational Electronics*, vol. 14, no. 2, pp. 820–827, July 2015.
71. Karmjit Singh and B. Raj, "Temperature dependent modeling and performance evaluation of multi-walled CNT and single-walled CNT as global interconnects", *Journal of Electronic Materials*, vol. 44, no. 12, pp 4825–4835, Dec. 2015.
72. V.K. Sharma, M. Pattanaik and B. Raj, "INDEP approach for leakage reduction in nanoscale CMOS circuits", *International Journal of Electronics*, vol. 102, no. 2, pp. 200–215, 2014.
73. Karmjit Singh and B. Raj, "Influence of temperature on MWCNT bundle, SWCNT bundle and copper interconnects for nanoscaled technology nodes", *Journal of Materials Science: Materials in Electronics*, vol. 26, no. 8, pp. 6134–6142, 2015.

74. Naveed Anjum, Tarun Bali and B. Raj, "Design and simulation of handwritten Gurumukhi and Devanagri numerical recognition", *International Journal of Computer Applications*, vol. 73(12), pp. 16–21, 2013.
75. S. Khandelwal, V. Gupta, B. Raj and R. D. Gupta, "Process variability aware low leakage reliable nano scale DG-FinFET SRAM cell design technique", *Journal of Nanoelectronics and Optoelectronics*, vol. 10, no. 6, pp. 810–817, Dec. 2015.
76. V.K. Sharma, M. Pattanaik and B. Raj, "ONOFIC approach: low power high speed nanoscale VLSI circuits design", *International Journal of Electronics*, vol. 101, no. 1, pp. 61–73, 2014.
77. S. Khandelwal, Balwinder Raj and R. D. Gupta, "FinFET based 6T SRAM cell design: analysis of performance metric, process variation and temperature effect", *Journal of Computational and Theoretical Nanoscience*, vol. 12, pp. 2500–2506, 2015.
78. Sumit Singh, Shekhar Yadav, Jagdeep Rahul, Anurag Srivastava and B. Raj, "Impact of HfO_2 in graded channel dual insulator double gate MOSFET", *Journal of Computational and Theoretical Nanoscience*, vol. 12, no. 6, pp. 950–953, April 2015.
79. Vijay Kumar Sharma, Manisha Pattanaik and B. Raj, "PVT variations aware low leakage INDEP approach for nanoscale CMOS Circuits", *Microelectronics Reliability*, vol. 54, pp 90–99, 2014.
80. B. Raj, A. K. Saxena and S. Dasgupta, "Quantum mechanical analytical modeling of nanoscale DG FinFET: evaluation of potential, threshold voltage and source/drain resistance", *Journal of Material Science in Semiconductor Processing*, vol. 16, no. 4, pp. 1131–1137, 2013.
81. Maisagalla Gopal, Siva Sankar D. Prasad and Balwinder Raj, "8T SRAM cell design for dynamic and leakage power reduction", *International Journal of Computer Applications*, vol. 71, no. 9, pp. 43–48, June 2013.
82. Manisha Pattanaik, B. Raj, Shashikant Sharma and Anjan Kumar, "Diode based trimode multi–threshold CMOS technique for ground bounce noise reduction in static CMOS adders", *Advanced Materials Research*, vol. 548, pp 885–889, 2012.
83. Balwinder Raj, A. K. Saxena and S. Dasgupta, "Nanoscale FinFET based SRAM cell design: analysis of performance metric, process variation, underlapped FinFET and temperature effect", *IEEE Circuits and System Magazine*, vol. 11, no. 2, pp. 38–50, 2011.
84. V. K. Sharma, M. Pattanaik and Balwinder Raj, "Leakage current ONOFIC approach for deep submicron VLSI circuit design", *International Journal of Electrical, Computer, Electronics and Communication Engineering*, vol. 7, no. 4, pp. 239–244, 2013.
85. Parminder Kaur, Sandeep Singh Gill and B. Raj, "Comparative analysis of OFETs materials and devices for sensor applications", *Journal of Silicon*, vol. 14, pp. 4463–4471, 2022.
86. Sanjeev Kumar Sharma, Parveen Kumar, Balwant Raj and B. Raj, "$In_{1-x}Ga_xAs$ double metal gate-stacking cylindrical nanowire MOSFET for highly sensitive photo detector", *Journal of Silicon*, vol. 14, pp. 3535–3541, 2022.

87. B. Raj, A. K. Saxena and S. Dasgupta, “Analytical modeling of quasi planar nanoscale double gate FinFET with source/drain resistance and field dependent carrier mobility: a quantum mechanical study”, *Journal of Computer*, vol. 4, no. 9, pp. 1–8, 2009.
88. S. Bhushan, S. Khandelwal and B. Raj, “Analyzing different mode FinFET based memory cell at different power supply for leakage reduction”, *Seventh International Conference on Bio-Inspired Computing: Theories and Application, (BIC-TA 2012) Advances in Intelligent Systems and Computing*, vol. 202, pp. 89–100, 2013.
89. Jeetendra Singh and B. Raj, “Temperature dependent analytical modeling and simulations of nanoscale memristor”, *Engineering Science and Technology, an International Journal*, vol. 21, pp. 862–868, Oct. 2018.
90. Shradhya Singh, Shashi Bala, Balwant Raj and B. Raj, “Improved sensitivity of dielectric modulated junctionless transistor for nanoscale biosensor design”, *Sensor Letter*, vol. 18, pp. 328–333, April 2020.
91. Vivek Kumar, Santosh Kumar Vishvakarma and B. Raj, “Design and performance analysis of ASIC for IoT applications”, *Sensor Letter*, vol. 18, pp. 31–38, Jan. 2020.
92. Akanksha Jaiswal, R. K. Sarin, B. Raj and Shikha Sukhija, “A novel circular slotted microstrip-fed patch antenna with three triangle shape defected ground structure for multiband applications”, *Advanced Electromagnetic*, vol. 7, no. 3, pp. 56–63, Aug. 2018.
93. Girish Wadhwa and B. Raj, “Label free detection of biomolecules using charge-plasma-based gate underlap dielectric modulated junctionless TFET”, *Journal of Electronic Materials*, vol. 47, no. 8, pp 4683–4693, Aug. 2018.
94. Gurmohan Singh, R. K. Sarin and B. Raj, “Design and performance analysis of a new efficient coplanar quantum-dot cellular automata adder”, *Indian Journal of Pure & Applied Physics*, vol. 55, pp. 97–103, Feb. 2017.
95. Parminder Kaur, Vikas Pandey and B. Raj, “Comparative study of efficient design, control and monitoring of solar power using IoT”, *Sensor Letter*, vol. 18, pp. 419–426, May 2020.
96. Anil Kumar Bhardwaj, Sumeet Gupta and B. Raj, “Development & analysis of compact model for double gate Schottky barrier CNTFET”, *Journal of Nanoelectronics and Optoelectronics*, vol. 15, pp. 1199–1208, Aug. 2020.
97. Girish Wadhwa, Priyanka Kamboj, Jeetendra Singh and B. Raj, “Design and investigation of junctionless DGTFET for biological molecule recognition”, *Transactions on Electrical and Electronic Materials*, vol. 22, pp. 282–289, 2021.
98. Sachin Kumar Verma, Shailendra Singh, Girish Wadhwa and B. Raj, “Detection of biomolecules using charge-plasma based gate underlap dielectric modulated dopingless TFET”, *Transactions on Electrical and Electronic Materials*, vol. 21, pp. 528–535, June 2020.
99. Neeraj Jain and B. Raj, “Impact of underlap spacer region variation on Electrostatic and analog/RF performance of symmetrical high–k SOI FinFET at 20 nm channel length”, *Journal of Semiconductors*, vol. 38, no. 12, pp. 122002, Dec. 2017.

100. Shailendra Singh and B. Raj, "Analytical modeling and simulation analysis of T-shaped III-V heterojunction vertical T-FET", *Superlattices and Microstructures*, vol. 147, pp. 106717, Nov. 2020.
101. Gurmohan Singh, R. K. Sarin and B. Raj, "Design and analysis of area efficient QCA based reversible logic gates", *Journal of Microprocessors and Microsystems*, vol. 52, pp. 59–68, May 2017.
102. Shailendra Singh and B. Raj, "Parametric variation analysis on hetero-junction vertical t-shape TFET for suppressing ambipolar conduction", *Indian Journal of Pure and Applied Physics*, vol. 58, pp. 478–485, June 2020.
103. Shailendra Singh Soniya, Girish Wadhwa and Balwinder Raj, "Design and analysis of dual source vertical tunnel field effect transistor for high performance", *Transactions on Electrical and Electronics Materials,* vol. 21, pp. 74–82, Oct. 2019.
104. Manjit Kaur, Neena Gupta, Sanjeev Kumar, B. Raj and Arun Kumar Singh, "RF performance analysis of intercalated graphene nanoribbon based global level interconnects", *Journal of Computational Electronics,* vol. 19, pp. 1002–1013, June 2020.
105. Girish Wadhwa and B. Raj, "Design and performance analysis of junctionless TFET biosensor for high sensitivity", *IEEE Nanotechnology*, vol. 18, pp. 567–574, 2019.
106. Jeetendra Singh and B. Raj, "Enhanced nonlinear memristor model encapsulating stochastic dopant drift", *Journal of Nanoelectronics and Optoelectronics*, vol. 14, pp. 958–963, 2019.

Chapter 7

FeFET in computing memory

Mandeep Singh, Mukesh Kumar, Bellamkonda Prudhvi, Pindi Vishal, Tarun Chaudhary, and Balwinder Raj,

7.1 INTRODUCTION

For a range of uses, ferroelectric materials are becoming more and more common in solid-state electronics. Ferroelectric materials are particularly appealing for non-volatile memory applications due to their intrinsic properties, which include their capacity to become polarized due to any applied electric field and to maintain that polarization when electric field is removed [1–5]. In certain operating regimes, ferroelectrics also show an effect called negative capacitance, a property that has been used in silicon-based MOS technologies to produce subthreshold slopes lower than the theoretical Boltzmann limit of 60 mV/dec. These films might enable ultra-low power operation in devices that incorporate them when biased suitably [6–9]. A crucial component of modern digital systems is data processing and evaluation, especially in light of the growing artificial intelligence (AI) paradigms. von Neumann computer architectures have demonstrated a significant bottleneck that restricts data throughput and reduces processing system power consumption: data transfer congestion is caused by the architecture's separation of processing and storage [10–13]. In von Neumann designs, data transfer is correlated with the component of computer systems that uses the greatest power. The amount of data being collected and analyzed is growing exponentially (by several zettabytes annually) due to the development of concepts like Internet of Things (IoT) and AI. As a result, the traditional computing paradigm is reaching its limit and requires investigation in new directions. The logic-in-memory (LiM) concept is a viable solution to this issue since it minimizes the amount of data transfers between the computational core and system memory [10]. Although the idea was first presented in 1970, it has recently attracted renewed attention as non-volatile memories – which integrate CMOS (complementary metal–oxide–semiconductor) transistors more deeply than previous memory technologies – have emerged [14–19]. The potential applications of LiM designs, such as AI cryptography or sensor computing for the Internet of Things, are another element driving interest in these concepts. Technologies for LiM applications

 DOI: 10.1201/9781003612797-7

include ferroelectric (FeRAM and FeFET), phase change memory, magnetic (MRAM), resistive (ReRAM), and phase change memories [20–25].

7.2 FERROELECTRIC MATERIALS

By definition, ferroelectric materials are those that, in the presence of an electric field, have the ability to spontaneously polarize. These materials' polarization vs voltage (P–V) characteristics, which display bistability and hysteretic activity even in the absence of bias, account for their appeal in the non-volatile memory space [26–31]. A ferroelectric material's hysteretic P–V characteristics are displayed in Figure 7.1 along with several labeled locations of interest [1]. On the y- and x-axes, respectively, are indicated the remaining polarization charges and the coercive fields.

The charge – positive or negative – that remains in the material after external bias has been eliminated is known as the remnant polarization charge [1, 2]. The electric field that the material must have in order to cause switching from +ve polarization to -ve polarization, or vice versa, is known as the coercive field connected to these loops [1, 3]. Coercive voltage values

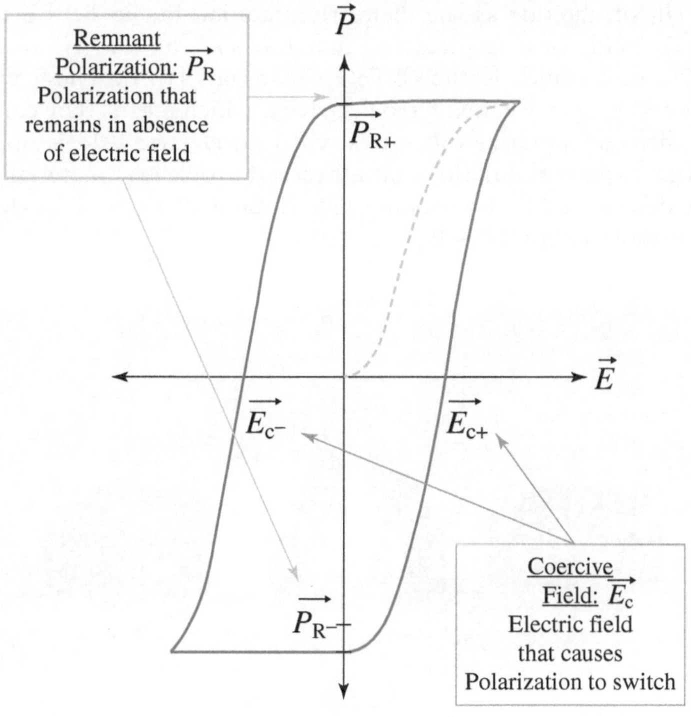

Figure 7.1 Characteristic curve for ferroelectric materials (polarization vs electric field) [1].

(Vc) can be obtained instead by multiplying the coercive field values by the ferroelectric film thickness, if that is known [32–37]. Despite being widely used as ferroelectric materials today, hafnium oxide doped with silicon or aluminum does not exhibit ferroelectric characteristics when deposited. The doped hafnium oxide needs to be in the orthorhombic phase, a certain non-centrosymmetric crystalline phase, for spontaneous polarization to happen. The material is often forced into this phase by a combination of continuous thermal processing and a TiN capping layer, which provides physical stress to the underlying hafnium oxide during cooling [1]. Figure 7.2 depicts this change in hafnium oxide to the intended crystalline phase. To preserve the integrity of the now ferroelectric film, the temperature budget of subsequent processing steps should be lowered once the hafnium oxide has been deposited and pushed into this crystalline phase [38–44].

7.3 CONSTRUCTION

FeFET has a structure similar to a traditional metal–oxide-semiconductor field-effect transistor (MOSFET); it has source, drain, and a channel in between them. The only difference lies in the gate structure. In MOSFETs, we have silicon dioxide as the dielectric material, but in FeFET the gate dielectric is made of a ferroelectric material such as lead zirconate titanate (PZT) or bismuth ferrite ($BiFeO_3$) materials [1, 4]. These materials have a special property called ferroelectricity, which means they can switch between different polarization states when an electric field is applied to them. After various simulations on all these ferroelectric materials, scientists have determined that hafnium oxide is the best material for designing logic-in-memory circuit [45–49].

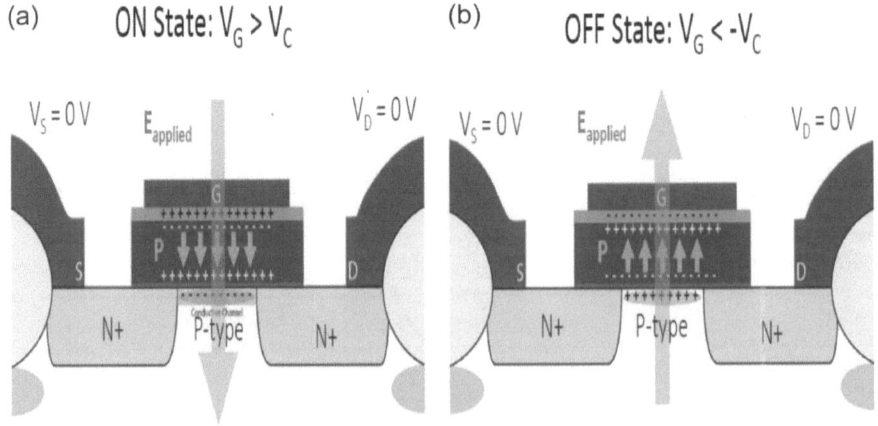

Figure 7.2 Depiction of (a) ON- and (b) OFF-state FeFET charging effects.

7.4 WORKING

Ferroelectric materials are inserted between the gate and the substrate to produce bistable transfer characteristics and demonstrate the bistability principle. FeFET has two operational states: ON and OFF. The source and drain of the device are grounded in order to achieve ON-state transfer characteristics, and the gate is pulsed with a high positive voltage that is greater than the ferroelectric film's coercive voltage [50–56]. This lowers the threshold voltage by presenting a positive polarization charge to the transistor channel's surface. The gate is now pulsed with a negative voltage for off-state operation, which is smaller than the ferroelectric film's coercive voltage. Therefore, the majority carriers from the substrate gather at the surface as a result of this negative polarization charge. Overall, this results in a rise in the threshold voltage.

7.5 EXISTING FEFET-BASED PIM (PROCESSING IN MEMORY) DESIGNS

One-time use of FeFET-based designs have been proposed; however, because they require additional circuits and have a single purpose, their usefulness is limited. Lately, novel FeFET-based multifunctional cell array designs have been presented by researchers. These designs combine more cells to perform different tasks. The structure of the attention in-memory (Aim) array is depicted in Figure 7.3(a).

The dotted wireframe shows the location of an Aim cell. Both storage and TCAM capabilities can be offered by the Aim array [57–62]. The layout of the FeFET-based memory architecture (FeMAT) is depicted in Figure 7.3(b) [5].

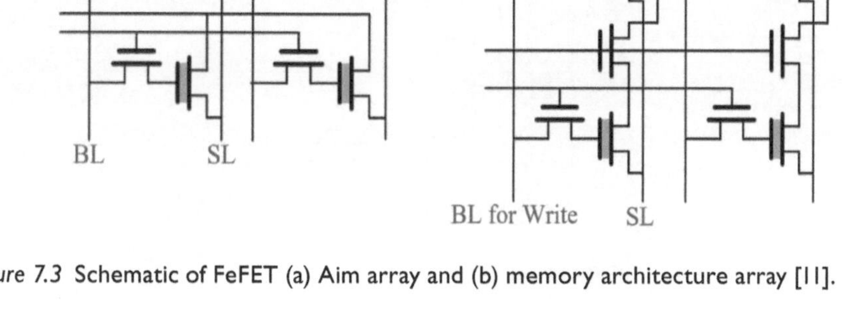

Figure 7.3 Schematic of FeFET (a) Aim array and (b) memory architecture array [11].

There is an FeMAT cell in the wireframe with dots. Storage, logic computation, and TCAM functions are implemented by the FeMAT. Furthermore, several configurations utilizing several FeFET devices were put out in an effort to boost the storage capacity of the cell [63–68].

7.6 FEMPIM CELL

The suggested multifunctional PIM cell based on FeFET, or FeMPIM, is schematically depicted in the image. Two CMOS (complementary metal–oxide–semiconductor) transistors and two FeFETs (FeFET1 and FeFET2) make up the FeMPIM. The source of FeFET1 is attached to ML2 by a transistor that WL2 controls. FeFET2's source is linked to ML1 via a transistor that WL1 controls [69–75] (Figure 7.4).

7.7 FEMPIM ARRAY STRUCTURE

The suggested FeMAT's array structure is depicted in Figure 7.5. The FeMPIM array structure will be similar to the FeMAT array structure, replacing unit cell with the FeMPIM cell.The address and input data are found in the array using the chosen SLs, WLs, and MLs [5]. To keep the other cells in their current state, the lines that are not selected are kept at the opposite level of the chosen lines. MLs and SAs for read operations are related [76–82]. Because reading and writing are separated, there are

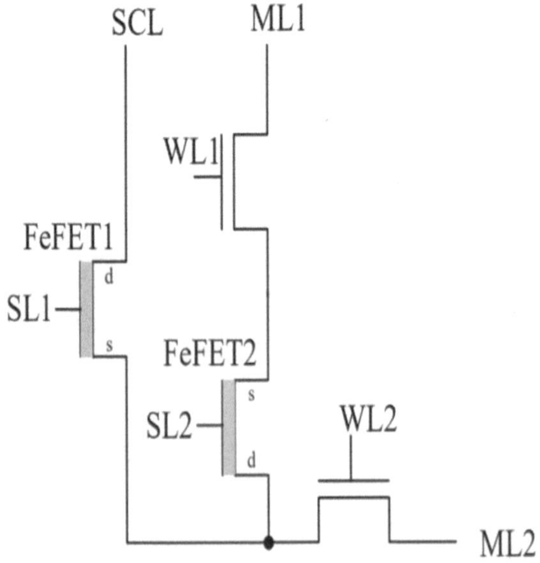

Figure 7.4 Schematic of FeMPIM cell [5].

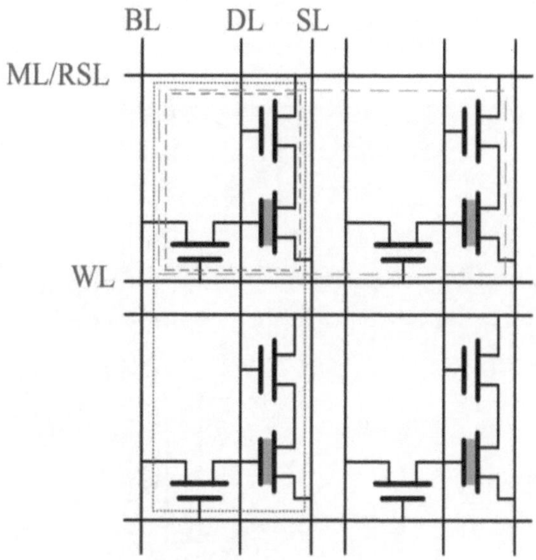

Figure 7.5 Array structure of FeMAT [11].

no data conflicts between SAs and MLs. The suggested FeMPIM offers various functions depending on SCLs and MLs, as explained below [5, 6]. The suggested FeMPIM array prevents negative voltages when compared to previous array designs, and the SAs can read the calculation results directly. Other cells on the same line are impacted when negative voltages are used in the array.

7.8 MEMORY MODE

The FeFET1 and FeFET2 operations are independent when in memory mode. The voltage that is above the V_{gs} threshold is known as the write voltage (V_w). Using this technique, we create V_{gs} by creating a voltage difference between the gate and source, and then we build the control signal [83–88].

7.9 WRITE OPERATION

The suggested FeMPIM's write operation is depicted in Figure 7.5. Figure 7.6 (a)–(b) shows how to write binary digits in the first FeFET. The write operation makes WL1 to have low value and WL2 to have a high value. SL1 and ML2 provide the proper V_{gs} to FeFET1. The purpose of SL2 is to shield FeFET2 from damage. The writing of 1 and 0 on FeFET2 is based on the same principle of working as FeFET1 and is depicted in Figure 7.6 (c)–(d) [5].

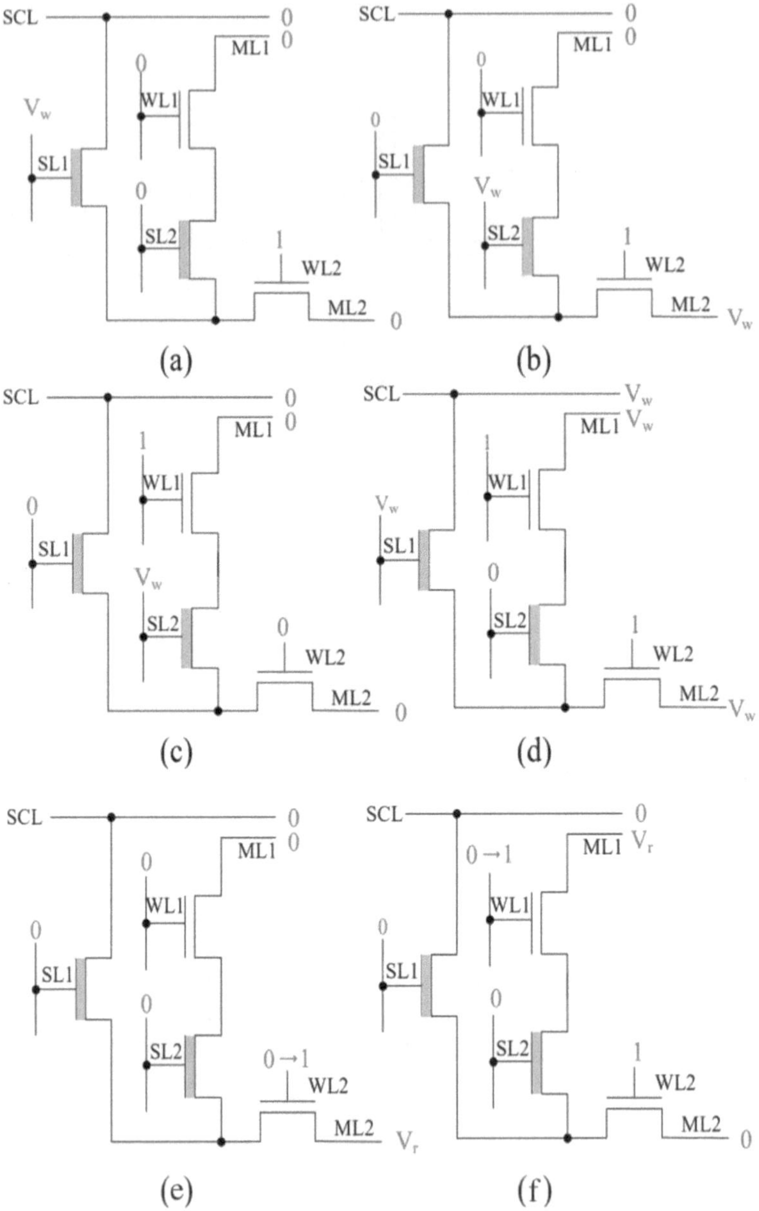

Figure 7.6 Writing operation (a) storing "1" in FeFET 1, (b) storing "0" in FeFET 1, (c) storing "1" in FeFET 2, (d) storing "0" in FeFET 2, (e) operation for reading FeFET 1, and (f) operation for reading FeFET 2 [5].

Turning off WLs and applying V_w at the other terminals of the FET devices in the array will keep the other cells in their original states [5]

7.10 READ OPERATION

The suggested FeMPIM's read procedure is depicted in Figure 7.8. Figure 7.6 (e) illustrates how FeFET1 reads data. SL1, SL2, and FeFET1 are read when WL1, ML1, and SCL are all low. First off, ML2 is already charged to Vr and WL2 is zero (low). WL2 then flips too high [5, 7]. Currently, when FeFET1 is in the "1" state, ML2's voltage rapidly decreases; while FeFET1 is in the "0" state, ML2's voltage remains constant. Ultimately, the voltage of ML2 is detected to determine the FeFET1. Figure 7.6(f) illustrates how FeFET2 reads data. ML1 is already charged to V_r for reading FeFET2, and subsequent operations are comparable to those of FeFET1. The ML1 determines the FeFET2 [89–93].

7.11 COMPUTING MODE

In contrast to logic gates in CMOS technology, logic operations using FeFET or any other non-volatile devices have low power dissipation, accurate readings, and high flexibility. For our suggested design, we have five logic operations in mind. Through ML1/ML2, all of these logic processes obtain their outcomes. The suggested cell's OR and AND operations are displayed in computing mode in Figure 7.7 [5, 8]. The inputs to the OR and AND operations are indicated by the states of FeFET1 and FeFET2 [94–99].

Figure 7.7 (a) illustrates the OR function. WL1 raises its hand and reads data from FeFET1. Only when both FeFET1 and FeFET2 are "0" does the outcome equal "0." The outcome in the remaining situations is "1". Figure 7.7(b) illustrates the AND function. WL2 reads data from FeFET2 while holding low [5]. Only when FeFET1 and FeFET2 are in the "1" state does their combined resistance become low. Because FeFET1 and FeFET2 are coupled in series, when any of them is "0", the read operation's result is "0". The NOT, NOR, and XOR operations are depicted in Figure 7.7, each of which requires FeFET devices to be in a particular state [100–106].

A state "1" FeFET device is needed for the NOT operation. Its source is the input, and its gate is low, as seen in Figure 7.7(c). The state of FeFET is the outcome. The NOR operation necessitates NOT operations on two inputs before performing an AND operation on them, as seen in Figure 7.7(d). ML1 is used to read the outcome [5]. A pair of FeFET devices in the state "0" are needed for the XOR function. In Figure 7.7, the input is displayed (e). FeFET devices stay in the "0" state when both inputs are the same; they transition to the "1" state when the inputs are different. After that, an OR operation is applied to them by the XOR procedure [107–110].

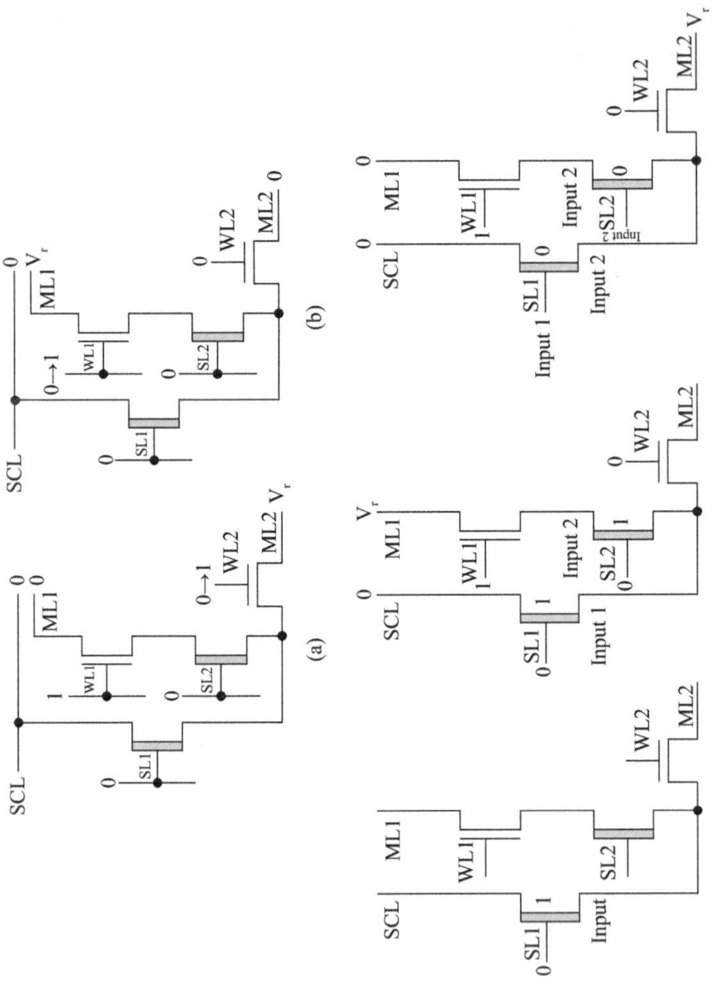

Figure 7.7 Computing mode: (a) OR gate, (b) AND gate, (c) NOT gate, (d) NOR gate, and (e) XOR gate [5].

7.12 TCAM MODE

The suggested FeMPIM's match operation in ternary content addressable memory (TCAM) mode is depicted from Figure 7.4. FeFET1 and FeFET2 have stored two complementary states (S and S′) prior to the match process. Y and Y′ are the complementary inputs of SCL and ML1. WL1 and WL2 are higher [5]. To VR, ML2 are already charged. The voltage at ML2 gives us the result of the match between S and Y. A FeFET device in a low state is linked to the low input if S and Y are the same. As a result, ML2's voltage won't decrease [5, 7]. The low input is attached to a FeFET device in a high state if S and Y are reversed, and the voltage of ML2, specifically, the states of S and S′ are both low.

In the TCAMs, it is "don't care" mode [5]. In this instance, the cell invariably yields the outcome that S and Y match. The address is returned with partially similar data in the "don't care" mode, which permits fuzzily found data. The match operation's efficacy and range will be greatly enhanced by this.

7.13 HYBRID MEMORY ELEMENT

Compact ternary content addressable memory (TCAM) has been designed using emerging technologies including ferroelectric (FeRAM and FeFET) and resistive RAM (ReRAM). These circuits can only be utilized as TCAMs, while combining memory functions (i.e., TCAM and RAM) will be intriguing for some applications [10]. CMOS technology has been used to suggest such a combination. Here, we suggest a brand-new hybrid memory component that combines TCAM and RAM functionality and is based on FeFETs.

7.14 THE TC-MEM CIRCUIT

Two modes of operation are required to make a 1-bit TC-MEM cell. The TC-MEM cell's first mode of operation is equivalent to the standard memory mode. As shown in Figure 7.8 (a), it is accomplished with only one FeFET [10]. Figure 7.8(b) shows the structure of another mode of operation, which is the TCAM mode, for which an XOR operation is used. By incorporating an NMOS transistor between the two FeFETs connected in parallel, as shown in the figure, it is possible to combine these two modes of operation. The NMOS is disconnected if the Mode input M is low. Because it is the memory mode, only

FeFET2 is able to discharge the output line designated MLWL for match/word line (TCAM/memory mode). In contrast, the NMOS is conducting if M is high (M = 1). FeFET1 and FeFET2 are linked; this indicates that the TCAM mode is being used. The MLWL line is precharged using the V_p input. Within the cell is stored a logical "0". When the right input is

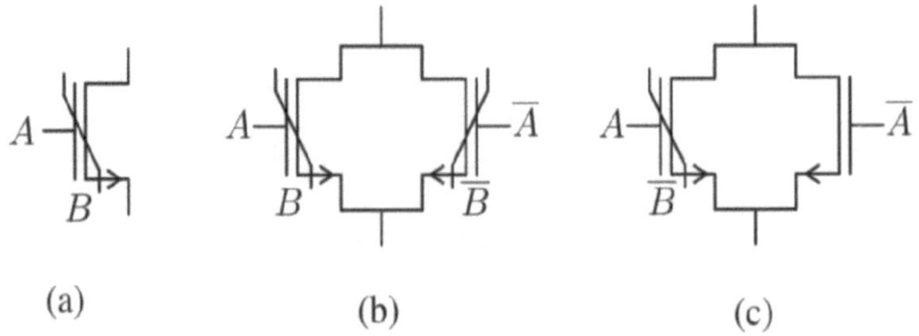

Figure 7.8 Logic gates: (a) NAND gate, (b) XOR gate, and (c) AND gate [10].

supplied, or when $V_i = 0$ and $V_i = 1$, the match line (ML) is discharged in TCAM mode. ML is high when $V_i = 1$ and $V_i = 0$. If $V_i = 1$, the word line (WL) in memory mode (i.e., M = 0) provides the stored value. If not, WL is elevated (Figure 7.9).

7.15 CONCLUSION

We used a multipurpose FET-based cell that is capable of carrying out several tasks, including TCAM, logic computation, and storage. Furthermore, our approach eliminates the need for negative voltages by offering independent read/write. When compared to similar designs, an evaluation using a calibrated FeFET model and the Cadence simulation tool has shown a moderate overhead [10].

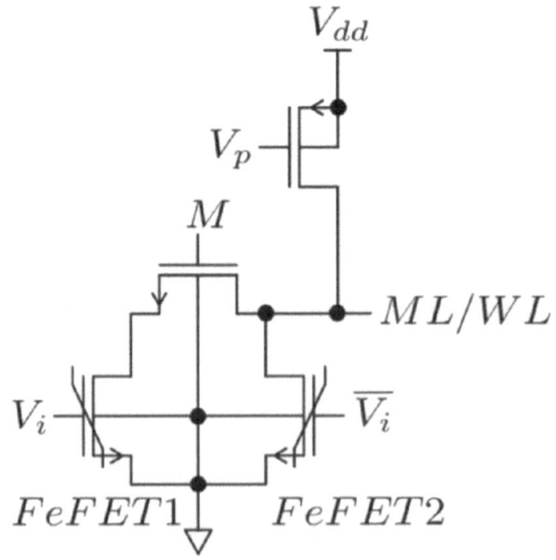

Figure 7.9 Circuit of a 1-bit TC-MEM cell [12].

REFERENCES

1. J. Merkel, "Fefet process integration and characterization", *Journal of the Microelectronic Engineering Conference,* vol. 25, no. 1, p. 20, 2019.
2. Foseph F. McGlone, "Ferroelectric HfO_2 thin films for memory devices", *Journal of the Microelectronic Engineering Conference*, Ghaziabad, India, Sept. 22–23, 2016.
3. Casey J. Gonta, "Atomic layer of deposition of ferroelectric HfO_2", *Journal of the Microelectronic Engineering Conference*, vol. 22, no. 1, Article 14, 2017.
4. Joshua L. Eschle, "Annealing study of ALD deposited ferroelectric aluminum-doped hafnium oxide". *Journal of the Microelectronic Engineering Conference*, pp. 1–8, 2018.
5. A. Yan, Y. Chen, Z. Gao, T. Ni, Z. Huang, J. Cui, ... and X. Wen, "FeMPIM: a FeFET-based multifunctional processing-in memory cell", *IEEE Transactions on Circuits and Systems II: Express Briefs,* vol. 12, pp. 143675–143694, 2023.
6. M. Gallo Sebastian and R. Khaddam-Aljameh, et al., "Memory devices and applications for in-memory computing", *Nature Nanotechnology,* vol. 15, no. 7, pp. 529–544, 2020.
7. Sharma and K. Roy, "1T non-volatile memory design using sub10nm ferroelectric FETs", *IEEE Electron Device Letters*, vol. 39, no. 3, pp. 359–362, 2018.
8. D. Reis, M. Niemier and X. S. Hu, "Computing in memory with FeFETs", *International Symposium on Low Power Electronics and Design*, pp. 1–6, 2018.
9. X. Zhang, R. Liu, T. Song, et al., "Re-FeMAT: a reconfigurable multifunctional FeFET-based memory architecture", *IEEE Transactions on Computer-Aided Design of Integrated Circuits and Systems*, early access, pp. 1–14, 2022.
10. C. Marchand, I. O'Connor, M. Cantan, E. T. Breyer, S. Slesazeck and T. Mikolajick, "FeFET based Logic-in-Memory: an overview", *2021 16th International Conference on Design and Technology of Integrated Systems in Nanoscale Era*, June 2021, pp. 1–6.
11. X. Zhang, X. Chen and Y. Han, "FeMAT: exploring in-memory processing in multifunctional FeFET-based memory array", *2019 IEEE 37th International Conference on Computer Design (ICCD)*, Abu Dhabi, United Arab Emirates, 2019, pp. 541–549.
12. C. Marchand, I. O'Connor, M. Cantan, E. T. Breyer, S. Slesazeck and T. Mikolajick, "A FeFET-based hybrid memory accessible by content and by address", *IEEE Journal on Exploratory Solid-State Computational Devices and Circuits*, vol. 8, no. 1, pp. 19–26, June 2022.
13. D. Wang, S. George, A. Aziz, S. Datta, V. Narayanan and S. K. Gupta, "Ferroelectric transistor based non-volatile flip-flop", *Proceedings of the 2016 International Symposium on Low Power Electronics and Design*, 2016, pp. 10–15.
14. I. Yoon, M. Chang, K. Ni, M. Jerry, S. Gangopadhyay, G. Smith, T. Hamam, V. Narayanan, J. Romberg, S. Lu, S. Datta and A. Ray Chowdhury, "A FeFET based processing-in-memory architecture for solving distributed least-square optimizations", *2018 76th Device Research Conference*, June 2018, pp. 1–2.

15. S. George, A. Aziz, X. Li, J. Sampson, S. Datta, S. Gupta and V. Narayanan, "NCFET based logic for energy harvesting systems", *SRC TECHCON*, Austin, TX, United States, Sep 20–22, 2015.
16. A. Aziz, S. Ghosh, S. Datta and S. K. Gupta, "Physics-based circuit compatible SPICE model for ferroelectric transistors", *IEEE Electron Device Letters*, vol. 37, no. 6, pp. 805–808, June 2016.
17. M. Courbariaux, I. Hubara, et al., "Binarized neural networks: training deep neural networks with weights and activations constrained to +1 or-1", *arXiv preprint arXiv:1602.02830*, 2016.
18. M. Rastegari, et al., "XNOR-net: imagenet classification using binary convolutional neural networks", *ECCV*, 2016, pp. 525–542.
19. R. Karam, R. Puri, S. Ghosh and S. Bhunia, "Emerging trends in design and applications of memory-based computing and content addressable memories", *Proceedings of the IEEE*, vol. 103, no. 8, pp. 1311–1330, Aug. 2015.
20. ASU, "Predictive technology model", 2011. [Online]. Available: http://ptm.asu.edu/.
21. Gurmohan Singh, R. K. Sarin and B. Raj, "Fault-tolerant design and analysis of quantum-dot cellular automata based circuits", *IEEE/IET Circuits, Devices & Systems*, vol. 12, pp. 638–64, 2018.
22. Jeetendra Singh and B. Raj, "Modeling of mean barrier height levying various image forces of metal insulator metal structure to enhance the performance of conductive filament based memristor model", *IEEE Nanotechnology*, vol. 17, no. 2, pp. 268–267, March 2018.
23. Aakash Jain, Sanjeev Sharma and B. Raj, "Analysis of triple metal surrounding gate (TM-SG) III–V nanowire MOSFET for photosensing application", *Opto-Electronics Journal*, vol. 26, no. 2, pp. 141–148, May 2018.
24. Aakash Jain, Sanjeev Sharma and B. Raj, "Design and analysis of high sensitivity photosensor using cylindrical surrounding gate MOSFET for low power sensor applications", *Engineering Science and Technology, an International Journal*, vol. 19, no. 4, pp. 1864–1870, Dec. 2016.
25. G. Saiphani Kumar, Amandeep Singh and B. Raj, "Design and analysis of gate all around CNTFET based SRAM cell design", *Journal of Computational Electronics*, vol. 17, no. 1, pp. 138–145, March 2018.
26. Gurinder pal Singh, B. S. Sohi and Balwinder Raj, "Material properties analysis of graphene base transistor (GBT) for VLSI analog circuits", *Indian Journal of Pure & Applied Physics*, vol. 55, pp. 896–902, Dec. 2017.
27. Sunil Kumar and B. Raj, "Estimation of stability and performance metric for inward access transistor based 6T SRAM cell design using n-type/p-type DMDG-GDOV TFET", *IEEE VLSI Circuits and Systems Letter*, vol. 3, no. 2, pp. 25–39, June2017
28. Shashikant Sharma, Anjan Kumar, Manisha Pattanaik and B. Raj, "Forward body biased multimode multi-threshold CMOS technique for ground bounce noise reduction in static CMOS adders", *International Journal of Information and Electronics Engineering*, vol. 3, no. 3, pp. 567–572, 2013.
29. Hamendra Singh, Pankaj Kumar and Balwinder Raj, "Performance analysis of majority gate SET based 1-bit full adder", *International Journal of Computer and Communication Engineering*, vol. 2, no. 4, pp. 1–8, 2013.

30. Anil Kumar Bhardwaj, Sumeet Gupta and B. Raj, "Investigation of parameters for Schottky Barrier (SB) height for Schottky Barrier based carbon nanotube field effect transistor device", *Journal of Nanoelectronics and Optoelectronics*, vol. 15, pp. 783–791, July 2020.
31. Priya Bansal and B. Raj, "Memristor: a versatile nonlinear model for dopant drift and boundary issues," *Journal of Computational and Theoretical Nanoscience*, vol. 14, no. 5, pp. 2319–2325, May 2017.
32. Neeraj Jain and B. Raj, "An analog and digital design perspective comprehensive approach on Fin-FET (Fin-field effect transistor) technology: a review", *Reviews in Advanced Sciences and Engineering*, vol. 5, pp. 1–14, 2016.
33. Balwinder Raj, A. K. Saxena and S. Dasgupta, "Analytical modeling for the estimation of leakage current and subthreshold swing factor of nanoscale double gate FinFET device", *Microelectronics International*, vol. 26, pp. 53–63, 2009.
34. Shailendra Singh Soniya, Girish Wadhwa and Balwinder Raj, "An analytical modeling for dual source vertical tunnel field effect transistor", *International Journal of Recent Technology and Engineering*, vol. 8 no. 2, pp. 603–608, July 2019.
35. Shailendra Singh and B. Raj, "Design and analysis of heterojunction vertical T-shaped tunnel field effect transistor", *Journal of Electronics Material*, vol. 48, no. 10, pp. 6253–6260, Oct. 2019.
36. Candy Goyal, Jagpal Singh Ubhi and B. Raj, "A low leakage CNTFET based inexact full adder for low power image processing applications", *International Journal of Circuit Theory and Applications*, vol. 47, no. 9, pp. 1446–1458, Sep. 2019.
37. B. Raj, A. K. Saxena and S. Dasgupta, "A compact drain current and threshold voltage quantum mechanical analytical modeling for FinFETs", *Journal of Nanoelectronics and Optoelectronics*, vol. 3, no. 2, pp. 163–170, 2008.
38. Girish Wadhwa and B. Raj, "An analytical modeling of charge plasma based tunnel field effect transistor with impacts of gate underlap region", *Superlattices and Microstructures*, vol. 142, pp. 106512, June 2020.
39. Shailendra Singh and B. Raj,"Modeling and simulation analysis of SiGe hetrojunction double GateVertical t-shaped tunnel FET", *Superlattices and Microstructures*, vol. 142, pp. 106496, June 2020.
40. Neeraj Jain and Balwinder Raj, "Device and circuit co-design perspective comprehensive approach on FinFET technology - a review", *Journal of Electron Devices*, vol. 23, no. 1, pp. 1890–1901, 2016.
41. Sunil Kumar and B. Raj, "Analysis of I_{ON} and ambipolar current for dual-material gate-drain overlapped DG-TFET", *Journal of Nanoelectronics and Optoelectronics*, vol. 11, pp. 323–333, June 2016.
42. Naveed Anjum, Tarun Bali and B. Raj, "Design and simulation of handwritten multiscript character recognition", *International Journal of Advanced Research in Computer and Communication Engineering*, vol. 2, no. 7, pp. 2544–2549, July 2013.
43. Karmjit Singh and B. Raj, "Performance and analysis of temperature dependent multi-walled carbon nanotubes as global interconnects at different technology nodes", *Journal of Computational Electronics*, vol. 14 no. 2, pp. 469–476, June 2015.

44. Sunil Kumar and B. Raj, "Compact channel potential analytical modeling of DG-TFET based on evanescent–mode approach", *Journal of Computational Electronics*, vol. 14, no. 2, pp. 820–827, July 2015.
45. Karmjit Singh and B. Raj, "Temperature dependent modeling and performance evaluation of multi-walled CNT and single-walled CNT as global interconnects", *Journal of Electronic Materials*, vol. 44, no. 12, pp. 4825–4835, Dec. 2015.
46. V. K. Sharma, M. Pattanaik and B. Raj, "INDEP approach for leakage reduction in nanoscale CMOS circuits", *International Journal of Electronics*, vol. 102, no. 2, pp. 200–215, 2014.
47. Karmjit Singh and B. Raj, "Influence of temperature on MWCNT bundle, SWCNT bundle and copper interconnects for nanoscaled technology nodes", *Journal of Materials Science: Materials in Electronics*, vol. 26, no. 8, pp. 6134–6142, 2015.
48. Naveed Anjum, Tarun Bali and B. Raj, "Design and simulation of handwritten Gurumukhi and devanagri numerical recognition", *International Journal of Computer Applications*, vol. 73, no. 12, pp. 16–21, 2013.
49. S. Khandelwal, V., Gupta, B. Raj and R. D. Gupta, "Process variability aware low leakage reliable nano scale DG-FinFET SRAM cell design technique", *Journal of Nanoelectronics and Optoelectronics*, vol. 10, no. 6, pp. 810–817, Dec. 2015.
50. V. K. Sharma, M. Pattanaik and B. Raj, "ONOFIC approach: low power high speed nanoscale VLSI circuits design", *International Journal of Electronics*, vol. 101, no. 1, pp. 61–73, 2014.
51. S. Khandelwal, Balwinder Raj and R. D. Gupta, "FinFET based 6T SRAM cell design: analysis of performance metric, process variation and temperature effect", *Journal of Computational and Theoretical Nanoscience*, vol. 12, pp. 2500–2506, 2015.
52. Sumit Singh, Shekhar Yadav, Jagdeep Rahul, Anurag Srivastava and B. Raj, "Impact of HfO_2 in graded channel dual insulator double gate MOSFET", *Journal of Computational and Theoretical Nanoscience*, vol. 12, no. 6, pp. 950–953, April 2015.
53. Vijay Kumar Sharma, Manisha Pattanaik and B. Raj, "PVT variations aware low leakage INDEP approach for nanoscale CMOS Circuits", *Microelectronics Reliability*, vol. 54, pp. 90–99, 2014.
54. B. Raj, A. K. Saxena and S. Dasgupta, "Quantum mechanical analytical modeling of nanoscale DG FinFET: evaluation of potential, threshold voltage and source/drain resistance", *Journal of Material Science in Semiconductor Processing*, vol. 16, no. 4, pp. 1131–1137, 2013.
55. Maisagalla Gopal, Siva Sankar D. Prasad and Balwinder Raj, "8T SRAM cell design for dynamic and leakage power reduction", *International Journal of Computer Applications*, vol. 71, no. 9, pp. 43–48, June 2013.
56. Manisha Pattanaik, B. Raj, Shashikant Sharma and Anjan Kumar, "Diode based trimode multi-threshold CMOS technique for ground bounce noise reduction in static CMOS adders", *Advanced Materials Research*, vol. 548, pp. 885–889, 2012.

57. Balwinder Raj, A. K. Saxena and S. Dasgupta, "Nanoscale FinFET based SRAM cell design: analysis of performance metric, process variation, underlapped FinFET and temperature effect", *IEEE Circuits and System Magazine*, vol. 11, no. 2, pp. 38–50, 2011.
58. V. K. Sharma, M. Pattanaik and Balwinder Raj, "Leakage current ONOFIC approach for deep submicron VLSI circuit design", *International Journal of Electrical, Computer, Electronics and Communication Engineering*, vol. 7, no. 4, pp. 239–244, 2013.
59. Parminder Kaur, Sandeep Singh Gill and B. Raj, "Comparative analysis of OFETs materials and devices for sensor applications", *Journal of Silicon,* vol. 14, pp. 4463–4471, 2022.
60. Sanjeev Kumar Sharma, Parveen Kumar, Balwant Raj and B. Raj, "$In_{1-x}Ga_xAs$ double metal gate-stacking cylindrical nanowire MOSFET for highly sensitive photo detector", *Journal of Silicon*, vol. 14, pp. 3535–3541, 2022.
61. B. Raj, A. K. Saxena and S. Dasgupta, "Analytical modeling of quasi planar nanoscale double gate FinFET with source/drain resistance and field dependent carrier mobility: a quantum mechanical study", *Journal of Computer*, vol. 4, no. 9, pp. 1–8, 2009.
62. S. Bhushan, S. Khandelwal and B. Raj, "Analyzing different mode FinFET based memory cell at different power supply for leakage reduction", *Seventh International Conference on Bio-Inspired Computing: Theories and Application* (BIC-TA 2012) Advances in Intelligent Systems and Computing, vol. 202, pp. 89–100, 2013.
63. Jeetendra Singh and B. Raj, "Temperature dependent analytical modeling and simulations of nanoscale memristor", *Engineering Science and Technology, an International Journal*, vol. 21, pp. 862–868, Oct. 2018.
64. Shradhya Singh, Shashi Bala, Balwant Raj and B. Raj, "Improved sensitivity of dielectric modulated junctionless transistor for nanoscale biosensor design", *Sensor Letter*, vol. 18, pp. 328–333, April 2020.
65. Vivek Kumar, Santosh Kumar Vishvakarma and B. Raj, "Design and performance analysis of ASIC for IoT applications", *Sensor Letter*, vol. 18, pp. 31–38, Jan. 2020.
66. Akanksha Jaiswal, R. K. Sarin, B. Raj and Shikha Sukhija, "A novel circular slotted microstrip-fed patch antenna with three triangle shape defected ground structure for multiband applications", *Advanced Electromagnetic*, vol. 7, no. 3, pp. 56–63, Aug. 2018.
67. Girish Wadhwa and B. Raj, "Label free detection of biomolecules using charge-plasma-based gate underlap dielectric modulated junctionless TFET", *Journal of Electronic Materials*, vol. 47, no. 8, pp. 4683–4693, Aug. 2018.
68. Gurmohan Singh, R. K. Sarin and B. Raj, "Design and performance analysis of a new efficient coplanar quantum-dot cellular automata adder", *Indian Journal of Pure & Applied Physics*, vol. 55, pp. 97–103, Feb. 2017.
69. Parminder Kaur, Vikas Pandey and B. Raj, "Comparative study of efficient design, control and monitoring of solar power using IoT", *Sensor Letter*, vol. 18, pp. 419–426, May 2020.
70. Anil Kumar Bhardwaj, Sumeet Gupta and B. Raj, "Development & analysis of compact model for double gate Schottky barrier CNTFET", *Journal of Nanoelectronics and Optoelectronics*, vol. 15, pp. 1199–1208, Aug. 2020.

71. Girish Wadhwa, Priyanka Kamboj, Jeetendra Singh and B. Raj, "Design and investigation of junctionless DGTFET for biological molecule recognition", *Transactions on Electrical and Electronic Materials,* vol. 22, pp. 282–289, 2021.
72. Sachin Kumar Verma, Shailendra Singh, Girish Wadhwa and B. Raj, "Detection of biomolecules using charge-plasma based gate underlap dielectric modulated dopingless TFET", *Transactions on Electrical and Electronic Materials*, vol. 21, pp. 528–535, June 2020.
73. Neeraj Jain and B. Raj, "Impact of underlap spacer region variation on electrostatic and analog/RF performance of symmetrical high-k SOI FinFET at 20 nm channel length", *Journal of Semiconductors*, vol. 38, no. 12, p. 122002, Dec. 2017.
74. Shailendra Singh and B. Raj, "Analytical modeling and simulation analysis of T-shaped III–V heterojunction vertical T-FET", *Superlattices and Microstructures*, vol. 147, p. 106717, Nov. 2020.
75. Gurmohan Singh, R. K. Sarin and B. Raj, "Design and analysis of area efficient QCA based reversible logic gates", *Journal of Microprocessors and Microsystems*, vol. 52, pp. 59–68, May 2017.
76. Shailendra Singh and B. Raj,"Parametric variation analysis on hetero-junction Vertical t-shape TFET for Supressing ambipolar conduction", *Indian Journal of Pure and Applied Physics*, vol. 58, pp. 478–485, June 2020.
77. Soniya, Shailendra Singh, Girish Wadhwa and Balwinder Raj, "Design and analysis of dual source vertical tunnel field effect transistor for high performance", *Transactions on Electrical and Electronics Materials*, vol. 21, pp. 74–82, Oct. 2019.
78. Manjit Kaur, Neena Gupta, Sanjeev Kumar, B. Raj and Arun Kumar Singh, "RF performance analysis of intercalated graphene nanoribbon based global level interconnects", *Journal of Computational Electronics,* vol. 19, pp. 1002–1013, June 2020
79. Girish Wadhwa and B. Raj, "Design and performance analysis of junctionless TFET biosensor for high sensitivity", *IEEE Nanotechnology*, vol. 18, pp. 567–574, 2019.
80. Jeetendra Singh and B. Raj, "Enhanced nonlinear memristor model encapsulating stochastic dopant drift, *Journal of Nanoelectronics and Optoelectronics*, vol. 14, pp. 958–963, 2019.
81. Anjana Bhardwaj, Pradeep Kumar, B. Raj and Sunny Anand, "Design and performance optimization of doping-less vertical nanowire TFET using gate stack technique", *Journal of Electronic Materials*, vol. 41, no. 7, pp. 4005–4013, 2022.
82. Jeetendra Singh and B. Raj, "Tunnel current model of asymmetric MIM structure levying various image forces to analyze the characteristics of filamentary memristor", *Applied Physics A*, vol. 125, no. 3, pp. 203.1–203.11, Feb. 2019.
83. Candy Goyal, Jagpal Singh Ubhi and B. Raj, "Low leakage zero ground noise nanoscale full adder using source biasing technique", *Journal of Nanoelectronics and Optoelectronics*, vol. 14, pp. 360–370, March 2019.

84. Gurmohan Singh, R. K. Sarin and B. Raj, "A novel robust exclusive-OR function implementation in QCA nanotechnology with energy dissipation analysis", *Journal of Computational Electronics*, vol. 15, no. 2, pp. 455–465, June 2016.
85. Girish Wadhwa, Priyanka Kamboj and Balwinder Raj, "Design optimisation of junctionless TFET biosensor for high sensitivity", *Advances in Natural Sciences: Nanoscience and Nanotechnology*, vol. 10, no. 7, p. 045001, 2019.
86. Priya Bansal and B. Raj, "Memristor modeling and analysis for linear dopant drift kinetics", *Journal of Nanoengineering and Nanomanufacturing*, vol. 6, pp. 1–7, 2016.
87. D. Vaithiyanathan Ashima and B. Raj, "Performance analysis of charge plasma induced graded channel Si nanotube", *Journal of Engineering Research*, EMSME Special no. pp. 146–154, Aug. 2021.
88. Abhishek Singh Tomar, Vijay Kumar Magraiya and B. Raj, "Scaling of access and data transistor for high performance DRAM cell design", *Quantum Matter*, vol. 2, pp. 412–416, Oct. 2013.
89. Neeraj Jain and B. Raj, "Parasitic capacitance and resistance model development and optimization of raised source/drain SOI FinFET structure for analog circuit applications", *Journal of Nanoelectronics and Optoelectronins*, vol. 13, pp. 531–539, April 2018.
90. Shradhya Singh, S. K. Vishvakarma and B. Raj, "Analytical modeling of split-gate junction-less transistor for a biosensor application", *Sensing and Bio-Sensing*, vol. 18, pp. 31–36, April 2018.
91. Maisagalla Gopal and Balwinder Raj, "Low power 8T SRAM cell design for high stability video applications", *ITSI Transaction on Electrical and Electronics Engineering*, vol. 1, no. 5, pp. 91–97, 2013.
92. Balwinder Raj, Jatin Mitra, Deepak Kumar Bihani, V. Rangharajan, A. K. Saxena and S. Dasgupta, "Analysis of noise margin, power and process variation for 32 nm FinFET based 6T SRAM cell", *Journal of Computer*, vol. 5, no. 6, pp. 1–8, 2010.
93. Divya Sharma, Rajesh Mehra and B. Raj, "Comparative analysis of photovoltaic technologies for high efficiency solar cell design", *Superlattices and Microstructures,* vol. 153, pp. 106861, May 2021.
94. Pawandeep Kaur, Avtar Singh Buttar and B. Raj, "A comprehensive analysis of nanoscale transistor based biosensor: a review", *Indian Journal of Pure and Applied Physics*, vol. 59, pp. 304–318, April 2021.
95. Divya Yadav, Balwant Raj and B. Raj, "Design and simulation of low power microcontroller for IoT applications", *Journal of Sensor Letters*, vol. 18, pp. 401–409, May 2020.
96. Shailendra Singh and B. Raj, "A 2-D analytical surface potential and drain current modeling of double-gate vertical t-shaped tunnel FET", *Journal of Computational Electronics*, vol. 19, pp. 1154–1163, April 2020.
97. Jeetendra Singh and B. Raj, "An accurate and generic window function for non-linear memristor model", *Journal of Computational Electronics*, vol. 18 no. 2, pp. 640–647, June 2019.
98. Manjit Kaur, Neena Gupta, Sanjeev Kumar, B. Raj and Arun Kumar Singh, "Comparative RF and crosstalk analysis of carbon based nano interconnects", *IET Circuits, Devices & Systems*, vol. 15, no. 6, pp. 493–503, Feb. 2021.

99. Nehru Kandasamy, Firdous Ahmad, D. Ajitha, B. Raj and Nagarjuna Telagam "Quantum dot cellular automata based scan flip flop and boundary scan register", *IETE Journal of Research,* vol. 66, pp. 535–548, 2020.
100. Jeetendra Singh and B. Raj, "Design and investigation of 7T2M NVSARM with enhanced stability and temperature impact on store/restore energy", *IEEE Transactions on Very Large Scale Integration Systems,* vol. 27, no. 6, pp. 1322–1328, June 2019.
101. Anil Kumar Bhardwaj, Sumeet Gupta, B. Raj and Amandeep Singh, "Impact of double gate geometry on the performance of carbon nanotube field effect transistor structures for low power digital design", *Computational and Theoretical Nanoscience*, vol. 16, pp. 1813–1820, 2019.
102. Neeraj Jain and B. Raj, "Thermal stability analysis and performance exploration of asymmetrical dual-k underlap spacer (ADKUS) SOI FinFET for security and privacy applications", *Indian Journal of Pure & Applied Physics*, vol. 57, pp. 352–360, May 2019.
103. Neeraj Jain and Balwinder Raj, "Dual-k spacer region variation at the drain side of asymmetric SOI FinFET structure: performance analysis towards the analog/RF design applications", *Journal of Nanoelectronics and Optoelectronics*, vol. 14, pp. 349–359, March 2019.
104. Jeetendra Singh, Sanjeev Sharma, B. Raj and Mamta Khosla, "Analysis of barrier layer thickness on performance of In1-xGaxAs based gate stack cylindrical gate nanowire MOSFET", *Journal of Nanoelectronics and Optoelectronics*, vol. 13, pp. 1473–1477, Oct. 2018.
105. Neeraj Jain and B. Raj, "Analysis and performance exploration of high-k SOI FinFETs over the conventional low-k SOI FinFET toward analog/RF design", *Journal of Semiconductors*, vol. 39, no. 12, pp. 124002-1–7, Dec. 2018.
106. Candy Goyal, Jagpal Singh Ubhi and B. Raj, "A reliable leakage reduction technique for approximate full adder with reduced ground bounce noise", *Journal of Mathematical Problems in Engineering*, vol. 2018, Article ID 3501041, p. 16, 15 Oct. 2018.
107. Anuradha, Jeetendra Singh, B. Raj and Mamta Khosla, "Design and performance analysis of nano-scale memristor-based nonvolatile SRAM", *Journal of Sensor Letter*, vol. 16, pp. 798–805, Oct. 2018.
108. Girish Wadhwa and B. Raj, "Parametric variation analysis of charge-plasma-based dielectric modulated JLTFET for biosensor application", *IEEE Sensor Journal*, vol. 18, no. 15, pp. 6070–6077, 1 Aug. 2018.
109. Jeetendra Singh and B. Raj, "Comparative analysis of memristor models for memories design", *Journal of Semiconductors*, vol. 39, no. 7, pp. 074006-1–12, July 2018.
110. Divya Yadav, Shailesh Singh Chouhan, Santosh Kumar Vishvakarma and B. Raj, "Application specific microcontroller design for IoT based WSN", *Sensor Letter*, vol. 16, pp. 374–385, May 2018.

Chapter 8

Advancements and implications of bioplastic development

From material synthesis to environmental impact

Sunil Sankathala, Yadnyesh Khapekar, Aryan Vasa, Aditya Shrivastav, Devanshi Soni, and Susanta Das

8.1 INTRODUCTION

The serious environmental problems caused by the extensive use and disposal of traditional plastics have come to light in recent years. Due to their affordability, adaptability, and durability, traditional plastics, which are manufactured from petrochemical sources, have long been a cornerstone of contemporary industry [1]. But because of their enduring presence in the environment – they can take decades to decompose – many ecosystems, including those of the air, soil, and water bodies, have been heavily contaminated. With its negative consequences on human health, marine ecosystems, and animals, plastic pollution has become a worldwide catastrophe [2, 3].

The creation and use of bioplastics have attracted a lot of attention and funding in an effort to alleviate these environmental issues. By utilizing biodegradable polymers and renewable resources, bioplastics present a viable substitute for traditional plastics [4]. Bioplastics may be made from natural resources such as plant starches, proteins, cellulose, and even waste biomass, in contrast to their petroleum-derived equivalents. They also demonstrate the vital benefit of biodegradability, disintegrating gradually into safe natural components in the presence of microorganisms [5, 6].

A major step in reducing plastic pollution and achieving sustainability goals is the move to bioplastics. Because of this, the subject of bioplastics has seen a tremendous expansion in study and innovation in recent years, embracing a variety of topics related to material science, polymer chemistry, and environmental engineering [7, 8]. Bioplastics provide a flexible way to minimize the environmental impact of plastic products and lessen dependency on fossil fuels, with uses ranging from packaging and medical equipment to agriculture [9].

But there are still obstacles in the way of expanding the use and manufacturing of bioplastics. It is necessary to overcome technical obstacles like improving process efficiency, guaranteeing cost competitiveness, and optimizing material properties [10]. Furthermore, there is a need for robust policies and regulations to promote the transition towards bioplastics and

DOI: 10.1201/9781003612797-8

encourage sustainable waste management practices. To fully utilize bioplastics and create a more sustainable future, cooperation between government, academic, and industry sectors will be crucial [11, 12].

By synthesizing insights from recent literature and highlighting key trends and developments, this review seeks to contribute to a deeper understanding of the opportunities and challenges associated with bioplastics and inform future research directions in this crucial field. In this context, the study aims to provide a comprehensive overview of the current state of bioplastics research, with a focus on their production, properties, applications, and environmental impact [13, 14].

Synthetic and organic plastics are vital in a wide range of sectors, including electronics, automotive, packaging, and medicines. However, the threats to the environment and human health stem from their non-biodegradable nature [15, 16].

Bioplastic technologies present a viable substitute for conventional petroleum-based plastics. They are mostly made from organic resources such as corn, potatoes, cassava starch, banana peels, and cellulose-based materials [17]. Approximately half of the bioplastics industry is made up of starch-based polymers, which are readily available, reasonably priced, and biodegradable. They have potential uses in the pharmaceutical, medical, and packaging industries. The creation and use of bioplastics are essential first steps toward a future with more sustainability [18].

The plastics industry struggles to strike the best possible balance between acceptable biodegradability, production costs, and mechanical qualities. Many synthesis techniques are being investigated as solutions to this problem. Using nanotechnology and agricultural waste to create competitive, biodegradable plastic products is one strategy.

Using agricultural waste: This entails turning organic waste from farming operations into raw materials for the manufacture of plastic. This procedure encourages sustainability by lowering dependency on fossil fuels.

Using nanotechnology: This entails adding nanoparticles to the process of making plastic. The mechanical qualities, manufacturing costs, and biodegradability of plastic items can all be improved by these nanoparticles. Enhancing plastic products entails making them more biodegradable, more cost-effective to produce, and have better mechanical qualities [19–25].

Plastics are renowned for being inexpensive, easy to manufacture, highly resistant to corrosion, and having a low density. They fall into two categories: thermosetting plastics and thermoplastic (TPS) plastics. When thermoplastics are molten, they can be reheated and molded into a variety of shapes, such as branching low density polyethylene (LDPE). Because of its resistance to germs, LDPE, which has a high molecular weight and hydrophobicity, is frequently utilized in the packaging business [26–31].

Even during the recession, over the past 20 years, there has been a considerable increase in the usage of plastic and in industrial demand. With an estimated annual production of over 100 million tons, plastics are now the preferred material for many different industries. But their inability to break down has grown to be a serious environmental problem, particularly in the packaging sector [32–37].

8.2 LITERATURE REVIEW

Jackfruit seeds are used as a source of starch because of their large availability, low price, and high starch capacity to produce bioplastic. The materials that are used are jackfruit seeds, glycerol, citric acid, and baking soda.

The raw jackfruit seed has a thick outer shell that is removed to reveal the hard shell and brown spermoderm. The deseeded seeds were then washed. Rotten seeds were discarded, and healthy seeds were soaked overnight to eliminate latex. After an overnight soak, the jackfruit seeds were rinsed with filtered water. Clean jackfruit seeds were mixed into a slurry. The slurry was filtered via cloth, producing a crude starch suspension. The filtrate was precipitated overnight to remove latex and minerals. The supernatant was removed, and the crude starch was rinsed with distilled water [38–45].

The method was performed three times, and the cake was dried either naturally or in the oven. This technique extracts starch. The plasticization technique required heating a liquid mixture containing components in a specific ratio above a magnetic stirrer. When the mixture is clear and thick, stop stirring. The plastic mixture was placed on trays lined with PE film. Following that, the viscous liquid bioplastic was flattened on a tray and dried in an oven until its weight was consistent [46–52].

At first, the higher the glycerol ratio, the softer and more flexible the bioplastic became; however, increasing the percentage of citric acid increased the hardness and decreased the plasticity of the bioplastic. At the same time, if the flattened viscous biopolymer is thick, it will crack due to shrinkage.

Microcracks may be caused by microbubbles that form during the gelatinization process. The lack of a vacuum stage in the bioplastic production process causes uncontrolled microbubbles in specimens. Furthermore, mixing and swirling generate micro air bubbles that cannot be entirely removed [53–58].

In jackfruit seed, there is a high starch content, and may be easily collected in large quantities for extraction. When manufacturing bioplastic, the strength is often proportional to the starch:glycerol ratio [1].

Three starch-based bioplastic films were created using lemon extract, water, and different cross-linkers: PV (polylactic acid and crude palm oil), AV (glycerol and crude palm oil), and PAV (polylactic acid, glycerol, and crude palm oil). During thermogravimetric analysis (TGA), all films released volatile compounds between 150°C and 220°C, followed by plasticizers and cross-linkers around 300°C. Carbonated compounds pyrolyzed above 500°C. These low-cost plant starch films have potential for food packaging use [59–66].

Bio-based plastics, derived from renewable sources like corn, potatoes, wheat, and vegetable oil, undergo processes such as hydrolysis, acidification, and microbial fermentation. Starch, comprising amylose (linear) and amylopectin (branched), offers gelatinization properties. Bioplastics from amylose tend to be stronger and more flexible, while those from amylopectin are less resistant to tension and elongation. Despite starch's advantages of low cost and renewability, its use in bioplastics is limited due to brittleness and hydrophilicity, exacerbated by environmental factors like humidity-induced hydrolysis. Plasticizers, like crude palm oil, enhance thermal stability, while glycerol strengthens the bioplastic's chemical bonds, improving tensile strength. Initial characterization of sago, jackfruit seed, and core includes determining moisture, ash, carbohydrate, protein, and starch content through standard laboratory procedures [67–73].

Ten grams of pearl corn starch were mixed with 95 ml of distilled water and 5% concentrated sulfuric acid, then autoclaved for hydrolysis, filtered to obtain a pure hydrolysate, which was neutralized with NaOH. Poly(lactic acid) (PLA) was synthesized in the laboratory via direct polycondensation without a catalyst, and freshly polymerized lactic acid was immediately utilized for film formulation [74–81].

In another process, 97 ml of sterile distilled water, 10 g of a starch mixture (sago, jackfruit seed, and core flours in a ratio of 9:1:0.5), and 3 ml of lemon extract were sonicated for 5 minutes to ensure proper blending and prepared in triplicates. Each formulation was heated on a hot plate until a thick gelatinized consistency was achieved, then poured onto aluminum foil and placed in a hot air oven at 45°C for two days.

All essential components of the minimal media broth were mixed thoroughly with distilled water, and the pH was adjusted to 7.0 (optimal for bacterial growth) and ~5.5 (optimal for fungal growth) before pouring into boiling tubes. A cocktail of bacterial cultures (*Escherichia coli, Pseudomonas aeruginosa*, *Staphylococcus aureus*, *Klebsiella pneumoniae*, *Streptococcus* spp.) and fungal cultures (*Aspergillus flavus*, *Penicillium chrysogenum*, *Candida albicans*) was inoculated aseptically into sterilized tubes separately.

In Kelling's test, a canary yellow color indicated iron lactate formation, while water showed an orange hue. Lactic acid purity via TLC was half of the standard (88%), with a colorimetric concentration of 0.65 g/g. Exposed hydroxyl groups in amylose and amylopectin not cross-linked with plasticizers swelled upon water exposure. Bioplastic films were less transparent due to fillers. Glycerol-based films showed lower solubility [82–87].

Palm oil enhanced tensile strength by improving PLA interaction, while jackfruit core acted as a filler, reducing elongation strength due to PLA's limited ductility. Glycerol improved film flexibility, and palm oil increased cross-linking. Plasticizer addition inversely affected tensile strength with fillers, and films softened in alkali due to acid neutralization.

The initial broad peak at 3296.71 cm^{-1} indicated the presence of carboxylic acid O–H stretch, confirming the interaction of different –OH groups in the starch and polylactic acid blends; the peak at 1735.62 cm^{-1} (C = O stretching) and the peak at 997.017 cm^{-1} indicated the presence of (R-CH=CH_2) in the sample. (b) AV film and (c) PAV films had more or less similar peaks in which the first broad peak at 3280.32 cm^{-1} (AV) and 3292.86 cm^{-1} (PAV) indicated the presence of carboxylic acid O–H stretch; a small peak at 2921.63 cm^{-1} in both the films showed good C–H stretch, while a sharp peak at 1743.3 cm^{-1} in PAV indicated triglyceride ester linkage [11]; other peaks of AV films 1647.88 cm^{-1}, 1362.46 cm^{-1}, and 1149.37 cm^{-1} showed the presence of substituted alkene, C–H rock with phenolic group, and C–H wag with halo compounds, respectively [88–93].

Starch plasticized with glycerol degraded at a faster rate. During thermogravimetric analysis, all bioplastic films exhibited gradual desorption between 150°C and 220°C, followed by the release of lower molecular weight compounds around 300°C, and pyrolysis of carbonated compounds above 500°C. The addition of crude palm oil improved thermal stability. Voids and microvoids formed due to hydrogen bonding breakdown during gelatinization, allowing water infiltration.

By day 15, bioplastic films changed in size and color, starting to degrade slightly. By day 30, PV film began disintegrating, while 50% of AV and 80% of PAV films degraded. Synthetic polythene remained unchanged.

Bacterial cultures utilized the carbon source in bioplastic films, leading to growth by day 7, whereas no growth occurred in control media or synthetic polythene. Fungal species also utilized bioplastic films as a carbon source by day 7, with growth observed in all tubes. The three bioplastic formulations showed better tensile strength, elongation at break, thermal stability, and biodegradability compared to previous studies on jackfruit seed flour-based biofilms [2].

Bioplastics, such as PHBs and PHAs, have the potential to replace conventional plastics derived from petrochemicals. These bioplastics are natural and biodegradable, with commercial qualities comparable to those of conventional plastics. Research focuses on using fruit waste, specifically banana peels, to synthesize bioplastic material. Glycerol and banana peels are investigated for creating bioplastics with qualities like strength, pliability, and user-friendliness.

To ensure the bioplastic materials meet commercial standards, tests like swelling and solubility studies are conducted. Land pollution is increasing, and plastic pollution is a major environmental issue. Biodegradable and biocompatible polymers, derived from renewable raw materials or produced through other sustainable methods, are gaining importance in various fields, including pharmacology, biomedicine, and environmental applications. Biodegradable plastics can be classified into aerobic and landfill environments. The research emphasizes the importance of developing bioplastic materials as alternatives to conventional plastics, as they are environmentally friendly, trendy, user-friendly, and capable [94–97].

The study aimed to synthesize bioplastic made from banana peel using two phases: hydrolysis duration and pH, and comparing widely available plasticizers. The bioplastic material was tested using standardized parameters. Infrared spectroscopy via Fourier transform (FTIR) and powder diffraction by X-ray (XRD) were used to examine the structure of the bioplastic. Absorbency tests were conducted on two 400 ml beakers filled with soil, and the bioplastic material was positioned 5 cm below the surface of the beaker. Solubility tests were performed on two different types of bioplastics, with the bioplastic made from banana peels becoming slightly darker and the bioplastic made from potato starch softening and losing its green.

The biodegradable nature of the bioplastic stored in the beaker was confirmed, and the integrity of the bioplastic material remained largely unchanged after soaking in solvents. The study aims to determine a potential solution to the environmental tractability of biomaterials and identify biodegradable tractable factors in the synthesis of bioplastic materials.

Bioplastics, made from renewable resources like potatoes, corn, and bananas, can be synthesized from agricultural waste to reduce environmental lethality and cost. Reprocessing takes three to six months, and bioplastics are less expensive than traditional plastics. They are also less harmful to the environment and can be used in everyday household and utility applications. The study used hydrochloric acid, glycerol, potatoes, water, and sodium chloride to extract the starch from potatoes. Bioplastics are environmentally friendly, cost-effective, and a viable replacement for standard plastics. They decompose in months and are less expensive than traditional plastics. The bioplastic was tested for solubility, which was found to be water insoluble in the sea. This suggests that bioplastics can be used as a viable alternative to traditional plastics [3, 4].

Bioplastics are eco-friendly, secure, and easy to recycle, reducing pollution and providing a respite from plastic overload. By switching from fossil fuels to renewable energy sources, we can reduce greenhouse gas emissions and contribute to a more sustainable future.

Prosopis juliflora is used to extract cellulose, and after that, bioplastic film samples were produced using it. To make the bioplastic, the required materials are citric acid, gelatin powder, the glycerol, and purified water [98–101].

The wood of *Prosopis juliflora* was cut and pulverized. Water was used to clean the powder. It was then treated with NaOH solution. The powder bleaching technique was repeated several times. The powder underwent a second washing before being treated with $C_2H_2O_4$. The cellulose was then removed mechanically and through acid hydrolysis. Filtered water was used. The required amounts of cellulose gelatin, glycerol, and citric acid were added to the purified water. The mixture was vigorously stirred. The mixture was placed on a hot plate and cooked. The mixture was stirred manually. The liquid was equally dispersed over a glass plate coated with Teflon. The mixture was allowed to dry for three or four days before the cast film was removed.

Following the preparation, many tests were performed on the bioplastic sheet. Biodegradability tests revealed that bioplastic films became more biodegradable as citric acid, gelatin, and glycerol concentrations increased. Furthermore, bioplastic films made of cellulose are more biodegradable than those made of starch. Bursting tests demonstrated that cellulose-based bioplastic films are appropriate for packaging purposes. Fourier transform infrared spectroscopy research revealed that the cellulose has substantial C–O stretching due to the presence of alcohol, as well as O–H stretching due to both primary and secondary alcohols.

Moisture content tests revealed that increasing the ratios of citric acid, gelatin, and glycerol in bioplastic sheets increases their moisture content. A scanning electron microscopy study revealed that microbial action affects their surface structure, causing disintegration, and the uneven film surface and surface flaws indicate that the bioplastics produced are biodegradable [102–106].

Tensile testing demonstrates that adding more citric acid, gelatin, or glycerol to cellulose diminishes its tensile strength. Thickness measurements demonstrate that bioplastics with more than 50 μm are stronger, more robust, and recyclable. The thickness values of the bioplastic film samples indicate that cellulose-based bioplastics are appropriate for packaging applications. Water contact angle measurements revealed that cellulose-based bioplastic sheets are more hydrophilic than regular plastic. Hydrophilic films are ideal for some medicinal applications. Thermal investigations demonstrated that the bioplastics might be employed in high-temperature packaging applications [5].

The potential of cassava starch as the primary ingredient in the production of bioplastics is highlighted. The effect of glycerol on microbial degradation was investigated. The utilization of cassava flour mixed with glycerol plasticizer at various concentrations (0%, 2%, 2.5%, 3%) in the synthesis of bioplastics was examined. The aspects studied included biodegradability, moisture absorption, shelf life, and morphological properties, as well as SEM analysis. It was observed that a higher concentration of glycerol would accelerate the microbial degradation process, increase moisture absorption, and extend the shelf life of bioplastics when stored in a dry place [107–111].

A 5% cassava starch (by volume) was added to four different amounts of distilled water/aquades (98.5 ml, 98 ml, 97.5 ml, and 97 ml), and the mixture was stirred using a magnetic stirrer for 5 minutes at 900 rpm. Subsequently, glycerol at various concentrations (1.5%, 2%, 2.5%, 3% by volume) was added and stirred for 5 minutes at 900 rpm. The mixture was then heated to approximately 80°C for 45 minutes at 900 rpm.

The resulting bioplastics exhibited a hard, smooth, and transparent appearance. Cassava starch was found to consist of semi-crystalline structures, which became fragile due to the disruption of granules and the conversion into amorphous forms resulting from the addition of specific heat and solvents. The water-swelling ratio increased with higher concentrations of glycerol, attributed to the hydrophilic properties of glycerol and starch, which contain hydroxyl (OH), carbonyl (CO), and ester (COOH) groups. The presence of high concentrations of hydrophilic starch and glycerol accelerated degradation in soil.

After burying the bioplastics for six days, more than 50% of their mass was reduced, with complete degradation occurring by the ninth day due to hydrolysis reactions caused by the high concentration of glycerol. The level of degradation of a compound depends not only on the durability of the molecules but also on environmental factors such as pH, temperature, humidity, and oxygen content. Soil degradation was facilitated by microorganisms such as *Pseudomonas* sp., *Streptococcus* sp., *Staphylococcus* sp., *Bacillus* sp., and *Moraxella* sp., with the process continuing through surface expansion via erosion and perforation.

Organic cassava starch exhibited a low level of tensile strength, resulting in a relatively short shelf life for the bioplastics and potential damage to packaged food due to the migration of spoilage bacteria. The bioplastics were covered with white fungi, leading to the splitting of polymer chains and subsequent depolymerization due to enzymatic activity.

Fungal growth in bioplastics, influenced by glycerol as a plasticizer, included species such as *Aspergillus niger*, *Aspergillus versicolor*, *Cladosporium* sp., *Fusarium* sp., *Penicillium* sp., *Trichoderma* sp., and *Verticillium* sp., and was affected by environmental conditions, bioplastic catabolism, and temperature. To assess shelf life, the bioplastics were placed in a plastic box with limited oxygen or humidity of 45–60% RH. Higher glycerol concentrations served as an anti-fungal agent, thereby prolonging shelf life [6].

The negative environmental effects of synthetic plastics have led to the development of biodegradable polymers for industrial and commercial applications. This study uses the soil burial test (SBT) method to examine the biodegradability of biopolymers based on starch. *Manihot esculenta* and *Triticum aestivum* biopolymer particulates were used to create biopolymeric films, with blends and additional additives created in different ratios for each particle size. Physical, physicochemical, thermal, and microstructural analyses were used to characterize the biopolymeric films. A 30-day soil burial test was used to assess the biodegradability of starch-based polymers. The bioplastic films were hard and brittle with visible surface cracks, shedding 46.55–63.77% of their total weight. Throughout the test, there was no indication of any weight loss or macrostructural changes in the LDPE film. According to the study's findings, bioplastic films performed better than regular plastic films since they can be effectively used in packaging applications and have been shown to be biodegradable.

The growing use of synthetic plastics derived from petrochemicals has raised significant concerns about economic, environmental, and health issues. Locally produced biosourced plastics, such as starch-based bioplastics, have made great strides in recent years. This study focuses on hybrid starch-based films made from *Triticum aestivum* (TA) and *Manihot esculenta* (ME), which are botanical representatives of wheat starch and cassava, respectively [7].

This study used gelatin and potato starch to generate biodegradable and compostable films. To make bioplastics out of gelatin and starch, this process only employs renewable raw materials. The chosen sample was first washed with water to eliminate impurities such as dirt, soil, tiny roots, and other undesired plant matter. The sample was weighed again to determine the difference in pre- and post-washing weights. The sample was then painstakingly peeled with a knife. The potato was meticulously peeled and hand-cut into small, equal cubes.

Using a blender, the leftover potato sample was filtered to remove the starch extract. The filtrate was combined with double-distilled water from a wash bottle to remove the starch from the residue. Two washes of the starch sediment were used to obtain pure starch. The sample should be air-dried or exposed to sunlight. After completing the process, gather the starch powder in a container.

Following that, various tests were conducted. They are as follows: (1) the FTIR test demonstrated the ranges of potato starch and gelatin mix bioplastics created during the study. (2) The solubility test revealed that it is not soluble in water. (3) The biodegradable test demonstrated a higher level of biodegradation based on weight loss. (4) The water uptake test demonstrated that starch-based bioplastics absorb water. Because of its great overall performance, the biodegradable starch-based film developed in this work could be used for environmentally friendly food packaging [8].

The study explores the use of bioplastics as an eco-friendly substitute for conventional plastics. Waste biomaterials were used to extract three basic materials: starch, agar, and gelatin. The bioplastics were insoluble in water and solvents, had an elongation strength of 17.64%, a folding endurance of 15 fold, a melting point of 90°C, 60% degradation in 25 days, and a hardness of 30. The bioplastic film was tested for solubility in water, ethanol, and chloroform after three hours of soaking. The bioplastic film was stretched until it broke, and the initial and final lengths were measured. The folding endurance test showed that the bioplastic could withstand 15 folds under standardized conditions.

After 25 days, the weight of the bioplastic was reduced from 2 g to 0.79 g, indicating a biodegradation of 1.21 g and a %degradation of 60.5%. Scanning electron microscopy (SEM) and energy-dispersive X-ray spectroscopy (EDS) analysis revealed a smooth surface with occasional cracks, possibly due to incomplete mixing, and good adhesion between components and plasticizer. Economic factors affect the commercialization process of the bioplastics industry, which has seen a boom in the use of renewable resources. This study highlights the importance of considering environmental contamination and the availability of affordable alternatives in the bioplastics industry [9].

Bioplastic from oil palm empty fruit bunches produced the most CO_2 on landfill soil, indicating the highest rate of biodegradation, followed by oxodegradable and conventional plastics. Microbes from landfill soil and top soil were used as inoculums in the biodegradation process.

Measuring C-organic and degrading media helped to define degrading media. Microorganisms were counted using the total plate count (TPC) method, and pH was measured in the media using a pH meter. These measurements were carried out on a daily basis. Bacteria in biodegradation media were measured using the total plate count method. Each soil sample was homogenized with aquadest and then diluted. At the fourth dilution, the sample was placed in a Petri dish with nutrient agar (NA) solid and flattened. Microorganism colonies were counted. To determine pH, water and homogenized. The pH meter was calibrated before being used to monitor pH. The pH meter electrodes were then dipped into the sample to obtain readings. After that, a test was performed using the respirometry method.

Landfill soil with bioplastics from oil palm empty fruit bunches produced the most CO_2, followed by oxodegradable plastics. Biodegradable materials, such as oil palm fruit bunch and cassava starch, can be consumed by microorganisms. In conclusion, we can say that bioplastic from oil palm empty fruit bunch produced the most CO_2 and biodegraded the fastest in landfill soil, followed by oxodegradable and conventional plastic [10].

Researchers developed a biodegradable plastic from potato starch mixed with various plasticizers, such as glycerol, sorbitol, and xylitol. The hydrogel was heated at 70–80°C for five minutes before adding the plasticizers. The bioplastic block served as the starch source, and the bioplastic film displayed high water vapor permeability. The bioplastic was tested using fungal diastase at room temperature and 50°C, with a solution containing 7.5 g of fungal α-amylase in 100 mL of water. An iodine reagent solution was prepared with 1 g of iodine and 10 g of KI in 200 mL.

The bioplastic film showed high water vapor permeability, while petroleum-based plastic showed significantly lower permeability. Samples with different plasticizers exhibited varying degrees of water solubility, with sample 3 showing the highest solubility at 0.84%. Thermal stability tests indicated that starch-based bioplastic was more heat-resistant than petroleum-based plastic, which melted completely at 150°C.

Bioplastics made with various plasticizers underwent complete degradation in both natural environments and through bacterial and enzymatic actions. Bacteria appeared on solid media within 48 hours, and bioplastic degradation started 24 hours after the initial time. Bioplastic degradation was quicker in compost than in regular soil, likely due to the higher bacterial presence in compost.

The breakdown of bioplastic by the fungal α-amylase enzyme is slower compared to free starch molecules, as the enzyme gradually breaks down amylose into glucose and maltose. Degradation of bioplastic was faster at 50°C, with turbidity appearing after 30 minutes of enzyme addition, indicating degradation. Undigested bioplastic turned dark blue-black in the presence of iodine solution, indicating the presence of amylose. Proper storage in a moisture-free environment away from direct sunlight ensures a good lifespan for bioplastic, as prolonged sunlight exposure can affect its mechanical and structural properties. Higher temperatures above 50°C inhibit biodegradation by starch-digesting bacteria, while excess moisture weakens bioplastic. In soil, complete degradation occurred within two months, while enzymatic breakdown took 36 to 48 hours [11].

This study focuses on the creation of biodegradable bioplastics as a solution to the environmental issue of plastics derived from fossil fuels. It centers on cellulolysis, which converts waste newspaper cellulose into a more environmentally friendly bioplastic made in a lab. Waste newspapers, high in cellulose, are suggested as a raw material for bioplastics, as they are high in cellulose and can be used in various ways such as bottles, cups, containers, and packaging.

The process of creating bioplastic involves extracting cellulose from pulp using a solvent called 1-butyl-3-methylimidazolium chloride. Cellulolysis is the process of hydrolyzing cellulose into smaller polysaccharides or glucose units, with enzymes like exo-acting glucosidases and endo-acting cellulases breaking the glycosidic bond in cellulose. A patented procedure consists of two stages: the preparation of PLA bioplastic involves extruding waste newspaper starch into dextrose, a plant-produced glucose that is then fermented to generate lactic acid and a long chain of bioplastic products.

The advantages of bioplastic over petroleum-based plastic include its reversible process when exposed to a compost heap, where the fundamental elements of PLA are broken down by bacteria and fungi in the soil. PLA composts like an organic material when given the right aerobic conditions, such as heat and moisture. By being consumed by microorganisms in compost, PLA turns into humus, a nutrient-rich material that feeds plants naturally [12].

By being made from carbon-absorbing, renewable plants, PLA is considered "carbon neutral" and helps to lower greenhouse gas emissions. Burning PLA does not create any harmful pollutants. Cellulose was derived from rice straw, which was obtained by removing the leaves of paddy rice. The straw was then cut and dried in the sun. As a catalyst, an ethanol and NaOH cooking solution was created, and rice straw was cooked in the digester. The digester residue and filtrate were separated with a sieve. The residue was rinsed with ethanol solution, then hot water, until the filtrate was clear and neutral. The "clean" residue was dried in an oven. After that, soak 10 g of dried pulp in NaOH solution in an Erlenmeyer flask for 1 minute. NaOH solution was added to the Erlenmeyer flask, which was then constantly agitated. NaOH solution was added to the mixture at regular intervals.

The amount of NaOH solution was added each time. The mixture was allowed to rest. After adding water, the liquid was filtered using Whatman filter paper. A precipitate was formed and washed with water. The precipitated solid that remained on the filter paper was transferred to another flask and washed with water. An acetic acid solution was added to the precipitated material and stirred. The substance was rinsed multiple times until it attained a neutral pH. The leftover precipitate was dried in an oven until its weight was consistent.

Drawing on the results of the research, the following conclusions may be made [13]:

- (i) Glycerol and CMC can be used to create bioplastics from rice straw cellulose.
- (ii) Bioplastics that have extra chemicals in them, including glycerol and CMC, are not waterproof.
- (iii) At the same CMC level, bioplastics with a greater glycerol content showed more oil resistance.
- (iv) Bioplastics that have lower glycerol and CMC content break down more quickly.

Polylactic acid (PLA), a bioplastic derived from plants, has the potential to replace 94% of toxic oil-based plastics. However, concerns have been raised about mixing PLA with ordinary plastic garbage as it becomes more prevalent.

The study demonstrates that near-infrared spectroscopy can accurately classify broken PLA in recycling procedures, indicating the feasibility of sensor-based quality management for PLA. Four primary categories of plastic packaging trash account for a large portion of Germany's Lightweight Packaging (LWP) waste stream: PP, HDPE, PET, and PS. To expedite PLA degradation, PLA samples were tested in a lab setting and exposed to an alkaline environment at a temperature of approximately 65°C. The study utilized a Helios-G2-320 NIR sensor to record spectra of samples between 930 and 1700 nm. The study found that NIR spectroscopy can effectively classify broken PLA in recycling procedures, with a classification accuracy of 97.4%. This suggests that sensor-based quality management for PLA can be feasible, allowing for effective detection and classification of the material's state through spectrum shifts.

In using NIR spectroscopy, one can efficiently separate ordinary plastics and bioplastics like PLA for recycling, which is crucial to preserving the quality of PLA and preventing deterioration. Additional sorting stages may be needed for more precise separation, and sorting equipment should be adjusted to prevent accidental sorting with normal plastics. Future research should examine PLA samples with varying colors, transparency levels, and deterioration stages for a more precise understanding.

Generally, over 1×10^8 bacterial cells are present in 1 g of soil. The commercial PBS, PBS-starch, and PLA were the bioplastics, while PA66 was the non-biodegradable petrochemical plastic used in this study. For a separate experiment, all three kinds of bioplastics (PBS, PBS starch, and PLA) were powdered using a power grinder [14].

Several observations were conducted to assess bioplastics' biodegradability. The short-term degradation of bioplastics and non-biodegradable plastic films reveals that bioplastics are degraded by various microorganisms in nature and turned into water and carbon dioxide through microbial metabolism. Non-biodegradable petrochemical plastics such as PA66, polypropylene, and polyethylene persist in the environment for a long time because they are resistant to decomposition by microbes, and degradation products from PBS and PBS-starch have little effect on microbial biomass and variety.

When we observed the short-term degradation of powdered bioplastics, we discovered that increasing the surface area improved the biodegradability of powdered bioplastics in the soil. Powdered bioplastics exhibited similar biodegradability to commercial bioplastics. When we examined the effect of bacterial biomass on the degradation of bioplastic films in soil, we discovered that PBS-starch degradation in environmental soil samples is closely related to bacterial biomass, and bioplastic degradation is dependent on bacterial biomass in the soil.

When we observed the long-term degradation of bioplastics and non-biodegradable plastics, we discovered that bioplastic degradation had no effect on soil bacterial biomass or diversity. When we examined the bioplastic surface using SEM, we discovered that the structures of the bioplastics were deteriorated and destroyed in the soil environment; however, the degradation rates varied amongst bioplastics. PBS and PBS-starch decomposed quickly, whereas PLA took a long time to degrade significantly.

When we investigated the effect of bioplastic degradation on nitrogen circulation activity in soil, we discovered that the biodegradation of PLA, PBS, and PBS-starch in agricultural soil had no significant effect on bacterial biomass, although bacterial diversity was somewhat altered. The breakdown of PBS and PBS-starch had no influence on nitrogen circulation activity, while PLA appeared to have a detrimental effect on the activities of both ammonium and nitrite-oxidizing bacteria.

The degradability of bioplastics in the soil environment varied greatly depending on the major polymer component. PBS-starch degraded faster than PLA or PBS. Although the decomposition of bioplastics in this study had no effect on bacterial diversity in the soil environment, the pace of deterioration was slow [15].

Three starch-based bioplastic films were produced using lemon extract, water, and cross-linkers – PV (polylactic acid and crude palm oil), AV (glycerol and crude palm oil), and PAV (polylactic acid, glycerol, and crude palm oil). These films emitted volatile chemicals during thermogravimetric analysis (TGA), with plasticizers and cross-linkers releasing at around 300°C.

The films, which are inexpensive and renewable, can be used in food packaging. Starch's gelatinization capabilities are provided by amylopectin and amylose, with amylose being stronger and more flexible. However, starch's brittleness and hydrophilicity, exacerbated by environmental factors like humidity-induced hydrolysis, limit its application in bioplastics. Plasticizers like crude palm oil and glycerol enhance the bioplastics' tensile strength and thermal stability. The first analysis of sago, jackfruit seed, and core involved calculating moisture, ash, carbohydrates, protein, and starch contents.

Iron lactate production was indicated by a canary yellow color in Kelling's test, whereas water had an orange tint. With a colorimetric concentration of 0.65 g/g, the lactic acid purity measured by TLC was half of the standard (88%). When exposed to water, the exposed hydroxyl groups in amylose and amylopectin that had not been cross-linked with plasticizers swelled. Fillers made bioplastic sheets less translucent. Films based on glycerol demonstrated reduced solubility. While PLA's poor ductility reduced elongation strength, jackfruit core served as a filler and improved tensile strength through improved PLA contact. Glycerol enhanced the flexibility of the film, whereas palm oil boosted the cross-linking. Tensile strength with fillers was negatively impacted by the use of plasticizers, and films softened in alkali as a result of acid neutralization [16].

Glycerol-plasticized starch broke down more quickly. All bioplastic films showed progressive desorption between 150°C and 220°C during thermogravimetric examination, followed by the release of compounds with lower molecular weights around 300°C and the pyrolysis of carbonated compounds beyond 500°C. Thermal stability was enhanced by the use of crude palm oil. During the gelatinization process, hydrogen bonds broke, creating gaps and microvoids that let water in. By day 15, bioplastic films had undergone size and color changes and had begun to deteriorate slightly. By day 30, 50% of AV and 80% of PAV films had deteriorated, while PV film had started to crumble. The synthetic polythene did not alter. The use of carbon sources in bioplastic films by bacterial cultures resulted in growth by day 7, but no growth was observed in control medium or synthetic polythene. By day 7, fungal species were using bioplastic sheets as a carbon source as well, and growth was seen in every tube. The three bioplastic formulations outperformed earlier research on jackfruit seed flour-based biofilms in terms of tensile strength, elongation at break, thermal stability, and biodegradability.

This study aimed to create a starch-based bio-coat for food products by combining different types of starch (corn, potato, and tapioca) with different plasticizers (glycerine, rice bran oil, sesame oil, castor oil, soy bean oil, and olive oil). The goal was to create a sustainable substitute for traditional petro-plastic polymers that could be used in the future as edible food wrappers.

The best bioplastic sheet was produced by combining corn, tapioca, and potato starch in equal amounts; vegetable oils produced brittle bioplastics, so glycerine was selected as the best plasticizer. The addition of red cabbage extract to this bioplastic formulation caused a color shift in response to pH changes, providing a straightforward solution for spoilage detection.

In the modern market, packaging has a wide range of applications from food products to commodities. Plastic is unquestionably the best material for packaging, but due to its inability to decompose, it is also the most feared environmental pollutant. The world urgently needs a synthetic polymer substitute that is completely degradable and sustainable. Bioplastics, which are sourced from microbes or vegetable stock, are the only viable solution in this situation. The next big thing in packaging is active and intelligent packaging, which has an inbuilt feature that gives the items packed extra protection and helps to indicate the quality of the items packed. One of the most intriguing developments in intelligent packaging is color-changing packaging, which makes it simple for consumers to judge the freshness of the product visually. Temperature changes (for thermochromic materials), light irradiation (for photochromic materials), the application of an electric potential (for electrochromic materials), or pH changes can trigger this response. This study attempted to create a proof of concept for the potential development of color-changing bio-wraps or bio-dips for edible items as a way to detect spoilage.

The main components of starch-based bioplastics were water, plasticizer, vinegar, and starch. Three distinct sources of starch were obtained: tapioca, corn, and potatoes. Eight trials were carried out, seven with different plasticizers and one without any. A variety of plasticizers, including castor oil, gelatin, glycerol, sesame oil, soybean oil, and olive oil, were employed in the investigation. The best plasticizer was chosen after analyzing the smoothness and elasticity of the plastic. The next step was to experiment with different sugar combinations to determine which would work for making the best bioplastics. Out of the 24 combinations tested using a single starch composition and various plasticizers, only the S-2 series (S-2–1 to S-2–4) produced sheets that were smooth.

When gelatin was used as the plasticizer, soft sheets of bioplastics were produced (S-3 series). The bioplastic sheets (S1 series) without a plasticizer were poor and brittle. It was discovered that glycerol was the best plasticizer, producing bioplastics that were smooth, stretchable, transparent, and of high quality. Corn and tapioca plastics looked the clearest or most transparent, after potato starch plastics. All of the D series displayed smooth, transparent bioplastics, but D5 (all sugars) and D4 (tapioca and potato combination) demonstrated the most consistent, stretchable, and wrappable qualities. The combination of D5 and red cabbage extracts subsequently displayed a discernible color shift.

In this study, we demonstrated the applicability of bioplastics in conjunction with natural pH indicators, providing a simple and clever way to let customers know how the quality of food is affected by pH changes [17].

The bags were composed of polyethylene with a pro-oxidant ingredient and HDPE combined with a degradable additive. In an effort to ascertain the properties of their degradability and biodegradability, four bags were certified as compostable.

Crude oil-derived synthetic polymers have transformed material sciences and chemistry by creating useful materials like polyethylene (PE). These materials, however, are renowned for their robustness and resistance to biodegradation, which pollutes the environment. Attempts are underway to create modified petroleum-based plastics with biodegradable additives or biodegradable polymers derived from agricultural resources in order to solve this. Composting aids in the process of biodegradation, which turns plastic waste into organic materials that may be used as fertilizer. But not all plastics can decompose entirely in natural settings, nor can all plastics be biodegradable.

Since they became accessible in 1990, biodegradable polymers have sparked questions about how truly biodegradable they are. National organizations like ASTM and ISO provide standards that assess compostability, disintegration, and safety in a range of settings. While compostable plastics disintegrate biologically during composting, leaving no harmful residues behind, biodegradable plastics break down because of microorganisms. Composting biodegradable plastics after use is one approach to addressing plastic waste.

In Eastern Europe, chain businesses provide compostable and biodegradable plastic bags as environmentally acceptable substitutes. Renewable resource-based compostable polymers are thought to be good for the environment. Over the course of 12 weeks, the degradability of single-use plastic bags labeled as compostable, biodegradable, and/or degradable was to be evaluated.

The disintegration of plastic parts was evaluated in the study using modified versions of CSN EN ISO 20200 and CSN EN 14806 Norm. To simulate real-world circumstances, a synthetic substance consisting of wood sawdust, compost, biodegradable municipal garbage, and glucose was combined with plastic pieces. For every material, two reactors were set up, and 90% of the reactor capacity was filled with the mixture. For 12 weeks, aerobic degradation was carried out at 58.0°C under periodic moisture, mixing, and aeration control. Plastic fragment variability and disintegration degree were computed. To evaluate the compost quality and biodegradability of plastic bags, it is essential to keep an eye on the composting processes.

The study measured variables including dry mass, volatile solids, and pH to evaluate a synthetic material's suitability for composting. The average volatile solids concentration was 88.57%, but the dry mass percentage was within the advised range. Temperature, odor, pH, and moisture content were all used to track the composting process. The temperature was maintained at 58.0°C, and the pH change was consistent with theory. Controlling moisture was essential for both the rate of breakdown and microbial activity. Odor observation was conducted according to CSN EN 14806 criteria, with microbial decomposition aided by starch's high biodegradability.

A test that demonstrated total deterioration in ten days on cellulose filter paper (CFP) validated the validity of the test. After 12 weeks, though, there was still no progress. Within six weeks, certified compostable bags began to disintegrate significantly, and by twelve weeks, four bags had completely disintegrated. The effects of composting conditions on the decomposition of biodegradable plastics should be investigated further [18].

The study looks at fillers in PVA matrices that are biodegradable and manufactured from pinewood nanocellulose. It shows that when compared to bioplastics without extract, orange peel extract improves tensile strength, elastic modulus, elongation, and degradability, indicating that it functions as a reinforcing agent.

The development of biodegradable bioplastics can help lessen the environmental effect of plastic waste, which is a global problem. Bioplastics can be blended with polymers and additives. They are made from cellulose and nanocellulose. These plentiful resources may be used to control biomass waste by Indonesia, a significant producer of orange fruits and wood. Waste orange peel may be utilized as a linker in bioplastic formulations, while pine wood waste can be hydrolyzed to produce nanocellulose. This strategy may encourage the bioplastics sector to flourish sustainably.

Using a variety of methods, the study examined the mechanical, morphological, and structural characteristics of cellulose, nanocellulose, and bioplastic samples. These included mechanical property investigation utilizing a tensile strength apparatus in accordance with the ASTM-D1708-18 standard procedure, X-ray diffraction to ascertain the crystallinity index, SEM to investigate surface morphology, and FTIR spectroscopy to detect functional groups. Tensile tests were performed on specimens shaped like dumbbells until they broke. The findings offer insightful information on how well certain materials work in different contexts. The study provides insightful information about these materials' characteristics and their uses.

Pinewood waste was delignified and acid pre-treated to extract cellulose and nanocellulose, producing a dark brown powder and white cellulose. White nanocellulose, a softer form, is used as a reinforcing filler in bioplastics due to its mechanical strength, low toxicity, and high degradability. FTIR analysis revealed distinctive absorption bands, while orange peel extract-based bioplastics have a faint yellow hue and high elastic modulus, tensile strength, and elongation at break.

The project investigates the creation of ecologically friendly bioplastics from pinewood waste. Bioplastics were produced by extracting cellulose and nanocellulose through hydrolysis and alkaline bleaching, then combining them with PVA. The tensile strength and flexibility of the bioplastic were enhanced by the addition of orange peel waste extract. Its potential to improve the material's long-term durability and environmental effect is indicated by its slower rate of deterioration [19].

In the plastic sector, there is a continuing need to strike the right balance between mechanical characteristics, production costs, and acceptable levels of biodegradability. This research focuses on two potential synthetic approaches for increasing plastic production: employing agricultural waste and nanotechnology.

The utilization of agricultural waste in plastic synthesis provides a more sustainable and cost-effective alternative to typical petroleum-based polymers. Waste materials such as corn starch, wheat straw, and rice husk can be used into the polymerization process to create biodegradable plastics with improved mechanical qualities.

These plastics have use in a variety of industries, including packaging, construction, and automobile components. Using nanotechnology in plastic synthesis can improve the qualities of plastic products, making them more durable, biodegradable, and cost effective.

The utilization of agricultural waste and nanotechnology in plastic synthesis provides promising solutions to the issues that the plastic industry faces. Waste materials such as corn starch, wheat straw, and rice husk can be used into the polymerization process to create biodegradable plastics with improved mechanical qualities. These plastics have use in a variety of industries, including packaging, construction, and automobile components.

Using nanotechnology in plastic synthesis can improve the qualities of plastic products, making them more durable, biodegradable, and cost-effective. This can be accomplished by adding nanoparticles to the polymerization process. These nanoparticles can improve the mechanical properties of the plastic, including tensile strength and impact resistance. Furthermore, nanoparticles can be employed to incorporate biodegradability into the plastic matrix, allowing the plastic.

The increased need for plastic, along with its negative environmental impact, has resulted in the creation of the bioplastic sector. Bioplastics are predicted to gradually replace conventional plastics on a large scale, increasing waste management alternatives and lowering reliance on fossil fuels. The bioplastic sector accounts for 1% of the 359 million tonnes of plastic produced globally, but this figure is predicted to rise to 2.43 million tonnes by 2024.

Bioplastics are employed in a variety of industries, including packaging, electronics, automotive, healthcare, and textiles. Packaging represents the largest market, accounting for more than 53% of total bioplastic production in 2019. The bioplastic market is rapidly expanding, with bio-based polyesters PLA and PHAs driving expansion due to their superior properties and adaptability [20].

Although nanotechnology has been successful in improving the qualities of bioplastics, its commercial use has been limited due to high costs. However, the use of nanotechnology has resulted in considerable advances in the mechanical, physical, and barrier properties of bioplastics, increasing their competitiveness in the plastics sector.

The source of nanoparticles is critical to the sustainability of the bioplastic industry. While metallic nanoparticles have shown promise, there has been a trend toward nanoparticle synthesis using renewable plant sources, particularly agricultural waste. Nanoparticles made from agricultural waste, including sesame husks, coconut shells, and oil palm ash, have been proven to improve the characteristics of bioplastics.

Several metallic nanoparticles (NPs) have been utilized to supplement polymer substances, particularly those used in the food packaging industry, to either improve the neatness of the polymers or to aid in packaging operations. The most common metallic nanoparticles are silver nanoparticles, aluminum nanoparticles, and zinc nanoparticles.

Silver nanoparticles (Ag NPs) have gained popularity as biocidal agents in food packaging materials due to their thermal resistance, safety, and broad-spectrum antibacterial action. The antibacterial capabilities of Ag NPs are determined by their size, concentration, composition, and form. When Ag NPs are introduced to polymeric substrates, they exhibit antibacterial properties against various bacteria and fungi.

A study that combined Ag NPs and PBAT (polybutylene adipate-*co*-terephthalate) in nanocomposite films discovered that the inclusion of Ag NPs boosted the material's tensile strength, water vapor permeability, and biocidal properties. Similarly, another study discovered that starch-based films coupled with Ag NPs have antibacterial properties against a variety of harmful bacteria and fungi. Copper nanoparticles (CuNPs) have also been used.

Plastic use is a major issue worldwide because of its low cost and availability, resulting in overconsumption and environmental impact. Plastic's non-biodegradable nature creates a significant load on the environment. Recycling plastic and developing sustainable, eco-friendly alternatives are critical to environmental conservation.

Agricultural waste products can be recycled into plastic, making it a more environmentally friendly option. Nanotechnology can also enhance plastic production and management. More research is needed to improve plastic management and develop sustainable plastic items for a healthy environment [21].

8.3 CONCLUSION

This study discusses the process of using jackfruit seeds as a source of starch to produce bioplastic. The seeds are deshelled, washed, and soaked to eliminate latex, then processed to extract crude starch, which is refined through filtration and precipitation to remove impurities. The starch is then mixed with glycerol, citric acid, and baking soda to create a plastic mixture, which is dried to form bioplastic. The ratio of glycerol affects the flexibility of the bioplastic, while citric acid impacts its hardness. Microcracks in the bioplastic are attributed to the presence of microbubbles formed during processing.

This study also discusses the characteristics of starch-based bioplastic films and their potential applications in food packaging. Various experiments and analyses are mentioned, including the synthesis process, characterization of the films, and their thermal properties.

The research aims to contribute to the development of eco-friendly and sustainable alternatives to conventional plastics, addressing the growing concerns related to plastic pollution and environmental degradation. The study involves the extraction and processing of raw materials, such as banana peels, cassava starch, potato starch, and cellulose extracted from *Prosopis juliflora* wood, to obtain the necessary components for bioplastic production. Various additives, such as glycerol, citric acid, and gelatin, are used to improve the mechanical properties and biodegradability of the bioplastics.

Tests and analyses are conducted to evaluate the properties of the synthesized bioplastics, including tensile strength, elongation at break, thermal stability, moisture absorption, and biodegradability. The results indicate that the synthesized bioplastics exhibit promising properties, such as good tensile strength, biodegradability, and suitability for packaging applications. The environmental advantages of using bioplastics over petroleum-based plastics, emphasizing the reduction of pollution and the conservation of resources, are also highlighted. Additionally, the potential of using locally available agricultural waste, such as banana peels and cassava starch, for bioplastic production is explored, offering a sustainable solution to plastic waste management.

The research aims to contribute to the development and adoption of eco-friendly alternatives to conventional plastics, addressing the environmental challenges associated with plastic pollution and waste management. Several key points are discussed in the research, including the microbial degradation of bioplastics, the development of starch-based bio-coats for food products, the evaluation of biodegradable plastic bags, and the use of orange peel extract in bioplastics. These studies highlight the importance of developing bioplastics with enhanced biodegradability and mechanical properties, as well as exploring innovative approaches such as intelligent packaging and utilizing waste-derived additives for sustainable plastic alternatives.

REFERENCES

1. Trieu Khoa Nguyen, Nguyen Thy Ton That, Nien Thi Nguyen and Ha Thi Nguyen, "Development of starch-based bioplastic from jackfruit seed", *Advances in Polymer Technology*, vol. 2022, pp. 9, 2022.
2. Akshaya Krishnamurthy and Pavithra Amritkumar, "Synthesis and characterization of eco-friendly bioplastic from low cost plant resources", *SN Applied Sciences*, vol. 1, p. 1432, 2019.
3. Jayachandra Yaradoddi, Vinay Patil, Sharanabasava Ganachari, Nagaraj Banapurmath, Anand Hunashyal and Ashok Shettar, "Biodegradable plastic production from fruit waste material and its sustainable use for green applications", *International Journal of Pharmaceutical Research & Allied Sciences*, vol. 5, no. 4, pp. 56–66, 2016.
4. Prachi Sonawane, Amisha Panchal, Shreeya Naik, Bhagyashree Mundaye and Unnati Padalia, "Potato starch based bioplastic as alternative packaging materials", vol. 3, no. 3, pp. 111–113, 2021, doi: 10.33472/AFJBS.3.3.2021.111–113.
5. M. K. Marichelvam, P. Manimaran, Anish Khan, M. Geetha, K. Kandakodeeswaran, Ahmed H. Abduljabbar, Edi Syafri, Mohammad A. Wazzan, Huda Wazzan and Waseem Khan, "Development and characterization of sustainable bioplastic films using cellulose extracted from prosopis juliflora", *Journal of Natural Fibers*, vol. 20, no. 2, p. 2231271, 2023.
6. Nanang Eko Wahyuningtiyas and Heru Suryanto, "Analysis of biodegradation of bioplastics made of cassava starch", *Journal of Mechanical Engineering Science and Technology*, vol. 1, no. 1, pp. 24–31, 2017.
7. Harrison O. Onovo, Theddeus T. Akano, Damian U. Onyegbule, Esther T. Towolawi and Tolulope S. Ajala, "A study of biodegradation of hybrid bioplastic films blend from manihot and triticum biopolymer", *European Journal of Engineering and Technology Research*, vol. 7, no. 3, pp. 30–38, 2022.
8. C. Ramesh, M. Premnath, S. V. Rakshan, N. Sai Sudharshan and R. Prakash, "Synthesis and characterization of biodegradable plastic from potato starch and gelatin", *Materials Science*, vol. 12, pp. 2327–2336, 2023.
9. S. Sharmila, E. Suganya Rajeswari, Neha Singh, R. Kamalambigeswari and E. Kowsalya, "Production of bioplastic from biowaste materials and its SEM-EDS report", *International Journal of Recent Technology and Engineering*, vol. 8, no. 3, pp. 8679–8682, 2019.
10. Anggun Rahman and Isroi Khaswar Syamsu, "Biodegradability of bioplastic from oil palm empty fruit bunch", *Journal of Natural Resources and Environmental Management*, vol. 9, no. 2, pp. 259–264, 2019.
11. Rajen Kundu and Priyanka Payal, "Biodegradation study of potato starch-based bioplastic", *Current Chinese Chemistry*, vol. 2, p. e190421192895, 2022.
12. Sudhanshu Joshi, Ujjawal Sharma and Garima Goswami, "Bio-plastic from waste newspaper", *International Journal of Engineering Research & Technology*, pp. 24–27, 1984, ISSN: 2278–0181.
13. S. Laya, Dr. S. Shamina and Moossa, *The Pharma Innovation Production of Bioplastic from Rice Straw Cellulose.* ISSN(E):2277–7695 ISSN(P):2349–8242. www.thepharmajournal.com. DOI: https://dx.doi.org/10.22271/tpi.

14. Xiaozheng Chen, Nils Kroell, Ke Li, Alexander Feil and Thomas Pretz, "Influences of bioplastic polylactic acid on near-infrared-based sorting of conventional plastic", *Waste Management & Research*, vol. 39, no. 9, pp. 1210–1213, 2021.
15. Dinesh Adhikari, Masaki Mukai, Kenzo Kubota, Takamitsu Kai, Nobuyuki Kaneko, Kiwako S. Araki and Motoki Kubo, "Degradation of bioplastics in soil and their degradation effects on environmental microorganisms", *Journal of Agricultural Chemistry and Environment*, vol. 5, pp. 23–34, 2016.
16. Rizka Wulandari Putri Rahmatullah, Sri Haryati, Harry Waristian, Aan Saputra and Robby Kurniawan, "Manufacture of bioplastic composite method of blending synthetic polymer (PP) with natural polymers from Kapok Fiber (Ceiba Pentandra)", *Chemical Engineering Transactions*, vol. 97, pp. 241–246, 2022.
17. Harsh Patel, Sudarshan Seshadri and J. R. Parvathi, "Edible bioplastic with natural pH indicators", *International Journal of Current Microbiology and Applied Sciences*, vol. 6, no. 7, 2017.
18. Magdalena Vaverková, František Toman, Dana Adamcová and Jana Kotovicov, "Study of the biodegrability of degradable/biodegradable plastic material in a controlled composting environment", *Ecological Chemistry and Engineering S,* vol. 19, no. 3, pp. 347–358, 2012.
19. Alyaa Farrah Dibha, Masruri Masruri and Arie Srihardyastutie, "Degradable bioplastic developed from pine-wood nanocellulose as a filler combined with orange peel extract", *Indonesian Journal of Chemistry*, vol. 23, no. 1, pp. 127–139, 2023.
20. R. K. Atchaya, U. Devipriya, K. Sreeja and P. Buvanesh, "Evaluation of bioplastic developed from corn starch and Euphorbia antiquorum Latex", *International Journal of Environment and Climate Change*, vol. 13, no. 11, pp. 1169–1177, 2023.
21. Abiodun A. Abioye, Sunday A. Afolalu, Olabisi O. Yusuf, Moses E. Emetere, Samson O. Ongbali, Udo Mfon, Olusegun D. Samuel and Ayotunde Olayinka, "Nanotechnology applications in synthesis of bioplastic", *IOP Conf. Series: Materials Science and Engineering*, vol. 1107, p. 012190, 2021.
22. Priya Bansal and B. Raj, "Memristor: a versatile nonlinear model for dopant drift and boundary issue", *Journal of Computational and Theoretical Nanoscience,* vol. 14, no. 5, pp. 2319–2325, May 2017.
23. Neeraj Jain and B. Raj, "An analog and digital design perspective comprehensive approach on Fin-FET (Fin-Field Effect transistor) technology – a review", *Reviews in Advanced Sciences and Engineering*, vol. 5, pp. 1–14, 2016.
24. Balwinder Raj, A. K. Saxena and S. Dasgupta, "Analytical modeling for the estimation of leakage current and subthreshold swing factor of nanoscale double gate FinFET device", *Microelectronics International,* vol. 26, pp. 53–63, 2009.
25. Shailendra Singh Soniya, Girish Wadhwa and Balwinder Raj, "An analytical modeling for dual source vertical tunnel field effect transistor", *International Journal of Recent Technology and Engineering,* vol. 8, no. 2, pp. 603–608, July 2019.
26. Shailendra Singh and B. Raj, "Design and analysis of heterojunction vertical T-shaped tunnel field effect transistor", *Journal of Electronics Material*, vol. 48, no. 10, pp. 6253–6260, Oct. 2019.

27. Candy Goyal, Jagpal Singh Ubhi and B. Raj, "A low leakage CNTFET based inexact full adder for low power image processing applications", *International Journal of Circuit Theory and Applications*, vol. 47, no. 9, pp. 1446–1458, Sep. 2019.
28. B. Raj, A. K. Saxena and S. Dasgupta, "A compact drain current and threshold voltage quantum mechanical analytical modeling for FinFETs", *Journal of Nanoelectronics and Optoelectronics*, vol. 3, no. 2, pp. 163–170, 2008.
29. Girish Wadhwa and B. Raj, "An analytical modeling of charge plasma based tunnel field effect transistor with impacts of gate underlap region", *Superlattices and Microstructures*, vol. 142, p. 106512, June 2020.
30. Shailendra Singh and B. Raj, "Modeling and simulation analysis of SiGe heterojunction double GateVertical t-shaped tunnel FET", *Superlattices and Microstructures*, vol. 142, p. 106496, June 2020.
31. Neeraj Jain and Balwinder Raj, "Device and circuit co-design perspective comprehensive approach on FinFET technology – a review", *Journal of Electron Devices*, vol. 23, no. 1, pp. 1890–1901, 2016.
32. Sunil Kumar and B. Raj, "Analysis of I_{ON} and ambipolar current for dual-material gate-drain overlapped DG-TFET", *Journal of Nanoelectronics and Optoelectronics*, vol. 11, pp. 323–333, June 2016.
33. Naveed Anjum, Tarun Bali and B. Raj, "Design and simulation of handwritten multiscript character recognition", *International Journal of Advanced Research in Computer and Communication Engineering*, vol. 2, no. 7, pp. 2544–2549, July 2013.
34. Karmjit Singh and B. Raj, "Performance and analysis of temperature dependent multi-walled carbon nanotubes as global interconnects at different technology nodes", *Journal of Computational Electronics*, vol. 14 no. 2, pp. 469–476, June 2015.
35. Sunil Kumar and B. Raj, "Compact channel potential analytical modeling of DG-TFET based on Evanescent–mode approach", *Journal of Computational Electronics*, vol. 14, no. 2, pp. 820–827, July 2015.
36. Karmjit Singh and B. Raj, "Temperature dependent modeling and performance evaluation of multi-walled CNT and single-walled CNT as global interconnects", *Journal of Electronic Materials*, vol. 44, no. 12, pp. 4825–4835, Dec. 2015.
37. V. K. Sharma, M. Pattanaik and B. Raj, "INDEP approach for leakage reduction in nanoscale CMOS circuits", *International Journal of Electronics*, vol. 102, no. 2, pp. 200–215, 2014.
38. Karmjit Singh and B. Raj, "Influence of temperature on MWCNT bundle, SWCNT bundle and copper interconnects for nanoscaled technology nodes", *Journal of Materials Science: Materials in Electronics*. vol. 26, no. 8, pp. 6134–6142, 2015.
39. Naveed Anjum, Tarun Bali and B. Raj, "Design and simulation of handwritten Gurumukhi and Devanagri numerical recognition", *International Journal of Computer Applications*, vol. 73, no. 12, pp. 16–21, 2013.
40. S. Khandelwal, V. Gupta, B. Raj and R. D. Gupta, "Process variability aware low leakage reliable nano scale DG-FinFET SRAM cell design technique", *Journal of Nanoelectronics and Optoelectronics,* vol. 10, no. 6, pp. 810–817, Dec. 2015.

41. V. K. Sharma, M. Pattanaik and B. Raj, "ONOFIC approach: low power high speed nanoscale VLSI circuits design", *International Journal of Electronics*, vol. 101, no. 1, pp. 61–73, 2014.
42. S. Khandelwal, Balwinder Raj, and R. D. Gupta, "FinFET based 6T SRAM cell design: analysis of performance metric, process variation and temperature effect", *Journal of Computational and Theoretical Nanoscience*, vol. 12, pp. 2500–2506, 2015.
43. Sumit Singh, Shekhar Yadav, Jagdeep Rahul, Anurag Srivastava and B. Raj, "Impact of HfO_2 in graded channel dual insulator double gate MOSFET", *Journal of Computational and Theoretical Nanoscience,* vol. 12, no. 6, pp. 950–953, April 2015.
44. Vijay Kumar Sharma, Manisha Pattanaik and B. Raj, "PVT variations aware low leakage INDEP approach for nanoscale CMOS circuits", *Microelectronics Reliability*, vol. 54, pp. 90–99, 2014.
45. B. Raj, A. K. Saxena and S. Dasgupta, "Quantum mechanical analytical modeling of nanoscale DG FinFET: evaluation of potential, threshold voltage and source/drain resistance", *Journal of Material Science in Semiconductor Processing*, vol. 16, no. 4, pp. 1131–1137, 2013.
46. Maisagalla Gopal, Siva Sankar D. Prasad and Balwinder Raj, "8T SRAM cell design for dynamic and leakage power reduction", *International Journal of Computer Applications*, vol. 71, no. 9, pp. 43–48, June 2013.
47. Manisha Pattanaik, B. Raj, Shashikant Sharma and Anjan Kumar, "Diode based trimode multi-threshold CMOS technique for ground bounce noise reduction in static CMOS adders", *Advanced Materials Research*, vol. 548, pp. 885–889, 2012.
48. Balwinder Raj, A. K. Saxena and S. Dasgupta, "Nanoscale FinFET based SRAM cell design: analysis of performance metric, process variation, underlapped FinFET and temperature effect", *IEEE Circuits and System Magazine*, vol. 11, no. 2, pp. 38–50, 2011.
49. V. K. Sharma, M. Pattanaik and Balwinder Raj, "Leakage current ONOFIC approach for deep submicron VLSI circuit design", *International Journal of Electrical, Computer, Electronics and Communication Engineering,* vol. 7, no. 4, pp. 239–244, 2013.
50. Parminder Kaur, Sandeep Singh Gill and B. Raj, "Comparative analysis of OFETs materials and devices for sensor applications", *Journal of Silicon*, vol. 14, pp. 4463–4471, 2022.
51. Sanjeev Kumar Sharma, Parveen Kumar, Balwant Raj and B. Raj, "$In_{1-x}Ga_xAs$ double metal gate-stacking cylindrical nanowire MOSFET for highly sensitive photo detector", *Journal of Silicon*, vol. 14, pp. 3535–3541, 2022.
52. B. Raj, A. K. Saxena and S. Dasgupta, "Analytical modeling of quasi planar nanoscale double gate FinFET with source/drain resistance and field dependent carrier mobility: a quantum mechanical study", *Journal of Computer*, vol. 4, no. 9, pp. 1–8, 2009.
53. S. Bhushan, S. Khandelwal and B. Raj, "Analyzing different mode FinFET based memory cell at different power supply for leakage reduction", *Seventh International Conference on Bio-Inspired Computing: Theories and Application* (BIC-TA 2012), Advances in Intelligent Systems and Computing, vol. 202, pp. 89–100, 2013.

54. Jeetendra Singh and B. Raj, "Temperature dependent analytical modeling and simulations of nanoscale memristor", *Journal: Engineering Science and Technology, an International Journal*, vol. 21, pp. 862–868, Oct. 2018.
55. Shradhya Singh, Shashi Bala, Balwant Raj and B. Raj, "Improved sensitivity of dielectric modulated junctionless transistor for nanoscale biosensor design", *Sensor Letter*, vol. 18, pp. 328–333, April 2020.
56. Vivek Kumar, Santosh Kumar Vishvakarma and B. Raj, "Design and performance analysis of ASIC for IoT applications", *Sensor Letter*, vol. 18, pp. 31–38, Jan. 2020.
57. Akanksha Jaiswal, R. K. Sarin, B. Raj and Shikha Sukhija, "A novel circular slotted microstrip-fed patch antenna with three triangle shape defected ground structure for multiband applications", *Advanced Electromagnetic*, vol. 7, no. 3, pp. 56–63, Aug. 2018.
58. Girish Wadhwa and B. Raj, "Label free detection of biomolecules using charge-plasma-based gate underlap dielectric modulated junctionless TFET", *Journal of Electronic Materials*, vol. 47, no. 8, pp. 4683–4693, Aug. 2018.
59. Gurmohan Singh, R. K. Sarin and B. Raj, "Design and performance analysis of a new efficient coplanar quantum-dot cellular automata adder", *Indian Journal of Pure & Applied Physics*, vol. 55, pp. 97–103, Feb. 2017.
60. Parminder Kaur, Vikas Pandey and B. Raj, "Comparative study of efficient design, control and monitoring of solar power using IoT", *Sensor Letter*, vol. 18, pp. 419–426, May 2020.
61. Anil Kumar Bhardwaj, Sumeet Gupta and B. Raj, "Development & analysis of compact model for double gate schottky barrier CNTFET", *Journal of Nanoelectronics and Optoelectronics*, vol. 15, pp. 1199–1208, Aug. 2020.
62. Girish Wadhwa, Priyanka Kamboj; Jeetendra Singh and B. Raj, "Design and investigation of junctionless DGTFET for biological molecule recognition", *Transactions on Electrical and Electronic Materials,* vol. 22, pp. 282–289, 2021.
63. Sachin Kumar Verma, Shailendra Singh, Girish Wadhwa and B. Raj, "Detection of biomolecules using charge-plasma based gate underlap dielectric modulated dopingless TFET", *Transactions on Electrical and Electronic Materials*, vol. 21, pp. 528–535, June 2020.
64. Neeraj Jain and B. Raj, "Impact of underlap spacer region variation on Electrostatic and analog/RF performance of symmetrical high-k SOI FinFET at 20 nm channel length", *Journal of Semiconductors*, vol. 38, no. 12, p. 122002, Dec. 2017.
65. Shailendra Singh and B. Raj, "Analytical modeling and simulation analysis of T-shaped III–V heterojunction vertical T-FET", *Superlattices and Microstructures*, vol. 147, pp. 106717, Nov. 2020.
66. Gurmohan Singh, R. K. Sarin and B. Raj, "Design and analysis of area efficient QCA based reversible logic gates", *Journal of Microprocessors and Microsystems*, vol. 52, pp. 59–68, May 2017.
67. Shailendra Singh and B. Raj, "Parametric variation analysis on hetero-junction vertical t-shape TFET for Supressing ambipolar conduction", *Indian Journal of Pure and Applied Physics*, vol. 58, pp. 478–485, June 2020.
68. Soniya, Shailendra Singh, Girish Wadhwa and Balwinder Raj, "Design and analysis of dual source vertical tunnel field effect transistor for high performance", *Transactions on Electrical and Electronics Materials,* vol. 21, pp. 74–82, Oct. 2019.

69. Manjit Kaur, Neena Gupta, Sanjeev Kumar, B. Raj and Arun Kumar Singh, "RF performance analysis of intercalated graphene nanoribbon based global level interconnects", *Journal of Computational Electronics,* vol. 19, pp. 1002–1013, June 2020
70. Girish Wadhwa and B. Raj, "Design and performance analysis of junctionless TFET biosensor for high sensitivity", *IEEE Nanotechnology,* vol. 18, pp. 567–574, 2019.
71. Jeetendra Singh and B. Raj, "Enhanced nonlinear memristor model encapsulating stochastic dopant drift, *Journal of Nanoelectronics and Optoelectronics,* vol. 14, pp. 958–963, 2019.
72. Anjana Bhardwaj, Pradeep Kumar, B. Raj and Sunny Anand, "Design and performance optimization of doping-less vertical nanowire TFET using gate stack technique", *Journal of Electronic Materials*, vol. 41, no. 7, pp. 4005–4013, 2022.
73. Jeetendra Singh and B. Raj, "Tunnel current model of asymmetric MIM structure levying various image forces to analyze the characteristics of filamentary memristor", *Applied Physics A*, vol. 125, no. 3, pp. 203 1–203.11, Feb. 2019.
74. Candy Goyal, Jagpal Singh Ubhi and B. Raj, "Low leakage zero ground noise nanoscale full adder using source biasing technique", *Journal of Nanoelectronics and Optoelectronics*, vol. 14, pp. 360–370, March 2019.
75. Gurmohan Singh, R. K. Sarin and B. Raj, "A novel robust exclusive-OR function implementation in QCA nanotechnology with energy dissipation analysis", *Journal of Computational Electronics*, vol. 15, no. 2, pp. 455–465, June 2016.
76. Girish Wadhwa, Priyanka Kamboj and Balwinder Raj, "Design optimisation of junctionless TFET biosensor for high sensitivity", *Advances in Natural Sciences: Nanoscience and Nanotechnology*, vol. 10, no. 7, p. 045001, 2019.
77. Priya Bansal and B. Raj, "Memristor modeling and analysis for linear dopant drift kinetics", *Journal of Nanoengineering and Nanomanufacturing*, vol. 6, pp. 1–7, 2016.
78. D. Vaithiyanathan Ashima and B. Raj, "Performance analysis of charge plasma induced graded channel Si nanotube", *Journal of Engineering Research*, Special Issue, pp. 146–154, Aug. 2021.
79. Abhishek Singh Tomar, Vijay Kumar Magraiya and B. Raj, "Scaling of access and data transistor for high performance DRAM cell design", *Quantum Matter*, vol. 2, pp. 412–416, Oct. 2013.
80. Neeraj Jain and B. Raj, "Parasitic capacitance and resistance model development and optimization of raised source/drain SOI FinFET structure for analog circuit applications", *Journal of Nanoelectronics and Optoelectronins*, vol. 13, pp. 531–539, April 2018.
81. Shradhya Singh, S. K. Vishvakarma and B. Raj, "Analytical modeling of split-gate junction-less transistor for a biosensor application", *Sensing and Bio-Sensing*, vol. 18, pp. 31–36, April 2018.
82. Maisagalla Gopal and Balwinder Raj, "Low power 8T SRAM cell design for high stability video applications", *ITSI Transaction on Electrical and Electronics Engineering*, vol. 1, no. 5, pp. 91–97, 2013.

83. Balwinder Raj, Jatin Mitra, Deepak Kumar Bihani, V. Rangharajan, A. K. Saxena and S. Dasgupta, "Analysis of noise margin, power and process variation for 32 nm FinFET Based 6T SRAM cell", *Journal of Computer*, vol. 5, no. 6, pp. 1–8, 2010.
84. Divya Sharma, Rajesh Mehra and B. Raj, "Comparative analysis of photovoltaic technologies for high efficiency solar cell design", *Superlattices and Microstructures*, vol. 153, p. 106861, May 2021.
85. Pawandeep Kaur, Avtar Singh Buttar and B. Raj, "A comprehensive analysis of nanoscale transistor based biosensor: a review", *Indian Journal of Pure and Applied Physics*, vol. 59, pp. 304–318, April 2021.
86. Divya Yadav, Balwant Raj and B. Raj, "Design and simulation of low power microcontroller for IoT applications", *Journal of Sensor Letters*, vol. 18, pp. 401–409, May 2020.
87. Shailendra Singh and B. Raj, "A 2-D analytical surface potential and drain current modeling of double-gate vertical t-shaped tunnel FET", *Journal of Computational Electronics*, vol. 19, pp. 1154–1163, April 2020.
88. Jeetendra Singh and B. Raj, "An accurate and generic window function for non-linear memristor model", *Journal of Computational Electronics*, vol. 18 no. 2, pp. 640–647, June 2019.
89. Manjit Kaur, Neena Gupta, Sanjeev Kumar, B. Raj and Arun Kumar Singh, "Comparative RF and crosstalk analysis of carbon based nano interconnects", *IET Circuits, Devices & Systems*, vol. 15, no. 6, pp. 493–503, Feb 2021.
90. Nehru Kandasamy, Firdous Ahmad, D. Ajitha, B. Raj and Nagarjuna Telagam, "Quantum dot cellular automata based scan flip flop and boundary scan register", *IETE Journal of Research*, vol. 66, pp. 535–548, 2020.
91. Jeetendra Singh and B. Raj, "Design and investigation of 7T2M NVSARM with enhanced stability and temperature impact on store/restore energy", *IEEE Transactions on Very Large Scale Integration Systems,* vol. 27, no. 6, pp. 1322–1328, June 2019.
92. Anil Kumar Bhardwaj, Sumeet Gupta, B. Raj and Amandeep Singh, "Impact of double gate geometry on the performance of carbon nanotube field effect transistor structures for low power digital design", *Computational and Theoretical Nanoscience*, vol. 16, pp. 1813–1820, 2019.
93. Neeraj Jain and B. Raj, "Thermal stability analysis and performance exploration of asymmetrical dual-k underlap spacer (ADKUS) SOI FinFET for security and privacy applications", *Indian Journal of Pure & Applied Physics*, vol. 57, pp. 352–360, May 2019.
94. Neeraj Jain and Balwinder Raj, "Dual-k spacer region variation at the drain side of asymmetric SOI FinFET structure: performance analysis towards the analog/RF design applications", *Journal of Nanoelectronics and Optoelectronics*, vol. 14, pp. 349–359, March 2019.
95. Jeetendra Singh, Sanjeev Sharma, B. Raj and Mamta Khosla, "Analysis of barrier layer thickness on performance of In1-xGaxAs based gate stack cylindrical gate nanowire MOSFET", *Journal of Nanoelectronics and Optoelectronics*, vol. 13, pp. 1473–1477, Oct. 2018.

96. Neeraj Jain and B. Raj, “Analysis and performance exploration of high-k SOI FinFETs over the conventional low-k SOI FinFET toward analog/RF design”, *Journal of Semiconductors*, vol. 39, no. 12, p. 124002-1–7, Dec. 2018.
97. Candy Goyal, Jagpal Singh Ubhi and B. Raj, “A reliable leakage reduction technique for approximate full adder with reduced ground bounce noise”, *Journal of Mathematical Problems in Engineering*, vol. 2018, p. 16, 15 Oct. 2018.
98. Anuradha, Jeetendra Singh, B. Raj and Mamta Khosla, “Design and performance analysis of nano-scale memristor-based nonvolatile SRAM”, *Journal of Sensor Letter*, vol. 16, pp. 798–805, Oct. 2018.
99. Girish Wadhwa and B. Raj, “Parametric variation analysis of charge-plasma-based dielectric modulated JLTFET for biosensor application”, *IEEE Sensor Journal*, vol. 18, no. 15, pp. 6070–6077, 1 Aug. 2018.
100. Jeetendra Singh and B. Raj, “Comparative analysis of memristor models for memories design”, *Journal of Semiconductors*, vol. 39, no. 7, pp. 074006-1–12, July 2018.
101. Divya Yadav, Shailesh Singh Chouhan, Santosh Kumar Vishvakarma and B. Raj, “Application specific microcontroller design for IoT based WSN”, *Sensor Letter*, vol. 16, pp. 374–385, May 2018.
102. Gurmohan Singh, R. K. Sarin and B. Raj, “Fault-Tolerant design and analysis of quantum-dot cellular automata based circuits”, *IEEE/IET Circuits, Devices & Systems*, vol. 12, pp. 638–64, 2018.
103. Jeetendra Singh and B. Raj, “Modeling of mean barrier height levying various image forces of metal insulator metal structure to enhance the performance of conductive filament based Memristor model”, *IEEE Nanotechnology*, vol. 17, no. 2, pp. 268–267, March 2018.
104. Aakash Jain, Sanjeev Sharma and B. Raj, “Analysis of triple metal surrounding gate (TM-SG) III–V nanowire MOSFET for photosensing application”, *Opto-Electronics Journal*, vol. 26, no. 2, pp. 141–148, May 2018.
105. Aakash Jain, Sanjeev Sharma and B. Raj, “Design and analysis of high sensitivity photosensor using cylindrical surrounding gate MOSFET for low power sensor applications”, *Engineering Science and Technology, an International Journal*, vol. 19, no. 4, pp. 1864–1870, Dec. 2016.
106. G. Saiphani Kumar, Amandeep Singh and B. Raj, “Design and analysis of gate all around CNTFET based SRAM cell design”, *Journal of Computational Electronics*, vol. 17, no. 1, pp. 138–145, March 2018.
107. Gurinder pal Singh, B. S. Sohi and Balwinder Raj, “Material properties analysis of graphene base transistor (GBT) for VLSI analog circuits”, *Indian Journal of Pure & Applied Physics*, vol. 55, pp. 896–902, Dec. 2017.
108. Sunil Kumar and B. Raj, “Estimation of stability and performance metric for inward access transistor based 6T SRAM cell design using n-type/p-type DMDG-GDOV TFET”, *IEEE VLSI Circuits and Systems Letter*, vol. 3, no. 2, pp. 25–39, June 2017.
109. Shashikant Sharma, Anjan Kumar, Manisha Pattanaik and B. Raj, “Forward body biased multimode multi-threshold CMOS technique for ground bounce noise reduction in static CMOS adders”, *International Journal of Information and Electronics Engineering*, vol. 3, no. 3, pp. 567–572, 2013.

110. Hamendra Singh, Pankaj Kumar and Balwinder Raj, "Performance analysis of majority gate SET based 1-bit full adder", *International Journal of Computer and Communication Engineering*, vol. 2, no. 4, pp. 1–8, 2013.
111. Anil Kumar Bhardwaj, Sumeet Gupta and B. Raj, "Investigation of parameters for Schottky Barrier (SB) height for Schottky barrier based carbon nanotube field effect transistor device", *Journal of Nanoelectronics and Optoelectronics*, vol. 15, pp. 783–791, July 2020.

Chapter 9

Variability in tunnel field-effect transistor

Ashish Maurya

9.1 INTRODUCTION

Tunnel field-effect transistor (TFET) is a gated p–i–n designed device, which is the most fundamental definition of its structure. The fact that the fabrication of TFETs is comparable to that of regular metal–oxide-semiconductor field-effect transistors (MOSFETs) gives them an edge over other newly developed devices [1]. With one key exception, the device architectures of a TFET and a MOSFET are similar. Unlike MOSFETs, which have both the source and the drain doped with the same sort of dopant, TFETs have their source and drain doped with distinct types of dopants [1–6]. Figures 9.1 and 9.2 depict potential n- and p-type single-gate TFETs (SG-TFETs).

TFETs operate by exploiting electron tunneling as a means of carrier transport as opposed to more traditional techniques. The idea of three-terminal devices originated with the band-to-band tunneling (BTBT) component included in a trench transistor cell. A three-terminal tunnel device utilizing this phenomenon was invented in 1987 by Sanjay Banerjee of Texas Instruments [4]. This device required a phenomenon called line tunneling or gate overlap of the source [7].

9.1.1 Tunnel FET

Because there is no temperature (kT) dependency on TFETs, the SS can be steeper than 60 mV/dec [7]. When functioning, TFETs employ the BTBT mechanism to establish the current flow. This leads to the production of a narrow energy barrier and a huge electric field. The TFET's sub-threshold behavior differs significantly from that of the MOSFET [1–4, 6].

The double-gate TFET (DG-TFET) offers effective gate control due to the presence of dual gates. The electric field acts to compress the gap between the source and the channel of a tunnel FET. This occurs as a result of the barrier between the source and the channel having less breadth. This leads to the tunneling effect, which enables the electrons to pass through the barrier and for current to pass through the transistor [7–9]. TFET in

 DOI: 10.1201/9781003612797-9

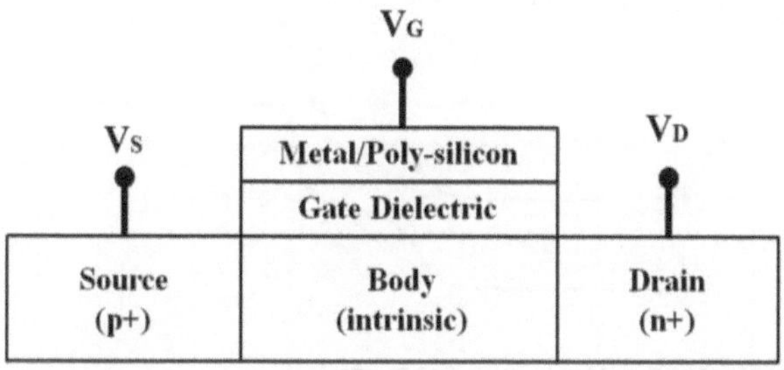

Figure 9.1 n-type TFET.

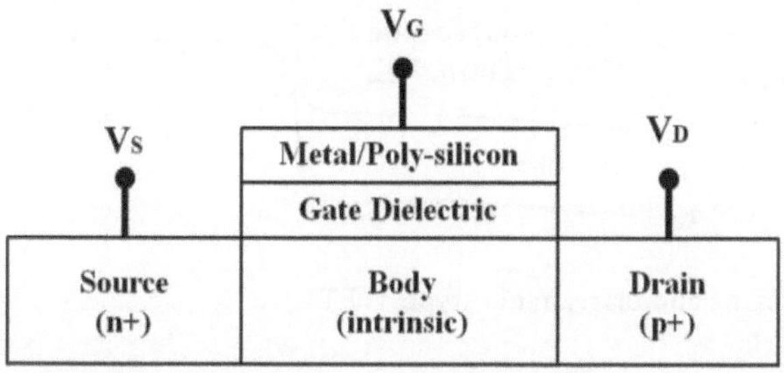

Figure 9.2 p-type TFET.

OFF-state, ON-state, and ambipolar state are shown in Figures 9.3, 9.4, and 9.5 respectively.

9.1.2 Characteristics of TFET

Figure 9.6 shows the drain current characteristics of MOSFET and TFET. It is clear from these characteristics that the TFET has a steeper sub-threshold slope than the MOSFET. But a TFET's I_{ON} is smaller than a MOSFET's [10].

9.1.3 Variability

Extrinsic sources of variability are mostly likely variations that show up from one chip or wafer to the next and are brought on by strain and modifications to the architecture of the device. Wafer-to-wafer, wafer-level, die-level, layout-dependent, and other variations are caused by significant process variances in the production of integrated circuits [11–14]. One important factor influencing the device's variability is doping variation or fluctuation, also referred to as random dopant fluctuation (RDF), which is produced by impurity occupancy occurring randomly in the device during manufacture [15–17]. A schematic diagram of randomly distributed dopant atoms is presented in Figure 9.7 along with the Gaussian features considered

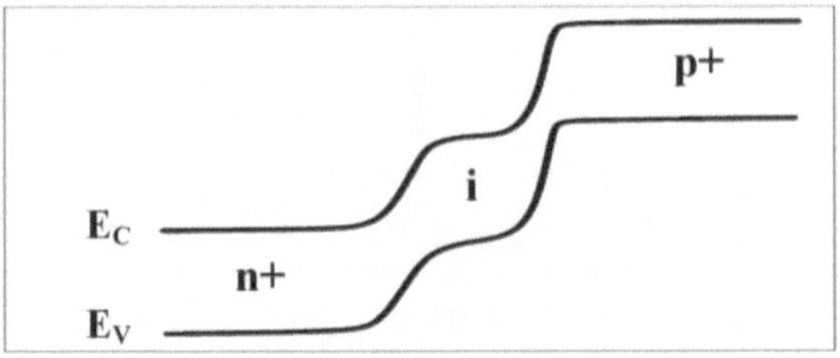

Figure 9.3 OFF-state band diagram in n-type TFET.

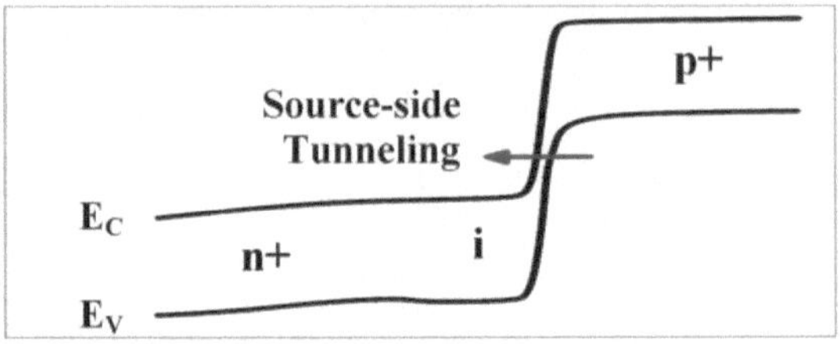

Figure 9.4 ON-state band diagram in n-type TFET.

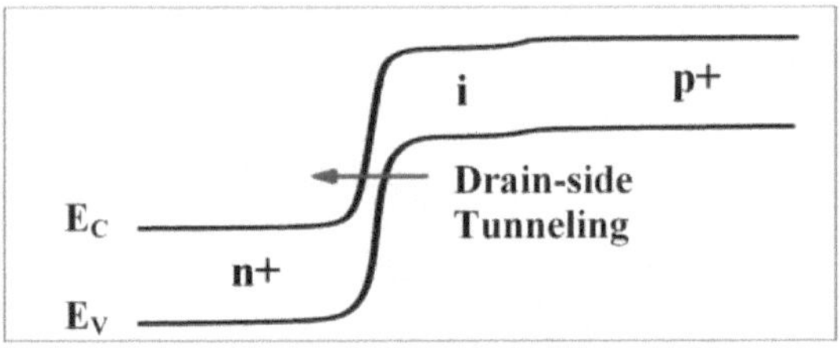

Figure 9.5 Ambipolar-state band diagram in n-type TFET.

for the atom distribution. RDF is produced at wafer surfaces that are not controllable during the substrate doping process via ion implantation [18, 19].

Kinetic Monte Carlo, an atomistic method, is the typical approach to analyze variability in semiconductor devices for examining the impact of RDF [21–23]. Furthermore, RDF is generated in devices to analyze variability using the analytical methodology [24] and the impedance field method (IFM) [25]. But for RDF modeling in devices, the Sano technique is more realistic since it takes into account the dopant atoms that generate long- and short-range Coulomb interactions, which result in screening between each dopant [26, 27].

It eliminates the singularity issue that develops when the drift-diffusion simulator's mesh element sizes are exceeded by the mean distance of the discrete dopants. It is important to choose the screening factor k_c so that

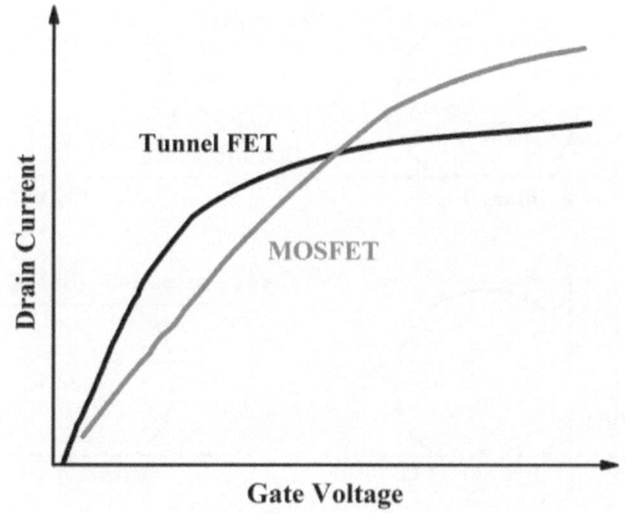

Figure 9.6 Drain current characteristics of TFET and MOSFET [26].

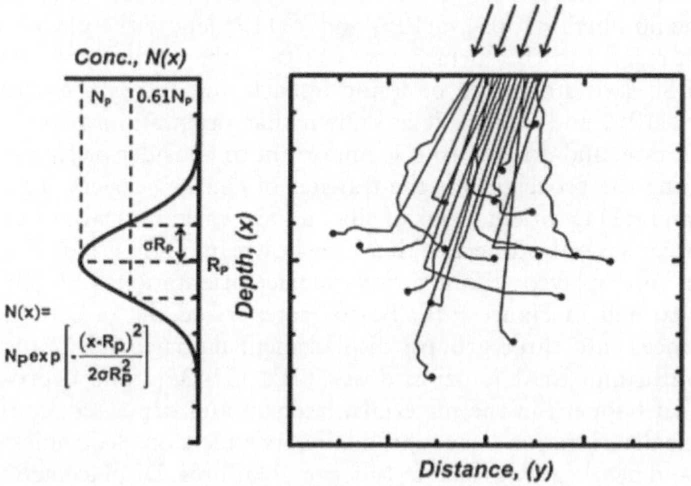

Figure 9.7 Random distributed dopant profile [20].

the carriers experience the charge density (N_a) from the dopants acting on their drift motion instead of scatterings [26–28].

$$k_c = 2 * N_a^{1/3} \tag{9.1}$$

Si wafers are typically used to describe single crystals of Si with specific orientations, dopant types, and resistivity (which are based on dopant

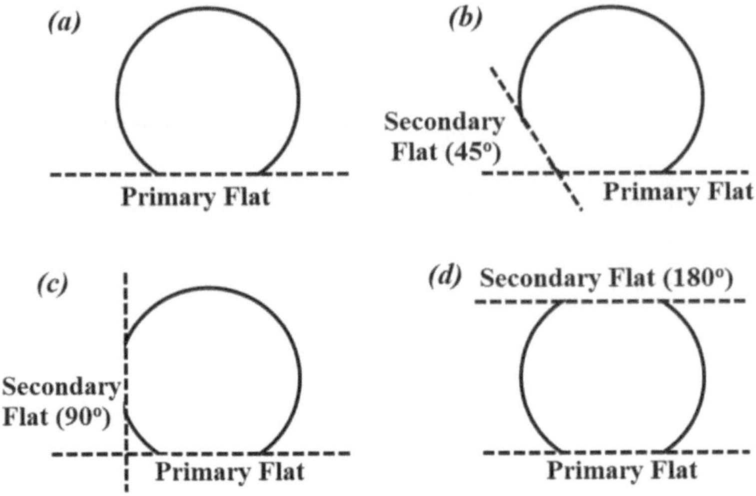

Figure 9.8 Wafer flats for (a) p<111>, (b) n<111>, (c) p<100>, and (d) n<100> [29].

concentrations). Most often, wafers composed of Si <100> or Si <111> are used. The numbers <100>, <110>, and <111> denote the plane's orientation with respect to the surface and <110> is the second potential orientation of Si. Two atoms can be found in each unit cell of the orientations <100>, <110>, and <111>. It is known that orientation affects etching, oxidation rate, and doping, so it is important to consider orientation when determining the processes for the transfer of charge carriers. The <100>, <110>, and <111> orientations of silicon have varied surface orientations, which affect several properties that are shown in Table 9.1. In Figure 9.9, length "a" is displayed within various surface orientations [29–33].

As illustrated in Figure 9.10, Bastos categorized the radiation-induced consequences into three groups: displacement damage (DD), single-event effect (SEE), and total ionizing dose (TID) [32–35]. The deprivation in energy that happens in the materials, built up and deposited, causes TID. While this deterioration occurs gradually, its effects on electronics are irreversible and nearly always lead to functional failures. Displacement damage (DD) is a type of gradual device deterioration that stems from an alternative

Table 9.1 Parameters for different surface orientations

Surface orientation	*Bond length (Å)*	*Silicon surface atom density (/cm²)*	*Effective mass*	*Tunneling mass*
<100>	1.36	6.78×10^{14}	$0.19m_0$	$0.087m_0$
<110>	1.92	9.59×10^{14}	$0.19m_0$	$0.087m_0$
<111>	0.78, 2.35	7.83×10^{14}	$0.26m_0$	$0.099m_0$

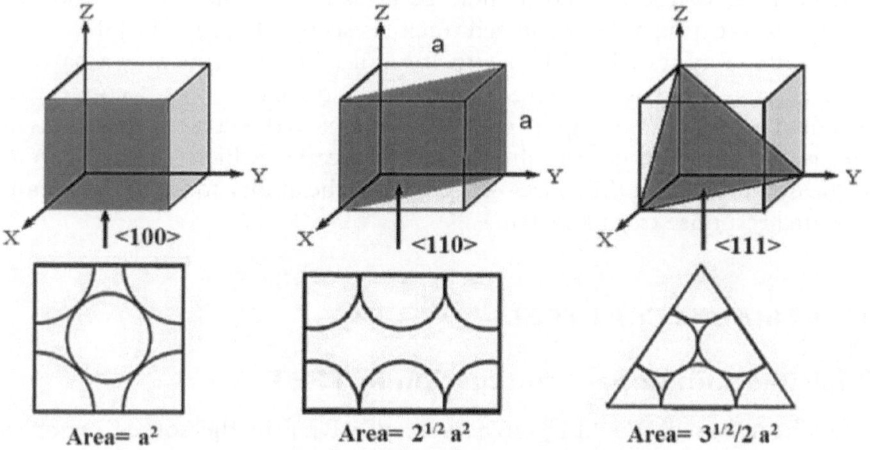

Figure 9.9 Unit cells of Si<100>, Si<110>, and Si<111> planes [30].

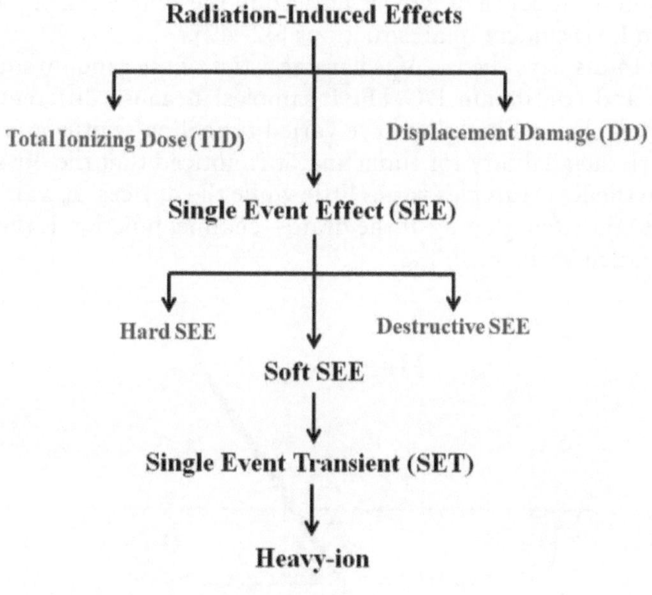

Figure 9.10 Radiation induction flow graph [31].

physical mechanism. SEE happens when a single ion strikes the device's material with enough energy to cause a systemic breakdown, as compared to TID and DD [35–37].

All ions with an atomic number more than two are referred to as heavy ions. The effect of a single heavy ion colliding with a silicon device is the loss of energy due to the creation of a stream of free electron–hole pairs.

The creation of these electron–hole pairs causes the depletion region to enlarge by creating a dense ionized track, as seen in Figure 9.11 [38].

Linear energy transfer (LET) is the term used to describe how a particle loses energy. A particle's charge deposition and LET are easily connected. Anytime a particle's energy is enough to increase the current in a transistor node, a transient current disturbance is created, which is referred to as single event transient (SET) [39–41]. SET has the ability to trigger an unanticipated response from a system.

9.2 VARIABILITY IN TFET

9.2.1 Random dopant fluctuation in TFET

The silicon-based DG-TFET structure with RDF in the source region is shown in Figure 9.12.

RDF is taken into account collectively in all regions and separately in the source, drain, and body regions. The bar chart in Figure 9.13 illustrates how the largest influence of RDF is noticed in the source region, where the variation in I_{on} is evident in all situations [42–47].

Figure 9.14 displays the $I_d - V_{gs}$ characteristics of the random source dopant (RSD) and continuum DG-TFET samples. Because different samples at the source–channel junction have varied tunneling widths and different charge carrier availability for tunneling, it is noticed that the off-state current (I_{off}) in the RSD samples varies little while the devices' I_{on} varies significantly. The consistent doping at the drain--channel junction is the cause of the insignificant variance in I_{off}.

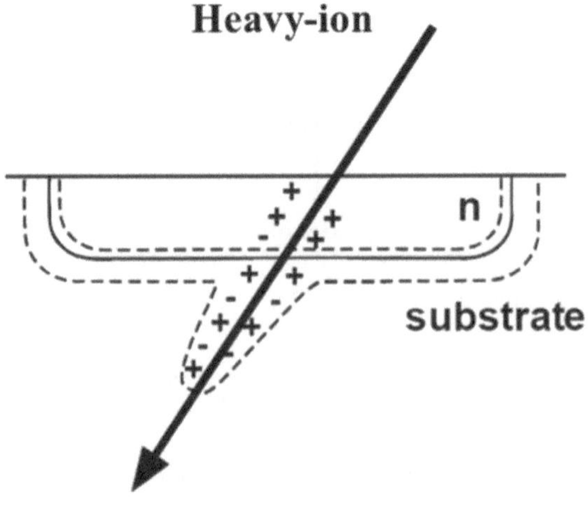

Figure 9.11 Heavy-ion strike [38].

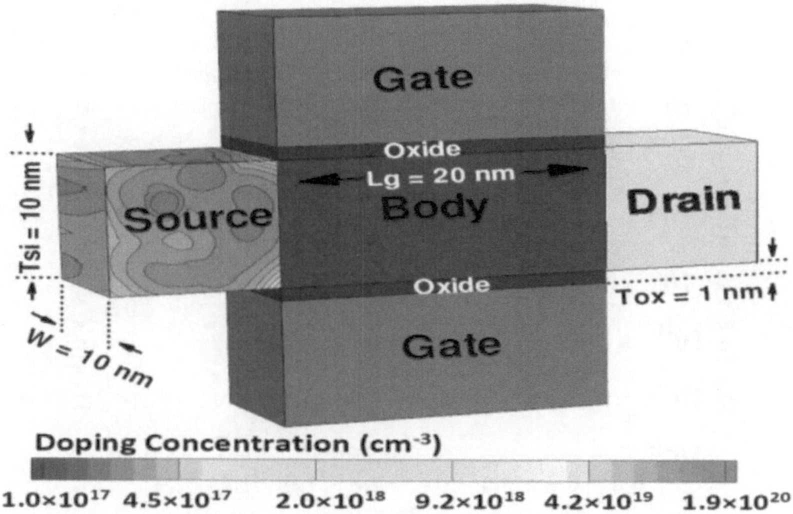

Figure 9.12 Sample of Si DG-TFET having RDF in the source region.

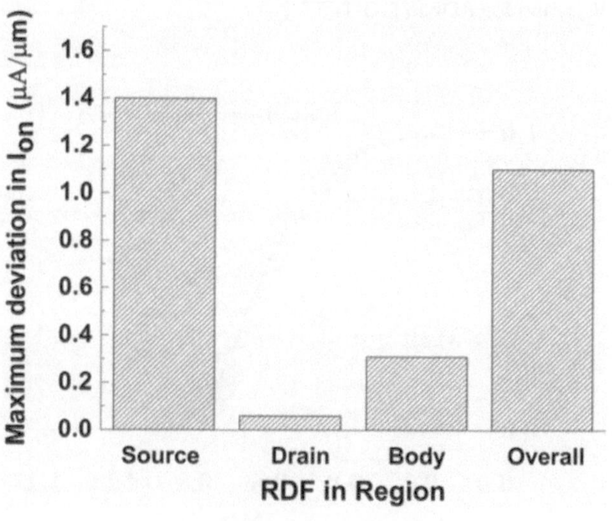

Figure 9.13 RDF-induced maximum deviation in I_{on}.

The analog and RF parameters as a result of RDF are used to study the common source (CS) amplifier circuit. nTFET and resistance R_d, which serves as a load, make up the CS amplifier circuit depicted in Figure 9.15. Complementary to the drain current variation (Figure 9.14) resulting from random source dopant, Figure 9.15 displays the variation in the VTC. This variance in VTC indicates that there may be significant variations in the circuit performance for various RDF device samples [48–55].

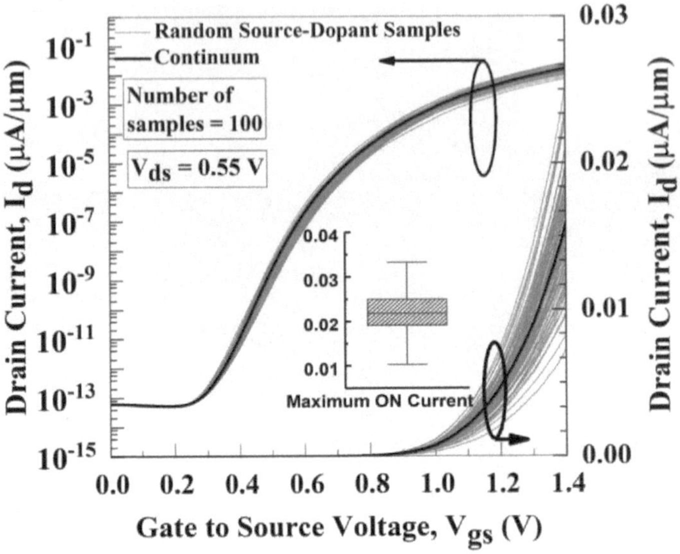

Figure 9.14 I_d–V_{gs} graph for RDF in DG-TFET.

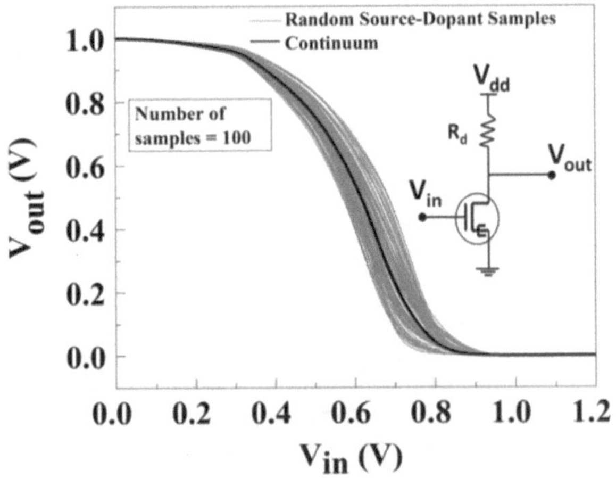

Figure 9.15 Voltage transfer characteristics of CS amplifier with resistive load.

The performance variation for the CS amplifier circuit by gain and gain-bandwidth product (GBWP) for all RDF device samples and the continuum device is displayed in Figure 9.16. Given the CS amplifier circuit [42], the transfer function is as follows:

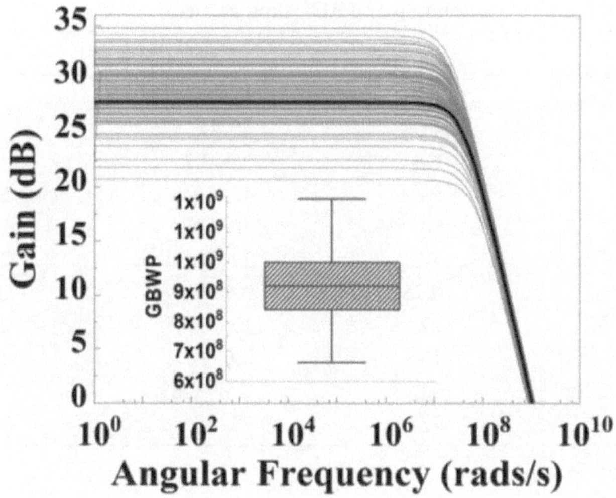

Figure 9.16 Gain and gain-bandwidth product.

$$\frac{V_{out}(s)}{V_{in}(s)} = \frac{(sC_{gd} - g_m)R_d}{R_sR_dC_{gs}C_{gd}s^2 + \left[R_s(1+g_mR_d)C_{gd} + R_sC_{gs} + R_dC_{gd}\right]s + 1} \tag{9.2}$$

Having dominant pole at:

$$\omega_{in} = \frac{1}{R_s\left[C_{gs} + (1+g_mR_d)C_{gd}\right]} \tag{9.3}$$

Table 9.2 displays the range of variability and the average value of the analyzed parameters for the RF and analogue FoMs. The computed average values for 100 RDF devices are displayed with the variation in values for RDF affected devices [56–62].

9.2.2 Orientation in TFET

The orientation of a surface influences the arrangement of atoms and lattices at its surface, which determines the direction of current flow. Analyzing various electrical and physical characteristics at the source–channel junction is crucial because tunneling in tunnel FETs occurs from the source to the channel area. Simulations are performed using Kane's model [43] for device performance analysis in order to calculate the generation rate (G_r), which is given by:

Table 9.2 Parameters for continuum and RDF simulations

	Continuum	*RDF (Variation)*	*RDF (Average)*
I_{on} ($\times10^{-2}$) (μA/μm)	1.9	1–3.4	2
SS_{min} (mV/dec)	41.5	40-47	44
g_m ($\times10^{-1}$) (μS/μm)	1.8	0.8–3.2	1.9
g_m/i_d (1/V)	54	48–56	51
R_0 (MΩ)	50	10–100	50
g_mR_0	27	22–28	25
C_{gg} (fF)	0.95	0.94–0.962	0.955
τ_m (nsec)	22	15–32	23
f_T (GHz)	1	0.6–4	1
V_{in} (V) at V_{out} = 0.5V	0.6	0.57–0.69	0.59
Gain (dB)	27	22–35	29

$$G_r = A_{ind}\left(\frac{E}{E_0}\right)^P \exp\left(-\frac{B_{ind}}{E}\right) \tag{9.4}$$

where E_0 = 1 V/cm and P = 2.5 for indirect BTBT.

The variance in the density of Si atoms on the surface in distinct cuts accounts for the observed variation in the G_r as shown in Figure 9.17. The fluctuation in the effective mass and decreased tunneling mass are caused by these variations in surface orientation. The effective mass (where m_0 is the free electron mass) for <100>, <110>, and <111> is computed to be 0.19 m_0 and 0.259 m_0, respectively. Because the atoms' surface space distribution is the same for square and rectangle, the effective mass for <100> and <110> is the same. Reduced tunneling mass (m_r), which is 0.087 m_0 for <100> and <110> and 0.099 m_0 for <111>, is a crucial BTBT parameter. Each of these variables has an impact on the BTB generation factor [63–67].

The drain current differs for <100>, <110>, and <111> as well because of the variations in the G_r for the various orientations. As seen in Figure 9.18, a larger G_r indicates the availability of more charge carriers, which in turn leads to a higher tunneling current. I_{on}/I_{off} on the displayed I_d–V_{gs} graph is of order 10^8, and I_d follows the G_r trend, meaning that I_{on} has order (<100>)> (<111>)> (<110>). About twice as much I_{on} is present for <100> to <110> and 1.5 times more to <111>.

The energy band diagram of Figure 9.19 shows the variation in tunneling barrier width (W_t), which explains the change in I_d. The device's tunneling probability, which is reliant on tunneling width, determines the I_{on}. The inset of Figure 9.19 illustrates how the variance of I_{on} is caused by the W_t being greater for <110> and smaller for <100>.

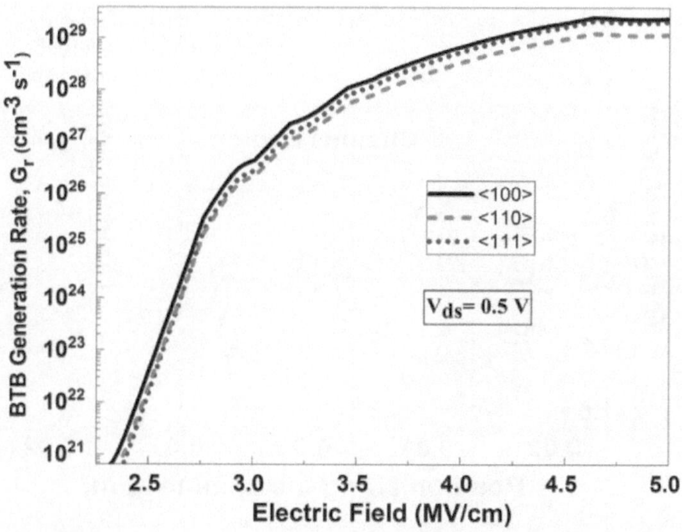

Figure 9.17 BTB generation rate for different orientations at V_{ds} = 0.5 V.

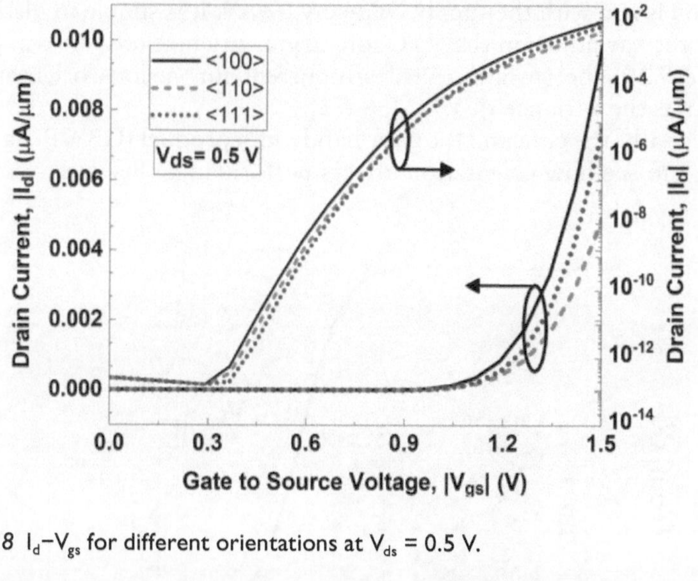

Figure 9.18 I_d–V_{gs} for different orientations at V_{ds} = 0.5 V.

Analyzing the device performance for orientations <100>, <110>, and <111> is done using RF and analog performance metrics in the preceding sections. Simulations of common source (CS) amplifier and single-ended output differential amplifier circuits are used to examine how orientation affects circuit-level performance. Resistance R, which serves as a load, and a double-gate pTFET (DG-pTFET) comprise the CS amplifier circuit seen

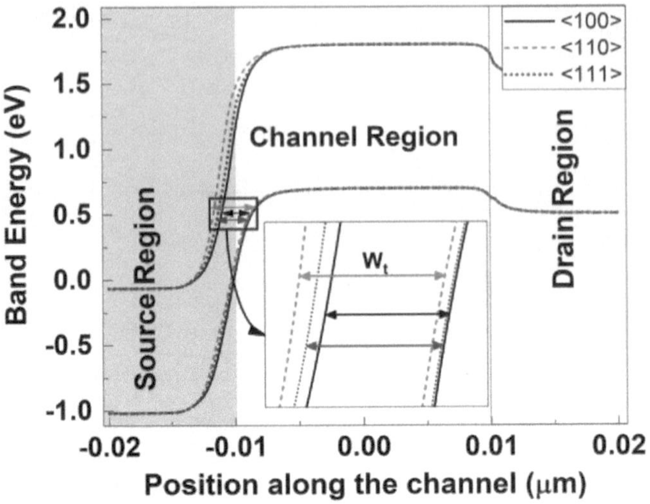

Figure 9.19 Energy band for different orientations.

in Figure 9.20. The input voltage V_{in} varies between 0 and 1V, and the simulation is run with the supply voltage V_{dd} = 1V. It is simple to distinguish between the variations in the VTC for various orientations. As can be seen in Figure 9.20, the variation in drain current for various orientations is what causes the variance in VTC [68–73].

Additionally, the gain and the gain-bandwidth product (GBWP) have been computed to see how orientation affects performance. Because orientation

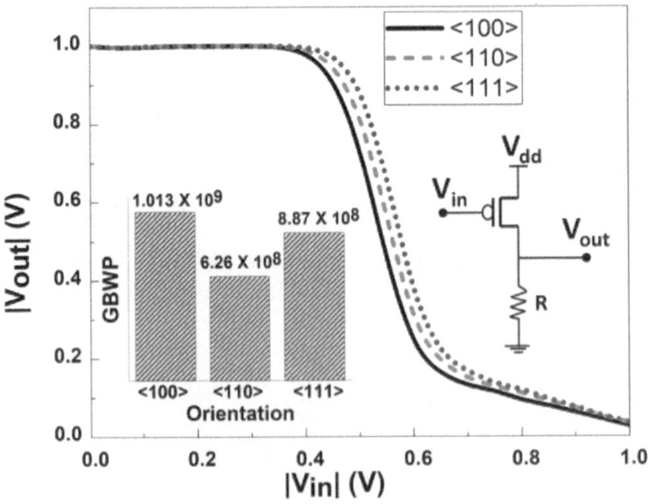

Figure 9.20 Voltage transfer characteristics of CS amplifier.

<100> has a larger 3dB bandwidth than the other simulated orientations, it gets the maximum value of GBWP [74–81]. According to the acquired data, orientation <100> performs better in the circuit than orientations <110> and <111>.

9.2.3 Radiation in TFET

The compatibility of the ion-track radius and the associated LET values should be confirmed while establishing the heavy-ion model [82–97]. Additionally, the radius should be defined so that it encompasses the device's sensitive region. The heavy-ion model of Sentaurus TCAD [43] incorporates the physics of the heavy-ion incident into the device in a way that makes it resemble genuine heavy ions in terms of characteristics. The heavy-ion striking position is indicated by arrows in Figure 9.21.

This chapter has examined the heavy-ion strike at five distinct positions: the source region, the source–body junction, the body region, the drain–body junction, and the drain region. As seen in Figure 9.21, the ions are aimed to strike vertically at x-coordinates of 5, 10, 20, 30, and 35 nm. To ensure correct generation, 3D device simulations are used to analyze the impact of heavy ions. The device's meshing is constructed so that the mesh spacing is modest relative to the ion-track width [98–117].

The drain current graph shown in Figure 9.22 shows that at LET = 20 MeV-cm^2/mg, the body is most susceptible to heavy ions [118–123]. After observation, it becomes crucial to examine how the drain current varies for various LETs at various places. The circled area indicates the increment in drain current with an increase in LET, which is larger at location x = 20 nm than it is at the other positions. In addition, heavy-ion production, charge

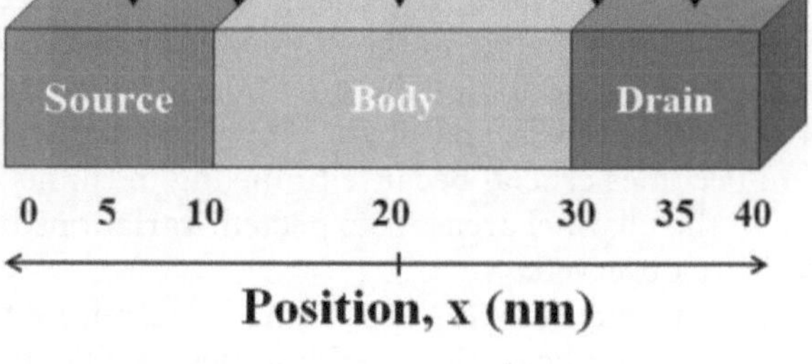

Figure 9.21 Heavy-ion strike tracks.

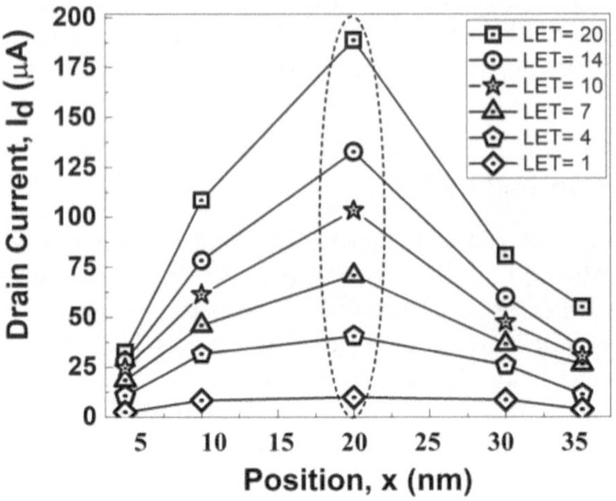

Figure 9.22 Variation in maximum drain current for different LETs.

density, transient drain current, collected charge, and bipolar gain can also be analyzed [124–133].

9.3 SUMMARY

Prior to releasing the device for commercial usage, the device's performance may be assessed at the fundamental levels: device and circuits. Within this framework, DG-TFET's behavior is examined that may have arisen during the fabrication processes. The substrate's orientation is crucial as it can influence how the device behaves differently. Based on observations for RDF, it appears that TFETs face a major difficulty with the RDF. After determining the ON-current deviation for each RDF separately in each region, the occurrence of RDF is limited to the source region. According to the source region RDF findings, the ion varies most (58%) and least (53%) in relation to the value of the uniformly doped device.

The orientation of a semiconductor surface influences the arrangement of atoms and lattices. Sometimes, the change in the arrangement of the lattice at its surface causes change in the physical parameters of materials. Since atom arrangement determines the direction of current flow as well, the analysis of various electrical and physical characteristics at the source–channel junction becomes crucial because tunneling in tunnel FETs occurs from the source to the channel area. As expected, variations in the characteristics of TFET are observed.

All ions with an atomic number more than 2 are referred to as heavy ions. The effect of a single heavy ion colliding with a silicon device is the loss of energy due to the creation of a stream of free electron–hole pairs.

Therefore, the radiation analysis is conducted by hitting the heavy ions vertically at five distinct sites. The effect of heavy ions on DG-TFET is examined, and the body area is determined to be the most sensitive of them all.

REFERENCES

1. A. Seabaugh and Q. Zhang, "Low-voltage tunnel transistors for beyond CMOS logic", *Proceedings of the IEEE*, vol. 98, no. 12, pp. 2095–110, 2010.
2. A. M. Ionescu and H. Riel, "Tunnel field-effect transistors as energy-efficient electronic switches", *Nature*, vol. 479, no. 7373, pp. 329–37, 2011.
3. H. Lu and A. Seabaugh, "Tunnel field-effect transistors: state-of-the-art", *IEEE Journal of Electron Devices Society*, vol. 2, no. 4, pp. 44–49, 2014.
4. D. Verreck, G. Groeseneken and A. Verhulst, "The tunnel field-effect transistor", *Wiley Encyclopedia of Electrical and Electronics Engineering*, International Journal of Engineering Research & Technology (IJERT), pp. 1–24, 2016
5. H. H. Radamson, *et al.*, "Miniaturization of CMOS", *Micromachines*. vol. 10, no. 5, 2019.
6. L. Zhang and M. Chan, *Tunneling Field Effect Transistor Technology*, ch. 2. pp. 33-60, Springer, 2016.
7. W. Reddick and G. Amaratunga, "Silicon surface tunnel transistor", *Applied Physics Letters*, vol. 67, no. 4, pp. 494–496, Jul. 1995.
8. P. F. Wang, K. Hilsenbeck, T. Nirschl, M. Oswald, C. Stepper, M. Weis, D. Schmitt-Landsiedel and W. Hansch, "Complementary tunneling transistor for low power application", *Solid-State Electronics*, vol. 48, no. 12, pp. 2281–2286, Dec. 2004.
9. W. Cheng, *et al.*, "Fabrication and characterization of a novel Si line tunneling TFET with high drive current", *IEEE Journal of the Electron Devices Society*, vol. 8, pp. 336–340, 2020.
10. D. S. Schulman, A. J. Arnold and S. Das, "Contact engineering for 2D materials and devices", *Chemical Society Reviews*, vol. 47, no. 9, pp. 3037–3058, 2018.
11. K. Takeuchi, A. Nishida and T. Hiramoto, "Random fluctuations in scaled MOS devices", *2009 International Conference on Simulation of Semiconductor Processes and Devices*, San Diego, CA, pp. 1–7, 2009.
12. N. Damrongplasit, C. Shin, S. H. Kim, R. A. Vega and T. J. K. Liu, "Study of random dopant fluctuation effects in germanium-source tunnel FETs", *IEEE Transactions on Electron Devices*, vol. 58, no. 10, pp. 3541–3548, 2011.
13. Y. Zhul, Y. Yel, Y. Caol, J. Hel, A. Zhangl, H. Hel, H. Wangl, C. Mal, Y. Hul, M. Chan and X. Zhu, "Numerical study on effects of random dopant fluctuation in double gate tunneling FET", *2013 IEEE International Conference of Electron Devices and Solid-state Circuits*, Hong Kong, pp. 1–2, 2013.
14. H. Nam, G. S. Lee, H. Lee, I. J. Park and C. Shin, "Analysis of random variations and variation-robust advanced device structures", *Journal of Semiconductor Technology and Science*, vol. 14, no. 1, pp. 8–22, 2014.
15. F. Yang, "Discrete dopant fluctuated 20nm/15nm-gate planar CMOS", *IEEE Symposium on VLSI Technology*, Kyoto, pp. 208–209, 2007.

16. A. Asenov, "Simulation of statistical variability in nano MOSFETs", *IEEE Symposium on VLSI Technology*, Kyoto, pp. 86–87, 2007.
17. K. Boucart and A. M. Ionescu, "Double-gate tunnel FET with high-k gate dielectric", *IEEE Transactions on Electron Devices*, vol. 54, no. 7, pp. 1725–1733, July 2007.
18. J. D. Plummer, M. D. Deal and P. B. Griffin, *Silicon VLSI Technology: Fundamentals, Practice and Modeling,* New Jersey: Prentice Hall, 2000.
19. R. C. Jaeger, *Introduction to Microelectronic Fabrication,* 2nd edn., New Jersey: Prentice Hall, 2002.
20. C. Shin, *Variation-Aware Advanced CMOS Devices and SRAM*, Springer Series in Advanced Microelectronics, Germany, 2016.
21. A. Asenov, "Random dopant induced threshold voltage lowering and fluctuations in sub-0.1 μm MOSFET's: a 3-D 'atomistic' simulation study", *IEEE Transactions on Electron Devices*, vol. 45, no. 12, pp. 2505–2513, Dec. 1998.
22. H. S. Wong and Y. Taur, "Three-dimensional "atomistic" simulation of discrete random dopant distribution effects in sub-0.1 μm MOSFET's", *Proceedings of IEEE International Electron Devices Meeting*, Washington, DC, USA, pp. 705–708, 1993.
23. D. J. Frank, Y. Taur, M. Ieong and H. -P. Wong, "Monte Carlo modeling of threshold variation due to dopant fluctuations", *1999 Symposium on VLSI Technology. Digest of Technical Papers*, Kyoto, Japan, pp. 169–170, 1999.
24. D. Reid, C. Millar, G. Roy, S. Roy and A. Asenov, "Analysis of threshold voltage distribution due to random dopants: a 100000-sample 3-D simulation study", *IEEE Transactions on Electron Devices*, vol. 56, no. 10, pp. 2255–2263, Oct. 2009.
25. G. Giusi and A. Lucibello, "Variability of the drain current in junctionless nanotransistors induced by random dopant fluctuation", *IEEE Transactions on Electron Devices*, vol. 61, no. 3, pp. 702–706, March 2014.
26. N. Sano, K. Matsuzawa, M. Mukai and N. Nakayama, "On discrete random dopant modeling in drift-diffusion simulations: physical meaning of 'atomistic' dopants", *Microelectronics Reliability*, vol. 42, no. 2, pp. 189–99, Feb. 2002.
27. N. Sano, K. Matsuzawa, M. Mukai and N. Nakayama, "Role of long-range and short-range Coulomb potentials in threshold characteristics under discrete dopants in sub-0.1 μm Si-MOSFETs", *International Electron Devices Meeting 2000*, San Francisco, CA, USA, pp. 275–278, 2000.
28. N. Sano and T. Fukui, "Fundamental aspects of semiconductor device modeling associated with discrete impurities: Monte Carlo simulation scheme", *IEEE Transactions on Electron Devices*, vol. 68, no. 11, pp. 5394–5399, Nov. 2021.
29. J. J. Quinn, G. Kawamoto and B. D. McCombe, "Subband spectroscopy by surface channel tunneling", *Surface Science*, vol. 73, pp. 190–196, May 1978.
30. C. Xiao, J. Guo, P. Zhang, *et al.*, "Effect of crystal plane orientation on tribochemical removal of monocrystalline silicon", *Scientific Reports* vol. 7, p. 40750, 2017.
31. S. Duzellier, "Radiation effects on electronic devices in space", *Aerospace Science and Technology,* vol. 9 no. 1, pp. 93–99, 2005.

32. National Aeronautics and Space Administration (NASA), Radiation effects and analysis. http://radhome.gsfc.nasa.gov/top.htm.
33. D. M. Fleetwood et al., "An overview of radiation effects on electronics in the space telecommunications environment", *Microelectronics Reliability*, vol. 40, no. 1, pp. 17–26, 2000.
34. C. Lazzari. "Transistor level automatic generation of radiation-hardened circuits. micro and nanotechnologies/microelectronics", *Institute National Polytechnique de Grenoble – INPG*, pp. 1–10, 2007.
35. K. Castellani-Coulie, D. Munteanu, V. Ferlet-Cavrois and J. Autran, "Simulation analysis of the bipolar amplification in fully-depleted SOI technologies under heavy-ion irradiations", *IEEE Transactions on Nuclear Science*, vol. 52, no. 5, pp. 1474–1479, Oct. 2005.
36. F. A. Ikraiam, "An analysis of radiation effects on electronics and SOI-MOS devices as an alternative", *Proceedings of International Conference on Sustainable Development through Nuclear Research and Education*, vol. 47, no. 1, pp. 6–14, 2013.
37. T. Colladant, O. Flament, A. L'Hoir, V. Ferlet-Cavrois, C. D'Hose and J. du Port de Potcharra, "Study of transient current induced by heavy-ion in NMOS/SOI transistors", *IEEE Transactions on Nuclear Science*, vol. 49, no. 6, pp. 2957–2964, Dec. 2002.
38. Q. Wang, H. Liu, S. Wang and S. Chen, "TCAD simulation of single-event-transient effects in L-shaped channel tunneling field-effect transistors", *IEEE Transactions on Nuclear Science*, vol. 65, no. 8, pp. 2250–2259, Aug. 2018.
39. Y. Wang, X. He and C. Shan, "A simulation study of SoI-like bulk silicon MOSFET with improved performance", *IEEE Transactions on Electron Devices*, vol. 61, no. 9, pp. 3339–3344, Sep. 2014.
40. G. Kaushal, S. S. Rathod, S. Maheshwaram, S. K. Manhas, A. K. Saxena and S. Dasgupta, "Radiation effects in Si-NW GAA FET and CMOS inverter: a TCAD simulation study", *IEEE Transactions on Electron Devices*, vol. 59, no. 5, pp. 1563–1566, May 2012.
41. P. Huang, S. Chen, J. Chen, B. Liang and Y. Chi, "Heavy-ion-induced charge sharing measurement with a novel uniform vertical inverter chains (UniVIC) SEMT test structure", *IEEE Transactions on Nuclear Science*, vol. 62, no. 6, pp. 3330–3338, Dec. 2015.
42. B. Razavi, *Design of Analog CMOS Integrated Circuits*, McGraw-Hill, New York, 2002.
43. *Sentaurus User's Manual*, Synopsys, Inc., Mountain View, CA, 2016
44. Abhishek Singh Tomar, Vijay Kumar Magraiya and B. Raj, "Scaling of access and data transistor for high performance DRAM cell design", *Quantum Matter*, vol. 2, pp. 412–416, Oct. 2013.
45. Neeraj Jain and B. Raj, "Parasitic capacitance and resistance model development and optimization of raised source/drain SOI FinFET structure for analog circuit applications", *Journal of Nanoelectronics and Optoelectronins*, vol. 13, pp. 531–539, April 2018.
46. Shradhya Singh, S. K. Vishvakarma and B. Raj, "Analytical modeling of split-gate junction-less transistor for a biosensor application", *Sensing and Bio-Sensing*, vol. 18, pp. 31–36, April 2018.

47. Maisagalla Gopal and Balwinder Raj, "Low power 8T SRAM cell design for high stability video applications", *ITSI Transaction on Electrical and Electronics Engineering*, vol. 1, no. 5, pp. 91–97, 2013.
48. Balwinder Raj, Jatin Mitra, Deepak Kumar Bihani, V. Rangharajan, A. K. Saxena and S. Dasgupta, "Analysis of noise margin, power and process variation for 32 nm FinFET based 6T SRAM cell", *Journal of Computer*, vol. 5, no. 6, pp. 1–8, 2010.
49. Divya Sharma, Rajesh Mehra and B. Raj, "Comparative analysis of photovoltaic technologies for high efficiency solar cell design", *Superlattices and Microstructures*, vol. 153, p. 106861, May 2021.
50. Pawandeep Kaur, Avtar Singh Buttar and B. Raj, "A comprehensive analysis of nanoscale transistor based biosensor: a review", *Indian Journal of Pure and Applied Physics*, vol. 59, pp. 304–318, April 2021.
51. Divya Yadav, Balwant Raj and B. Raj, "Design and simulation of low power microcontroller for IoT applications", *Journal of Sensor Letters*, vol. 18, pp. 401–409, May 2020.
52. Shailendra Singh and B. Raj, "A 2-D analytical surface potential and drain current modeling of double-gate vertical t-shaped tunnel FET", *Journal of Computational Electronics*, vol. 19, pp. 1154–1163, April 2020.
53. Jeetendra Singh and B. Raj, "An accurate and generic window function for non-linear Memristor model", *Journal of Computational Electronics*, vol. 18 no. 2, pp. 640–647, June 2019.
54. Manjit Kaur, Neena Gupta, Sanjeev Kumar, B. Raj and Arun Kumar Singh, "Comparative RF and crosstalk analysis of carbon based nano interconnects", *IET Circuits, Devices & Systems*, vol. 15, no. 6, pp. 493–503, Feb. 2021.
55. Nehru Kandasamy, Firdous Ahmad, D. Ajitha, B. Raj and Nagarjuna Telagam, "Quantum dot cellular automata based scan flip flop and boundary scan register", *IETE Journal of Research*, vol. 66, pp. 535–548, 2020.
56. Jeetendra Singh and B. Raj, "Design and investigation of 7T2M NVSARM with enhanced stability and temperature impact on store/restore energy", *IEEE Transactions on Very Large Scale Integration Systems*, vol. 27, no. 6, pp. 1322–1328, June 2019.
57. Anil Kumar Bhardwaj, Sumeet Gupta, B. Raj and Amandeep Singh, "Impact of double gate geometry on the performance of carbon nanotube field effect transistor structures for low power digital design", *Computational and Theoretical Nanoscience*, vol. 16, pp. 1813–1820, 2019.
58. Neeraj Jain and B. Raj, "Thermal stability analysis and performance exploration of asymmetrical dual-k underlap spacer (ADKUS) SOI FinFET for security and privacy applications", *Indian Journal of Pure & Applied Physics*, vol. 57, pp. 352–360, May 2019.
59. Neeraj Jain and Balwinder Raj, "Dual-k spacer region variation at the drain side of asymmetric SOI FinFET structure: performance analysis towards the analog/RF design applications", *Journal of Nanoelectronics and Optoelectronics*, vol. 14, pp. 349–359, March 2019.
60. Jeetendra Singh, Sanjeev Sharma, B. Raj and Mamta Khosla, "Analysis of barrier layer thickness on performance of In1-xGaxAs based gate stack cylindrical gate nanowire MOSFET", *Journal of Nanoelectronics and Optoelectronics*, vol. 13, pp. 1473–1477, Oct. 2018.

61. Neeraj Jain and B. Raj, "Analysis and performance exploration of high-k SOI FinFETs over the conventional low-k SOI FinFET toward analog/RF design", *Journal of Semiconductors*, vol. 39, no. 12, p. 124002-1–7, Dec. 2018.
62. Candy Goyal, Jagpal Singh Ubhi and B. Raj, "A reliable leakage reduction technique for approximate full adder with reduced ground bounce noise", *Journal of Mathematical Problems in Engineering*, vol. 2018, p. 16, 15 Oct. 2018.
63. Jeetendra Singh Anuradha, B. Raj and Mamta Khosla, "Design and performance analysis of nano-scale memristor-based nonvolatile SRAM", *Journal of Sensor Letter*, vol. 16, pp. 798–805, Oct. 2018.
64. Girish Wadhwa and B. Raj, "Parametric variation analysis of charge-plasma-based dielectric modulated JLTFET for biosensor application", *IEEE Sensor Journal*, vol. 18, no. 15, pp. 6070–6077, 1 Aug. 2018.
65. Jeetendra Singh and B. Raj, "Comparative analysis of Memristor models for memories design", *Journal of Semiconductors*, vol. 39, no. 7, p. 074006-1–12, July 2018.
66. Divya Yadav, Shailesh Singh Chouhan, Santosh Kumar Vishvakarma and B. Raj, "Application specific microcontroller design for IoT based WSN", *Sensor Letter*, vol. 16, pp. 374–385, May 2018.
67. Gurmohan Singh, R. K. Sarin and B. Raj, "Fault-tolerant design and analysis of quantum-dot cellular automata based circuits", *IEEE/IET Circuits, Devices & Systems*, vol. 12, pp. 638–64, 2018.
68. Jeetendra Singh and B. Raj, "Modeling of mean barrier height levying various image forces of metal insulator metal structure to enhance the performance of conductive filament based Memristor model", *IEEE Nanotechnology*, vol. 17, no. 2, pp. 268–267, March 2018.
69. Aakash Jain, Sanjeev Sharma and B. Raj, "Analysis of triple metal surrounding gate (TM-SG) III–V nanowire MOSFET for photosensing application", *Opto-Electronics Journal*, vol. 26, no. 2, pp. 141–148, May 2018.
70. Aakash Jain, Sanjeev Sharma and B. Raj, "Design and analysis of high sensitivity photosensor using cylindrical surrounding gate MOSFET for low power sensor applications", *Engineering Science and Technology, an International Journal*, vol. 19, no. 4, pp. 1864–1870, Dec. 2016.
71. G. Saiphani Kumar, Amandeep Singh and B. Raj, "Design and analysis of gate all around CNTFET based SRAM cell design", *Journal of Computational Electronics*, vol. 17, no. 1, pp. 138–145, March 2018.
72. Gurinder pal Singh, B. S. Sohi and Balwinder Raj, "Material properties analysis of graphene base transistor (GBT) for VLSI analog circuits", *Indian Journal of Pure & Applied Physics*, vol. 55, pp. 896–902, Dec. 2017.
73. Sunil Kumar and B. Raj, "Estimation of stability and performance metric for inward access transistor based 6T SRAM cell design using n-type/p-type DMDG-GDOV TFET", *IEEE VLSI Circuits and Systems Letter*, vol. 3, no. 2, pp. 25–39, June 2017.
74. Shashikant Sharma, Anjan Kumar, Manisha Pattanaik and B. Raj, "Forward body biased multimode multi-threshold CMOS technique for ground bounce noise reduction in static CMOS adders", *International Journal of Information and Electronics Engineering*, vol. 3, no. 3, pp. 567–572, 2013.

75. Hamendra Singh, Pankaj Kumar and Balwinder Raj, "Performance analysis of majority gate SET based 1-bit full adder", *International Journal of Computer and Communication Engineering*, vol. 2, no. 4, pp. 1–8, 2013.
76. Anil Kumar Bhardwaj, Sumeet Gupta and B. Raj, "Investigation of parameters for Schottky barrier (SB) height for Schottky Barrier based carbon nanotube field effect transistor device", *Journal of Nanoelectronics and Optoelectronics*, vol. 15, pp. 783–791, July 2020.
77. Priya Bansal and B. Raj, "Memristor: a versatile nonlinear model for dopant drift and boundary issues", *Journal of Computational and Theoretical Nanoscience*, vol. 14, no. 5, pp. 2319–2325, May 2017.
78. Neeraj Jain and B. Raj, "An analog and digital design perspective comprehensive approach on Fin-FET (Fin-Field Effect transistor) technology – a review", *Reviews in Advanced Sciences and Engineering*, vol. 5, pp. 1–14, 2016.
79. Balwinder Raj, A. K. Saxena and S. Dasgupta, "Analytical modeling for the estimation of leakage current and subthreshold swing factor of nanoscale double gate FinFET device", *Microelectronics International*, vol. 26, pp. 53–63, 2009.
80. Shailendra Singh Soniya, Girish Wadhwa and Balwinder Raj, "An analytical modeling for dual source vertical tunnel field effect transistor", *International Journal of Recent Technology and Engineering*, vol. 8 no. 2, pp. 603–608, July 2019.
81. Shailendra Singh and B. Raj, "Design and analysis of heterojunction vertical T-shaped tunnel field effect transistor", *Journal of Electronics Material*, vol. 48, no. 10, pp. 6253–6260, Oct. 2019.
82. Candy Goyal, Jagpal Singh Ubhi and B. Raj, "A low leakage CNTFET based inexact full adder for low power image processing applications", *International Journal of Circuit Theory and Applications*, vol. 47, no. 9, pp. 1446–1458, Sep. 2019.
83. B. Raj, A. K. Saxena and S. Dasgupta, "A compact drain current and threshold voltage quantum mechanical analytical modeling for FinFETs", *Journal of Nanoelectronics and Optoelectronics*, vol. 3, no. 2, pp. 163–170, 2008.
84. Girish Wadhwa and B. Raj, "An Analytical modeling of charge plasma based tunnel field effect transistor with impacts of gate underlap region", *Superlattices and Microstructures*, vol. 142, p. 106512, June 2020.
85. Shailendra Singh and B. Raj, "Modeling and simulation analysis of SiGe heterojunction double GateVertical t-shaped tunnel FET", *Superlattices and Microstructures*, vol. 142, p. 106496, June 2020.
86. Neeraj Jain and Balwinder Raj, "Device and circuit co-design perspective comprehensive approach on FinFET technology – a review", *Journal of Electron Devices*, vol. 23, no. 1, pp. 1890–1901, 2016.
87. Sunil Kumar and B. Raj, "Analysis of I_{ON} and ambipolar current for dual-material gate-drain overlapped DG-TFET", *Journal of Nanoelectronics and Optoelectronics*, vol. 11, pp. 323–333, June 2016.
88. Naveed Anjum, Tarun Bali and B. Raj, "Design and simulation of handwritten multiscript character recognition", *International Journal of Advanced Research in Computer and Communication Engineering*, vol. 2, no. 7, pp. 2544–2549, July 2013.

89. Karmjit Singh and B. Raj, "Performance and analysis of temperature dependent multi-walled carbon nanotubes as global interconnects at different technology nodes", *Journal of Computational Electronics*, vol. 14 no. 2, pp. 469–476, June 2015.
90. Sunil Kumar and B. Raj, "Compact channel potential analytical modeling of DG-TFET based on evanescent-mode approach", *Journal of Computational Electronics*, vol. 14, no. 2, pp. 820–827, July 2015.
91. Karmjit Singh and B. Raj, "Temperature dependent modeling and performance evaluation of multi-walled CNT and single-walled CNT as global interconnects", *Journal of Electronic Materials*, vol. 44, no. 12, pp. 4825–4835, Dec. 2015.
92. V. K. Sharma, M. Pattanaik and B. Raj, "INDEP approach for leakage reduction in nanoscale CMOS circuits", *International Journal of Electronics*, vol. 102, no. 2, pp. 200–215, 2014.
93. Karmjit Singh and B. Raj, "Influence of temperature on MWCNT bundle, SWCNT bundle and copper interconnects for nanoscaled technology nodes", *Journal of Materials Science: Materials in Electronics*, vol. 26, no. 8, pp. 6134–6142, 2015.
94. Naveed Anjum, Tarun Bali and B. Raj, "Design and simulation of handwritten Gurumukhi and Devanagri numerical recognition", *International Journal of Computer Applications*, vol. 73, no. 12, pp. 16–21, 2013.
95. S. Khandelwal, V. Gupta, B. Raj and R. D. Gupta, "Process variability aware low leakage reliable nano scale DG-FinFET SRAM cell design technique", *Journal of Nanoelectronics and Optoelectronics*, vol. 10, no. 6, pp. 810–817, Dec. 2015.
96. V. K. Sharma, M. Pattanaik and B. Raj, "ONOFIC approach: low power high speed nanoscale VLSI circuits design", *International Journal of Electronics*, vol. 101, no. 1, pp. 61–73, 2014.
97. S. Khandelwal, Balwinder Raj and R. D. Gupta, "FinFET based 6T SRAM cell design: analysis of performance metric, process variation and temperature effect", *Journal of Computational and Theoretical Nanoscience*, vol. 12, pp. 2500–2506, 2015.
98. Sumit Singh, Shekhar Yadav, Jagdeep Rahul, Anurag Srivastava and B. Raj, "Impact of HfO_2 in graded channel dual insulator double gate MOSFET", *Journal of Computational and Theoretical Nanoscience*, vol. 12, no. 6, pp. 950–953, April 2015.
99. Vijay Kumar Sharma, Manisha Pattanaik and B. Raj, "PVT variations aware low leakage INDEP approach for nanoscale CMOS Circuits", *Microelectronics Reliability*, vol. 54, pp. 90–99, 2014.
100. B. Raj, A. K. Saxena and S. Dasgupta, "Quantum mechanical analytical modeling of nanoscale DG FinFET: evaluation of potential, threshold voltage and source/drain resistance", *Journal of Material Science in Semiconductor Processing*, vol. 16, no. 4, pp. 1131–1137, 2013.
101. Maisagalla Gopal, Siva Sankar D. Prasad and Balwinder Raj, "8T SRAM cell design for dynamic and leakage power reduction", *International Journal of Computer Applications*, vol. 71, no. 9, pp. 43–48, June 2013.

102. Manisha Pattanaik, B. Raj, Shashikant Sharma and Anjan Kumar, "Diode based trimode multi-threshold CMOS technique for ground bounce noise reduction in static CMOS adders", *Advanced Materials Research,* vol. 548, pp. 885–889, 2012.
103. Balwinder Raj, A. K. Saxena and S. Dasgupta, "Nanoscale FinFET based SRAM cell design: analysis of performance metric, process variation, underlapped FinFET and temperature effect", *IEEE Circuits and System Magazine*, vol. 11, no. 2, pp. 38–50, 2011.
104. V. K. Sharma, M. Pattanaik and Balwinder Raj, "Leakage current ONOFIC approach for deep submicron VLSI circuit design", *International Journal of Electrical, Computer, Electronics and Communication Engineering,* vol. 7, no. 4, pp. 239–244, 2013.
105. Parminder Kaur, Sandeep Singh Gill and B. Raj, "Comparative analysis of OFETs materials and devices for sensor applications", *Journal of Silicon*, vol. 14, pp. 4463–4471, 2022.
106. Sanjeev Kumar Sharma, Parveen Kumar, Balwant Raj and B. Raj, "$In_{1-x}Ga_xAs$ double metal gate-stacking cylindrical nanowire MOSFET for highly sensitive photo detector", *Journal of Silicon,* vol. 14, pp. 3535–3541, 2022.
107. B. Raj, A. K. Saxena and S. Dasgupta, "Analytical modeling of quasi planar nanoscale double gate FinFET with source/drain resistance and field dependent carrier mobility: a quantum mechanical study", *Journal of Computer,* vol. 4, no. 9, pp. 1–8, 2009.
108. S. Bhushan, S. Khandelwal and B. Raj, "Analyzing different mode FinFET based memory cell at different power supply for leakage reduction", *Seventh International Conference on Bio-Inspired Computing: Theories and Application* (BIC-TA 2012), Advances in Intelligent Systems and Computing, vol. 202, pp. 89–100, 2013.
109. Jeetendra Singh and B. Raj, "Temperature dependent analytical modeling and simulations of nanoscale memristor", *Engineering Science and Technology, an International Journal*, vol. 21, pp. 862–868, Oct. 2018.
110. Shradhya Singh, Shashi Bala, Balwant Raj and B. Raj, "Improved sensitivity of dielectric modulated junctionless transistor for nanoscale biosensor design", *Sensor Letter*, vol. 18, pp. 328–333, April 2020.
111. Vivek Kumar, Santosh Kumar Vishvakarma and B. Raj, "Design and performance analysis of ASIC for IoT applications", *Sensor Letter*, vol. 18, pp. 31–38, Jan. 2020.
112. Akanksha Jaiswal, R. K. Sarin, B. Raj and Shikha Sukhija, "A novel circular slotted microstrip-fed patch antenna with three triangle shape defected ground structure for multiband applications", *Advanced Electromagnetic*, vol. 7, no. 3, pp. 56–63, Aug. 2018.
113. Girish Wadhwa and B. Raj, "Label free detection of biomolecules using charge-plasma-based gate underlap dielectric modulated junctionless TFET", *Journal of Electronic Materials*, vol. 47, no. 8, pp. 4683–4693, Aug. 2018.
114. Gurmohan Singh, R. K. Sarin and B. Raj, "Design and performance analysis of a new efficient coplanar quantum-dot cellular automata adder", *Indian Journal of Pure & Applied Physics*, vol. 55, pp. 97–103, Feb. 2017.
115. Parminder Kaur, Vikas Pandey and B. Raj, "Comparative study of efficient design, control and monitoring of solar power using IoT", *Sensor Letter*, vol. 18, pp. 419–426, May 2020.

116. Anil Kumar Bhardwaj, Sumeet Gupta and B. Raj, "Development & analysis of compact model for double gate Schottky Barrier CNTFET", *Journal of Nanoelectronics and Optoelectronics*, vol. 15, pp. 1199–1208, Aug. 2020.
117. Girish Wadhwa, Priyanka Kamboj, Jeetendra Singh and B. Raj, "Design and investigation of junctionless DGTFET for biological molecule recognition", *Transactions on Electrical and Electronic Materials*, vol. 22, pp. 282–289, 2021.
118. Sachin Kumar Verma, Shailendra Singh, Girish Wadhwa and B. Raj, "Detection of biomolecules using charge-plasma based gate underlap dielectric modulated dopingless TFET", *Transactions on Electrical and Electronic Materials*, vol. 21, pp. 528–535, June 2020.
119. Neeraj Jain and B. Raj, "Impact of underlap spacer region variation on electrostatic and analog/RF performance of symmetrical high-k SOI FinFET at 20 nm channel length", *Journal of Semiconductors*, vol. 38, no. 12, p. 122002, Dec. 2017.
120. Shailendra Singh and B. Raj," Analytical modeling and simulation analysis of T-shaped III–V heterojunction vertical T-FET", *Superlattices and Microstructures*, vol. 147, pp. 106717, Nov. 2020.
121. Gurmohan Singh, R. K. Sarin and B. Raj, "Design and analysis of area efficient QCA based reversible logic gates", *Journal of Microprocessors and Microsystems*, vol. 52, pp. 59–68, May 2017.
122. Shailendra Singh and B. Raj, "Parametric variation analysis on hetero-junction vertical t-shape TFET for suppressing ambipolar conduction", *Indian Journal of Pure and Applied Physics*, vol. 58, pp. 478–485, June 2020.
123. Shailendra Singh Soniya, Girish Wadhwa and Balwinder Raj, "Design and analysis of dual source vertical tunnel field effect transistor for high performance", *Transactions on Electrical and Electronics Materials,* vol. 21, pp. 74–82, Oct. 2019.
124. Manjit Kaur, Neena Gupta, Sanjeev Kumar, B. Raj and Arun Kumar Singh, "RF performance analysis of intercalated graphene nanoribbon based global level interconnects", *Journal of Computational Electronics,* vol. 19, pp. 1002–1013, June 2020.
125. Girish Wadhwa and B. Raj, "Design and performance analysis of junctionless TFET biosensor for high sensitivity", *IEEE Nanotechnology*, vol. 18, pp. 567–574, 2019.
126. Jeetendra Singh and B. Raj, "Enhanced nonlinear memristor model encapsulating stochastic dopant drift", *Journal of Nanoelectronics and Optoelectronics*, vol. 14, pp. 958–963, 2019.
127. Anjana Bhardwaj, Pradeep Kumar, B. Raj and Sunny Anand, "Design and performance optimization of doping-less vertical nanowire TFET using gate stack technique", *Journal of Electronic Materials*, vol. 41, no. 7, pp. 4005–4013, 2022.
128. Jeetendra Singh and B. Raj, "Tunnel current model of asymmetric MIM structure levying various image forces to analyze the characteristics of filamentary memristor", *Applied Physics A*, vol. 125, no. 3, pp. 203.1–203.11, Feb. 2019.
129. Candy Goyal, Jagpal Singh Ubhi and B. Raj, "Low leakage zero ground noise nanoscale full adder using source biasing technique", *Journal of Nanoelectronics and Optoelectronics*, vol. 14, pp. 360–370, March 2019.

130. Gurmohan Singh, R. K. Sarin and B. Raj, "A novel robust exclusive-OR function implementation in QCA nanotechnology with energy dissipation analysis", *Journal of Computational Electronics*, vol. 15, no. 2, pp. 455–465, June 2016.
131. Girish Wadhwa, Priyanka Kamboj and Balwinder Raj, "Design optimisation of junctionless TFET biosensor for high sensitivity", *Advances in Natural Sciences: Nanoscience and Nanotechnology*, vol. 10, no. 7, p. 045001, 2019.
132. Priya Bansal and B. Raj, "Memristor modeling and analysis for linear dopant drift kinetics", *Journal of Nanoengineering and Nanomanufacturing*, vol. 6, pp. 1–7, 2016.
133. D. Vaithiyanathan Ashima and B. Raj, "Performance analysis of charge plasma induced graded channel Si nanotube", *Journal of Engineering Research*, EMSME Special issue, pp. 146–154, Aug. 2021.

Chapter 10

Enhancing 5G network connectivity with fractal-inspired five-port MIMO antenna

Material-based analysis for satellite communication systems

Arun Raj and Durbadal Mandal

10.1 INTRODUCTION

In the realm of "wireless technology", antennas play a pivotal role, prompting researchers to innovate for compactness and reliability [1, 2]. The market now demands easily fabricated, low-profile antennas with small dimensions due to the advent of fifth-generation communication, driven by the need for high data rates in applications like machine-to-machine and vehicle-to-vehicle communication [3, 4]. To meet these demands, conventional techniques are evolving to align with the requirements of next-generation communications.

The global shift to the fifth generation of communication is a pivotal research focus, highlighting technology's evolution to meet high-speed connectivity and the burgeoning realm of Internet of Things (IoT) applications. The dynamic landscape prompts consistent updates in 5G technology standards by the 3GPP (Third Generation Partnership Project). This ongoing transformation ensures researchers can effectively pursue their goals and propel further advancements in the field [5–12].

The small size of the antenna enables resonance at higher frequencies, facilitating communication in 5G mm-wave and sub-5G (FR1) ranges. Previous studies explored MIMO (multiple input, multiple output) configurations with expansive coverage, addressing certain challenges. Notably, some antennas adopted a single input and single output to resolve specific issues. To enhance signal quality and achieve low-latency communication, diverse techniques have been employed. These advancements contribute to the optimization of antennas for efficient 5G communication across different frequency ranges, showcasing progress in resolving technical challenges and improving overall performance [13–15].

Researchers are currently focused on MIMO antennas, addressing issues like weak signals and low signal strength. The multi-patch property enhances communication between transmitters and receivers. In substrate cavity MIMO antenna design, multiple radiating elements are integrated

DOI: 10.1201/9781003612797-10

into a dielectric substrate with strategically placed voids. These voids serve as electromagnetic barriers, minimizing mutual coupling and interference among antennas. This approach significantly enhances antenna performance [16–26].

The advancement of MIMO systems hinges on leveraging spatial diversity for enhanced data throughput and reliability. By employing substrate cavities, these systems ensure improved spatial separation among antenna elements, resulting in heightened diversity gain and overall system efficiency. In a novel design, the integration of fractal-inspired elements and a unique substrate divider cylindrical cavity significantly reduces MIMO coupling. This innovative approach yields an envelope correlation coefficient below 0.0012 and isolation exceeding 20 dB [27–29]. Developed using CST software, the compact size of the antenna minimizes mutual coupling, offering improved performance. This design proves invaluable for achieving elevated data rates and superior signal quality in the realm of next-generation wireless devices. The proposed antenna explores diverse substrates and fractal shapes within a five-port congruent ring MIMO design, specifically tailored for individual applications in compact and efficient advanced 5G mm-wave systems. The substrate selection involves a comparative analysis between FR4 and Roger substrates. Roger substrate is an optimal choice due to its compact dimensions, high gain, and efficiency [30–36]. Notably, it exhibits an acceptable envelope correlation coefficient and diversity gain. These characteristics make the antenna well-suited for advanced 5G mm-wave applications, aligning with the targeted resonant frequencies. Focusing on compact size ensures practicality and adaptability to various contexts while emphasizing high gain and efficiency, which enhances overall performance. This investigation underscores the antenna's versatility, making it a promising candidate for diverse applications within the evolving landscape of advanced 5G communication systems [37–43].

10.2 ANTENNA DESIGN

For microstrip patch, design parameters are calculated using the following mathematical formulae's [21]:

Width (W): The "width of the patch" is given by:

$$W = \frac{C}{2f_r\sqrt{\frac{\varepsilon r + 1}{2}}} \tag{10.1}$$

where

c – "Free space velocity of light" = 3×10^8 m/s

f_r – "Frequency of operation", and ε_r – "Dielectric constant"

Calculation of "Effective dielectric constant" (ε_{reff}): The "effective dielectric constant" ε_r is considered by [21]:

$$W = \frac{1}{2fr\sqrt{\mu\varepsilon}}\sqrt{\frac{2}{\epsilon r+1}} \tag{10.2}$$

$$\varepsilon_{reff} = \frac{\epsilon r+1}{2}+\frac{\varepsilon r-1}{2}\left(1+12\frac{h}{w}\right)^{\frac{-1}{2}} \tag{10.3}$$

where h – "height of dielectric substrate" in mm
w – "Width of patch", and ε_{reff} – "Dielectric constant"
The "effective length" is given by:

$$L_{reff} =- \frac{c}{2f\sqrt{\varepsilon r}} - 2\Delta L \tag{10.4}$$

Once W is found using equation (10.2), determine the "extension of length"

$$\Delta L = h\frac{(\varepsilon_{reff}+0.3)\left(\frac{W}{h}+0.264\right)}{(\varepsilon_{reff}-0.258)\left(\frac{W}{h}+0.8\right)} \tag{10.5}$$

"Actual length of patch":

$$L = \left(\frac{1}{2fr\sqrt{\varepsilon}}\times\frac{1}{\sqrt{\mu\varepsilon}}\right) - 2\Delta L \tag{10.6}$$

"Effective length of ground" $L^1 = L + 6h$, "effective width" $W^1 = W + 6h$,

$$\text{"Band width" } \% = \frac{2(f_2-f_1)}{f_1+f_2} * 100 \tag{10.7}$$

The "feed point Position" for "50 Ohms" can be calculated as:

$$R_{in}(y = y_0) = R_{in}(y = 0)\cos\left(\frac{\pi}{L}y0\right)^2 \tag{10.8}$$

Where $R_{in}(y=y_0)$ is "50Ohms" and $R_{in}(y = 0)$ is roughly given as:
$Zin = 1/Y_{in} = R_{in} = 1/2G_1$

$$G_1 = \begin{cases} \frac{1}{90}\left(\frac{W}{\lambda}\right)^2 & W \ll \lambda \\ \frac{1}{120}\left(\frac{W}{\lambda}\right) & W \gg \lambda \end{cases} \tag{10.9}$$

10.3 DESIGN AND SIMULATION

10.3.1 Simulation and analysis of a simple patch antenna

Figure 10.1 and Tables 10.1, 10.2, and 10.3 outline the essential notations for design parameters crucial in fabricating and optimizing a simple patch antenna. Ground dimensions, denoted as Wg (22 mm) and Lg (22 mm), along with patch dimensions represented by Patch Lp (12.5 mm) and Patch Wp (12.5 mm), play a pivotal role in achieving precise geometric features [44–49]. Feed dimensions, expressed as feed length "Lf" (2.5 mm) and width "Wf" (1 mm), contribute to the antenna's accurate specification. The antenna, detailed in Figure 10.2 and Table 10.4, operates within the frequency range of f_1 (24 GHz) to f_2 (40 GHz), offering a bandwidth of 14 GHz. The return loss, measured at 45 dB, signifies efficient signal reflection, while the peak efficiency is recorded at -3.1 dB. These parameters collectively enhance the antenna's performance, positioning it as a viable choice

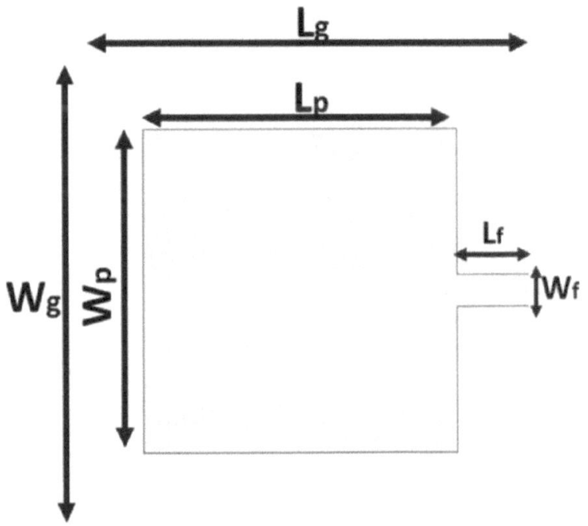

Figure 10.1 Simple patch antenna with mentioned dimensions concerning line feed.

Table 10.1 Design parameters for Simple patch and Proposed MIMO antenna

Antenna description	*Simple patch antenna*	*Prop. MIMO antenna*
Patch	12.5 × 12.5 mm^2	25 × 25 mm^2
Height of substrate (h)	1.6 mm	1.6 mm
Loss tangent (Tan δ)	0.002	0.002
Frequency range (f_r)	24–40 GHz	24–40 GHz
Patch used	Line feed	Line feed
Shape	Square shape	Square shape

Table 10.2 Different notations of design parameters for designed MIMO antenna

L_{mDGS}	W_{mDGS}	F_{pl}	$Fp2$	$Fp3$	$Fp4$	$Fp5$	$R2$
50 mm	50 mm	16.6 × 1 mm^2	16.6 × 1 mm^2	16.6 × 1 mm^2	16.6 × 1 mm^2	16.6 × 1 mm^2	8.5 mm
R1	**X**	$\mathbf{X_H}$	$\mathbf{X_P}$	$\mathbf{X_T}$	$\mathbf{X_S}$	$\mathbf{T_{DGS}}$**%**	$\mathbf{M_{PATCH}}$**%**
8	(H+P+T+S)	Hexagon	Pentagon	Triangle	Square	50%	50%

Table 10.3 Different notations of design parameters for simple patch antenna

Notations	*Ground Wg*	*Ground Lg*	*Patch Lp*	*Patch Wp*	*Feed length "Lf" and width "Wf"*
Dimension "mm"	22	22	12.5	12.5	2.5 and 1

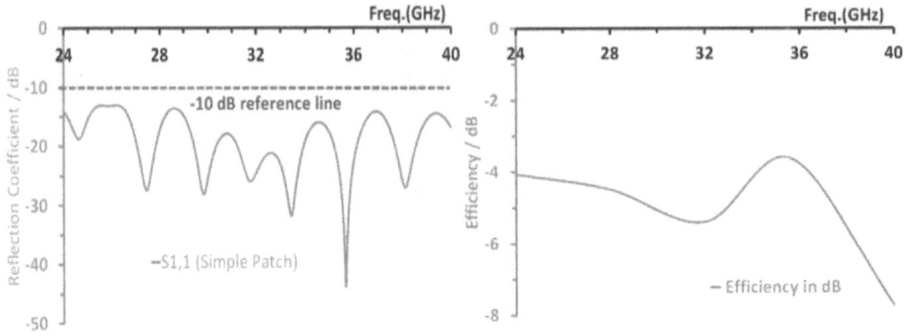

Figure 10.2 S_{11}, efficiency vs frequency graph for simple patch antenna.

Table 10.4 Study of different parameters of simple patch

Figure 1	f_1 *(GHz)*	f_2 *(GHz)*	*Bandwidth (GHz)*	*Return loss (dB)*	*Peak efficiency (dB)*
Simple patch	24	40	14	45	−3.1

for applications requiring a frequency range between 24 and 40 GHz. The study emphasizes the antenna's characteristics, highlighting its bandwidth, return loss, and peak efficiency, providing valuable insights for practical implementations in communication systems or relevant domains [50–57].

10.3.2 *Simulation and analysis of fractal antenna with carpet structure*

Figure 10.3 illustrates the simple patch antenna as sub-figure "a", alongside the triangular slot Sierpinski carpet and square pentagon-inspired patch antennas. The figure also showcases the intermediate stages of forming the Sierpinski carpet antenna [58–66]. Moving to Figure 10.4, the results for triangular and square-pentagon-inspired antennas are presented, including reflection coefficients and gains. The corresponding efficiency values are detailed in Table 10.5. These visual representations and data tables offer a comprehensive insight into the performance of the antennas at various stages of development. The comparative analysis in Table 10.5 allows for

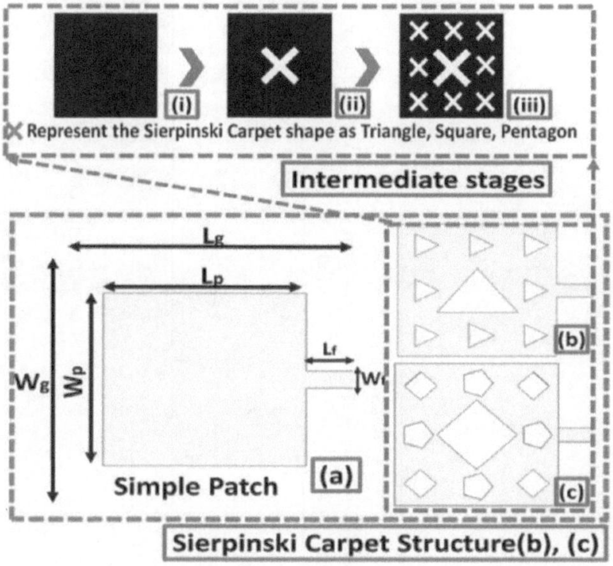

Figure 10.3 Fractal element antenna concerning Sierpinski carpet geometry with various slots type "X".

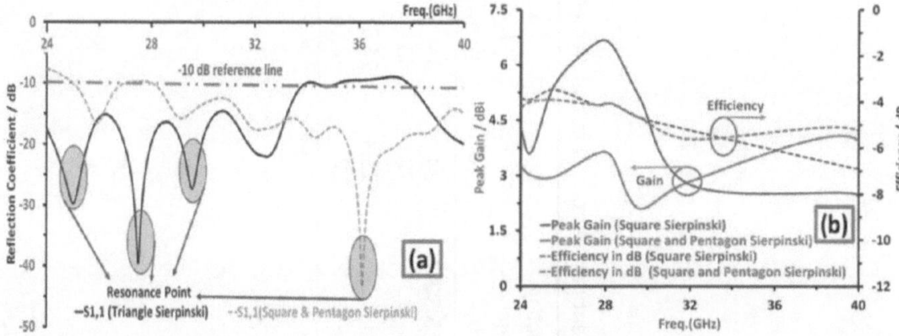

Figure 10.4 S_{11}, gain, efficiency, vs frequency graphs concerning fractal element antennas.

a clear assessment of the reflection coefficients, gains, and efficiencies of triangular and square-pentagon-inspired patch antennas, aiding in understanding their suitability for specific applications [67–74].

The figures and tables collectively offer a valuable resource for understanding the characteristics and effectiveness of these innovative patch antenna designs. Table 10.5 presents a comparative study of two types of designs, Antenna "b" (Triangle) and Antenna 'c' (Square), along with their corresponding parameters. For Antenna 'b,' resonating notches are observed at 25, 27.6, 29.5, 32.3, and 40 GHz, with resonant frequencies f1 and f2 at 24 and 38 GHz, respectively. The bandwidth is 10 GHz, return

Table 10.5 Comparative study with different parameters of fractal element antennas

Type of designs	*Resonating notches (GHz)*	F_1 *(GHz)*	F_2 *(GHz)*	*Bandwidth (GHz)*	*Return loss (dB)*	*Peak gain (dBi)*	*Peak efficiency (dB)*
Antenna "b" (Triangle)	**25, 27.6, 29.5 32.3,**	24	34	10	30, 40	6.5	–3.5
	40	38	40	2	26, 19.5		
Antenna "c" (Square)	**26**	25.4	26.5	1.1	17.5	4.5	–4
	36.1	28	40	12	44		

loss is 30, 40 dB, peak gain is 6.5 dBi, and peak efficiency is -3.5 dB. On the other hand, Antenna "c" exhibits resonating notches at 26 and 36.1 GHz, f1 and f2 at 25.4 and 28 GHz, and a bandwidth of 1.1 and 12 GHz, respectively [75–81]. The return loss is 17.4 and 44 dB, the peak gain is 4.5 dBi, and the peak efficiency is –4 dB. These findings offer insights into the performance variations between the two antenna designs concerning resonant frequencies, bandwidth, return loss, peak gain, and peak efficiency.

10.3.3 Simulation and analysis of 5 port MIMO antenna with simple ground

Figure 10.5 illustrates a MIMO (multiple input, multiple output) antenna configuration featuring five congruent rings in the presented study. Each ring incorporates a Sierpinski carpet patch antenna within, and they are connected to line feeding designated as portl, port2, port3, port4, and port5, with corresponding feeding labeled as Fpl, Fp2, Fp3, Fp4, and Fp5. Notably, the Sierpinski carpet is represented by X, XH, XP, XT, and XS, where XH, XP, ST, and XS denote different shapes within the Sierpinski carpet design. The antenna design is intended for applications in the 4G to advanced 5G ranges, showcasing its versatility across various communication needs [82–87]. Figures 10.6 and 10.7 present a thorough analysis of different parameters. The reflection coefficient and isolation metrics exhibit favorable performance, indicating effective signal transmission and reduced interference. However, the study notes that radiation efficiency and gain are somewhat marginal, primarily attributed to the use of FR4 substrate. Incorporating Sierpinski carpet patches within the rings adds a dimension of complexity and likely contributes to the antenna's enhanced capabilities. The connectivity to multiple ports allows for diverse applications and improved network performance [88–93]. Despite the marginal drawbacks

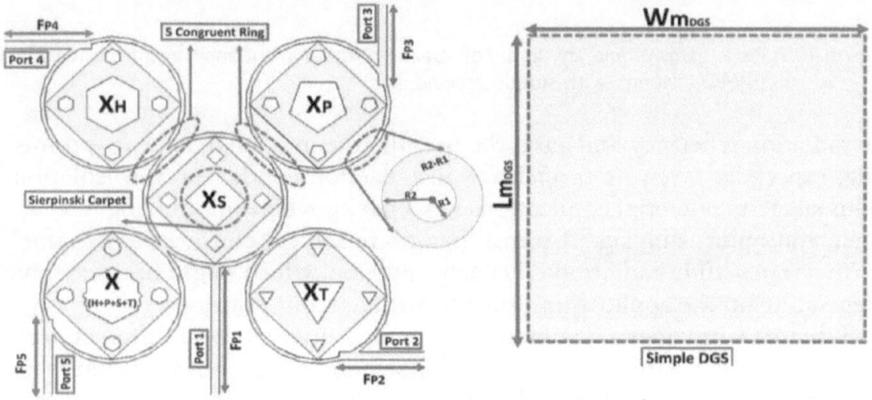

Figure 10.5 S11 vs frequency graph concerning MIMO antenna with simple.

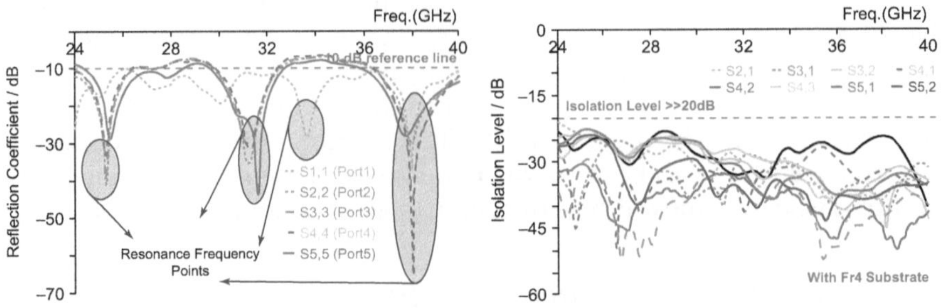

Figure 10.6 S1,1 and isolation vs frequency graph concerning MIMO antenna with simple ground.

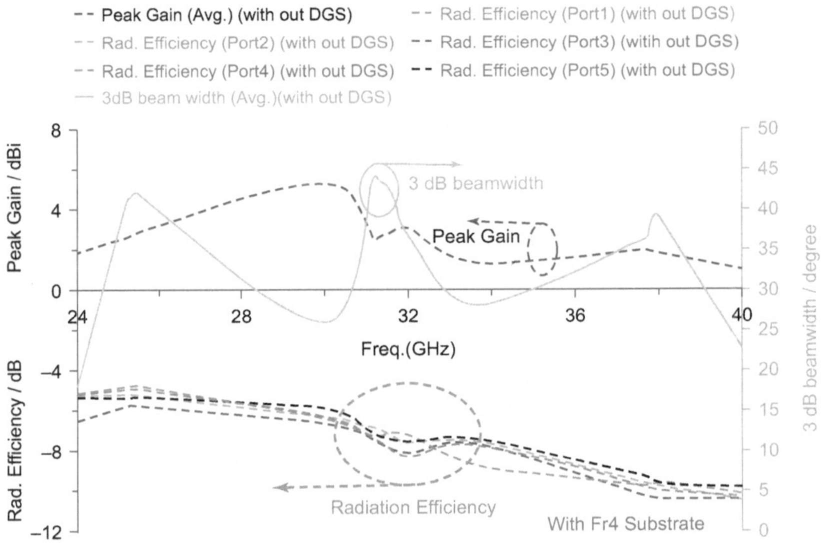

Figure 10.7 Peak gain, efficiency, and 3dB beamwidth vs frequency graph concerning MIMO antenna with simple ground.

in radiation efficiency and gain, the overall antenna design appears promising, especially given its proficiency in reflection coefficient and isolation. This suggests potential applications in scenarios where minimizing interference and optimizing signal transmission are critical considerations, albeit with a trade-off in radiation efficiency and gain, which might be acceptable depending on the application's specific use case and priorities.

Table 10.6 provides a comprehensive comparative study of different ports 1, 2, 3, 4, and 5 parameters [94–101]. For port 1, resonant frequencies (fr) are observed at 29.8, 33.6, and 38 GHz, with corresponding frequencies f1 and f2 at 24 and 40 GHz. The bandwidth is 16 GHz, and the return loss

Table 10.6 Comparative study with different parameters of MIMO antenna with simple ground

Type of designs	f_r *(GHz)*	f_l *(GHz)*	f_2 *(GHz)*	*Bandwidth (GHz)*	*Return loss (dB)*
Port 1	29.8, 33.6, 38	24	40	16	20, 29.5, 40
Port 2	25	24	26.4	2.4	40
	31.6	28.8	32.4	3.6	33
	38	36.1	40	3.9	29.4
Port 3	25	24	26.4	2.4	40
	31.6	28.8	32.4	3.6	30
	38	36.1	40	3.9	44
Port 4	25	24	26.4	2.4	32
	31.6	28.8	32.4	3.6	30
	38	36.1	40	3.9	64
Port 5	25.4	24.3	28.1	3.8	29.5
	31.8	29.5	32.5	3	43
	37.8	36	40	4	28

values are 20, 29.5, and 40 dB. Port 2 exhibits resonating notches at 25, 31.6, and 38 GHz, with f1 and f2 at 24 and 40 GHz.

$$R(c) = \frac{Z(1) - Z(0)}{Z(1) + Z(0)} \tag{10.10}$$

$$R(c) = \frac{VSWR - 1}{VSWR + 1}$$

$$R(L) = -20 * \log_{10}(R(c)) \tag{10.11}$$

where $R(c)$ is the reflection coefficient and $R(L)$ is return loss.

The bandwidths are 2.4, 3.6, 3.9 GHz, while return loss values are 40, 33, and 29.4 dB. similarly, port 3, port 4, and port 5 display resonant frequencies and corresponding parameters. Port 3 has resonant frequencies at 25, 31.6, and 38 GHz, with f1 and f2 at 24 and 40 GHz, a bandwidth of 2.4, 3.6, 3.9 GHz, and return loss values of 30, 44, and 40 dB. For port 4, resonant frequencies are 25, 31.6, and 38 GHz, f1 and f2 at 24 and 40 GHz, bandwidth of 2.4, 3.6, 3.9 GHz, and return loss values of 30, 64, and 32 dB. Lastly, port 5 has resonant frequencies at 25.4, 31.8, and 37.8 GHz, with f1 and f2 at 24.3 and 40 GHz, a bandwidth of 3.8, 3, 4 GHz, and return loss values of 29.5, 43, and 28 dB, respectively [102–107].

This comprehensive analysis highlights the differences in resonant frequencies, bandwidth, and return loss among the various ports, providing

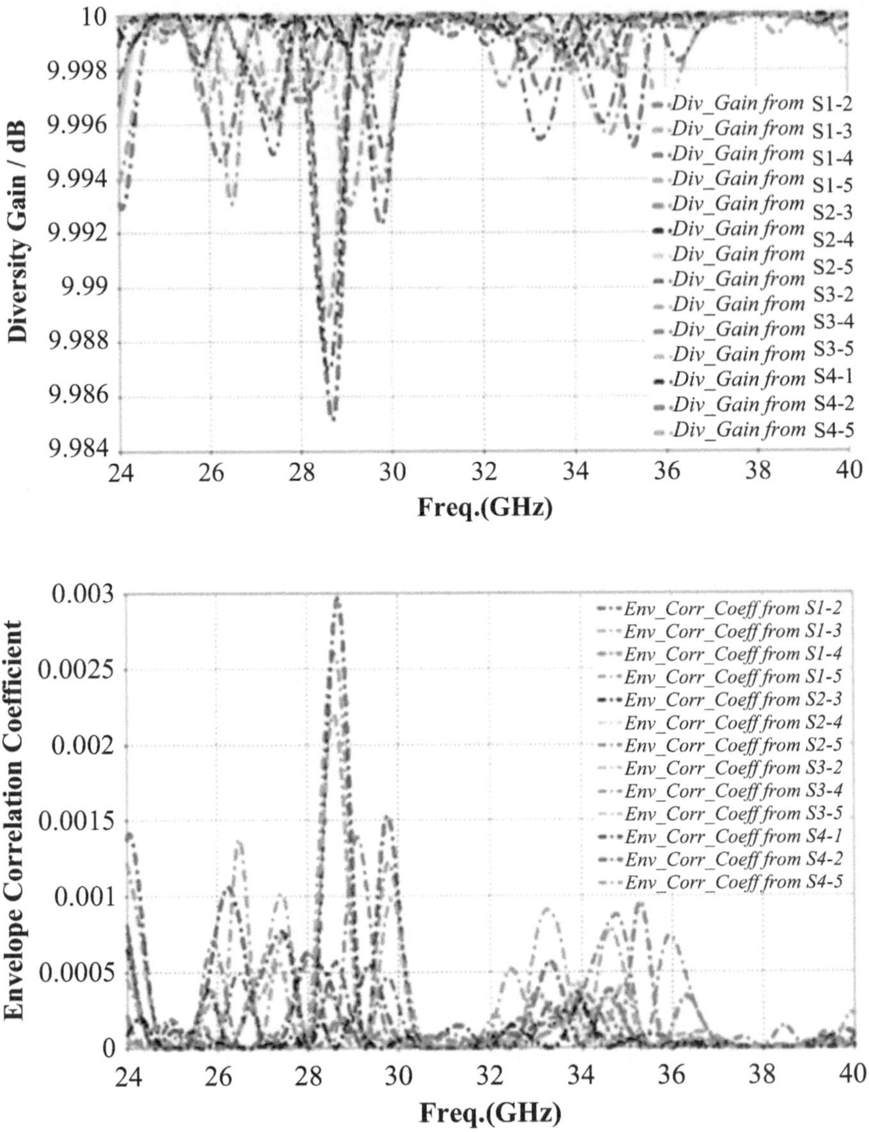

Figure 10.8 Diversity gain, ECC vs frequency graph concerning MIMO antenna with simple ground.

valuable insights for designing and optimizing the corresponding antenna system. Furthermore, Table 10.7 summarizes the comparative study of proposed MIMO antennas, showcasing their key parameters. The antennas exhibit high performance with a peak isolation level exceeding 20 dB, peak gain above 5.9 dBi, and peak efficiency below –5.6 dB. Notably,

Table 10.7 Comparative study with different parameters of MIMO antenna with simple ground

Peak isolation level (dB)	*Peak gain (dBi)*	*Peak efficiency (dB)*	*Minimum envelope correlation coefficient (ECC)*	*Diversity gain (dB)*	*3dB beam width (degree)*
>> 20	Near 5.9	<< −5.6	<< 0.003	>> 9.984	<< 45

the minimum envelope correlation coefficient (ECC) is impressively low at 0.003, signifying minimal mutual coupling. The diversity gain is substantial, surpassing 9.984 dB, underscoring the antennas' effectiveness in enhancing signal diversity [108–112]. Additionally, the 3 dB beamwidth is wide, exceeding 45 degrees, emphasizing the broad coverage of the MIMO antenna system. These findings underscore the antennas' capability to achieve significant isolation, gain, and diversity for improved wireless communication performance.

10.3.4 Simulation and analysis of 5 port MIMO antenna with defected ground

The installation process of the MIMO antenna underwent a significant enhancement by incorporating a fragment-inspired ground-defected structure, as depicted in Figure 10.9. This structural modification involved integrating four circular stubs strategically positioned at 45-degree intervals from the center. The iterative application of the T scaling factor was employed to boost gain while minimizing coupling between individual ports. This meticulous approach aimed to augment the antenna's diversity properties [113–119].

The comparative study of MIMO antennas with a defected ground depicted in Figures 10.9, 10.10, 10.11, and 10.12 reveals distinct performance

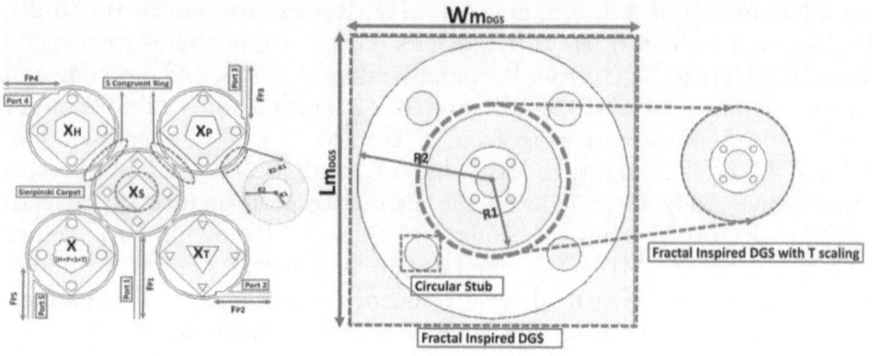

Figure 10.9 MIMO antenna with defected ground structure.

Table 10.8 Comparative study with different parameters of MIMO antenna with defected ground

Ports	f_r *(GHz)*	f_l *(GHz)*	f_2 *(GHz)*	*Bandwidth (GHz)*	*Return loss (dB)*
Port 1	25.5	24	28.8	4.8	18
	32.4	30.1	33.7	3.6	14
	39	35.6	40	4.4	13
Port 2	26.6	25	27.9	2.9	25
	32	29	33.2	4.2	27
	39	36.2	40	3.8	30
Port 3	26.6	25	27.9	2.9	25
	32	29	33.2	4.2	35
	39	36.2	40	3.8	43
Port 4	26.6	25	27.9	2.9	25
	32	29	33.2	4.2	27
	39	36.2	40	3.8	48
Port 5	27	25	33.4	8.4	24
	32.1, 38.8	35.6	40	4.4	32, 24

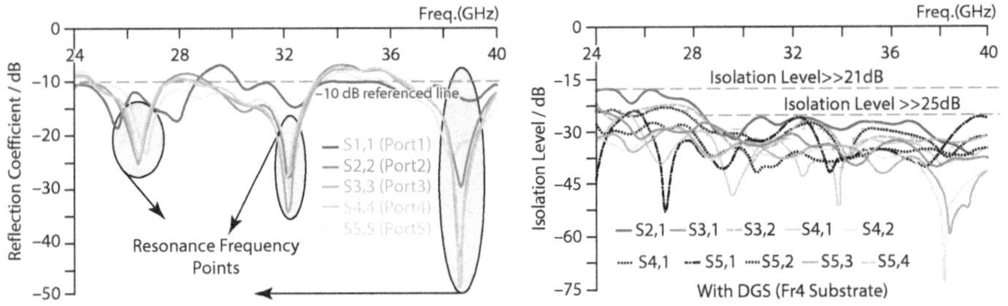

Figure 10.10 Ref. coefficient, isolation vs frequency graph concerning MIMO antenna with DGS.

parameters. For port 1, frequencies range from 25.5 GHz to 39 GHz, achieving a bandwidth of 4.8, 3.6, and 4.4 GHz. Return loss values are 18 dB, 14 dB, and 13 dB at different frequencies. Port 2 demonstrates frequencies from 26.6 GHz to 39 GHz with a bandwidth of 2.9, 4.2, and 3.8 GHz and return loss values of 25 dB, 27 dB, and 30 dB. Ports 3 and 4 exhibit similar trends. Port 5 offers frequencies from 27 GHz to 38.8 GHz, a bandwidth of 8.4 and 4.4 GHz, and return loss values of 24 dB, 32 dB, and 24 dB. This comparative study sheds light on the distinctive performance characteristics of each port in terms of resonant frequencies, bandwidth, and return loss. The proposed MIMO antenna design demonstrates versatility across multiple ports, catering to diverse frequency requirements. The findings underscore the potential for this antenna system to facilitate effective communication in MIMO applications with varying frequency demands. The

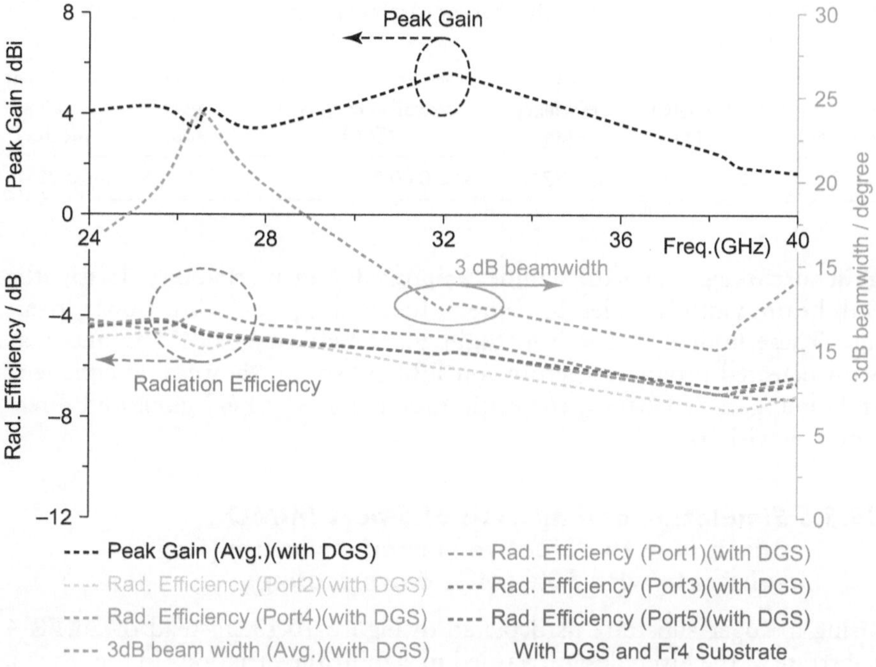

Figure 10.11 **Peak gain, efficiency, and 3dB beamwidth vs frequency graph concerning MIMO antenna with defected ground.**

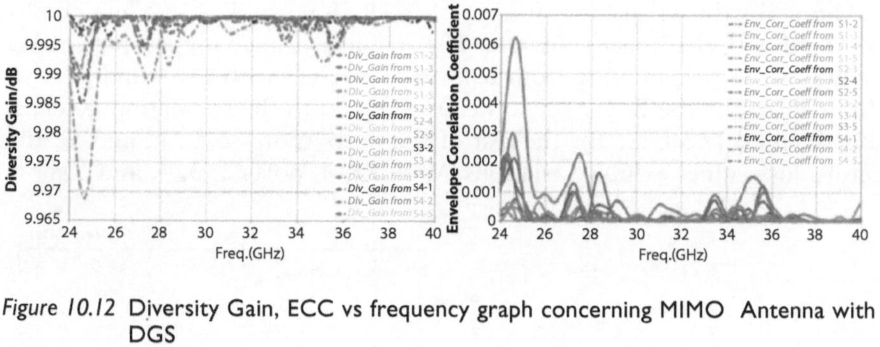

Figure 10.12 **Diversity Gain, ECC vs frequency graph concerning MIMO Antenna with DGS**

comparative analysis in Table 10.9 evaluates MIMO antennas with defective ground structures on FR4 substrate. Noteworthy results include a peak isolation level exceeding 25 dB, demonstrating strong signal separation. The peak gain reaches nearly 5.9 dBi, indicating favorable signal strength. A high peak efficiency of –3.8 dB underscores effective signal utilization. The minimum envelope correlation coefficient (ECC) is impressively low at 0.007, affirming minimal correlation between antenna elements. A diversity

Table 10.9 Comparative study with different parameters of MIMO antenna with defected ground

Peak isolation level (dB)	*Peak gain (dBi)*	*Peak efficiency (dB)*	*Minimum envelope correlation coefficient (ECC)*	*Diversity gain (dB)*	*3dB beam width (degree)*
>> 25	Near 5.9	<< −3.8	<< 0.007	>> 9.965	<< 25

gain surpassing 9.965 dB signifies improved system reliability. Lastly, the 3dB beam width is under 25 degrees, indicating precise directional coverage. These findings collectively highlight the efficacy of MIMO antennas with defected ground structures on FR4 substrate, showcasing enhanced isolation, gain, efficiency, correlation reduction, diversity gain, and directional precision.

10.3.5 Simulation and analysis of 5 port MIMO antenna with defective ground structure (Roger substrate with 50% (M%) down scaling)

Using a Roger substrate as depicted in Figure 10.13, instead of an FR-4 substrate in the given design, has led to significant variations in the results, rendering them unacceptable. Specifically, examining the details related to ports, we observe noteworthy discrepancies.

In the case of port 1, the resonating frequency is reported as 30 GHz when FR-4 is used, but with the Roger substrate, it increases to 35.8 GHz. Additionally, the return loss for port 1 is 11 dB with FR-4, while it climbs to 13 dB with the Roger substrate. The frequency range for resonance also shows disparities, ranging from 29.5 to 30.3 GHz with FR-4 and 34.6 to 36.3 GHz with the Roger substrate. Moving on to the resonating notches, identified as 37.3 GHz for the rest of the ports (ports 2, 3, 4, and 5), the return loss values exhibit variations. With FR-4, ports 2, 3, 4, and 5 show

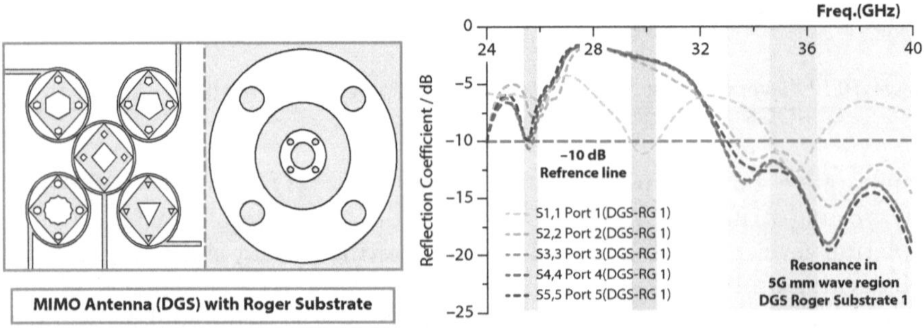

Figure 10.13 S1,1 vs frequency graph concerning MIMO antenna with DGS (Roger substrate).

16, 19, 19, and 20 dB return losses, respectively. However, when the Roger substrate is employed, these values may differ. These discrepancies are critical, as they can impact the overall performance and reliability of the design. Therefore, carefully considering substrate materials is imperative to ensure consistent and acceptable results across all ports and frequencies.

As illustrated in Figure 10.14, the antenna design utilizes a congruent ring implemented within a defective ground structure. To ensure the overall compactness of the proposed antenna without compromising results, a scaling-down approach was employed on the selected Rogers substrate, reducing it to "m" scaling. This reduction in scale aims to maintain the antenna's efficiency while adhering to space constraints. Four congruent rings were installed to enhance performance, four of which are in Figure 10.14. The fractal scaling was maintained in conjunction with the ground-defected structure to optimize the current distribution. Efficient current distribution is crucial for the movement of electrons, facilitating the radiation process in MIMO antennas. Resonating points were identified at 24.25 GHz and 34 GHz for port 1 and port 4 and 32 GHz for port 2, port 3, and port 5.

Notably, the proposed antenna significantly improved in isolation, achieving a 20 dB increase. This heightened isolation is crucial for the antenna's functionality in diverse applications. The resonance frequencies and enhanced isolation, as highlighted in Figure 10.15, contribute to the antenna's effectiveness in supporting various applications, aligning with the specific requirements of the intended use cases. The meticulous design, incorporating congruent rings and a defected ground structure, showcases a balanced approach, successfully achieving both compactness and optimal performance in resonance and isolation.

The proposed design introduces a novel approach to antenna construction, focusing on size reduction and optimization for MIMO applications. The total antenna size has been significantly scaled down compared to the initial design. This reduction is achieved by carefully considering the element antenna within the five-port congruent ring antenna. The optimization process involves transforming the element antenna into a carpet with iteration, as outlined in Figure 10.20. Additionally, a defective ground

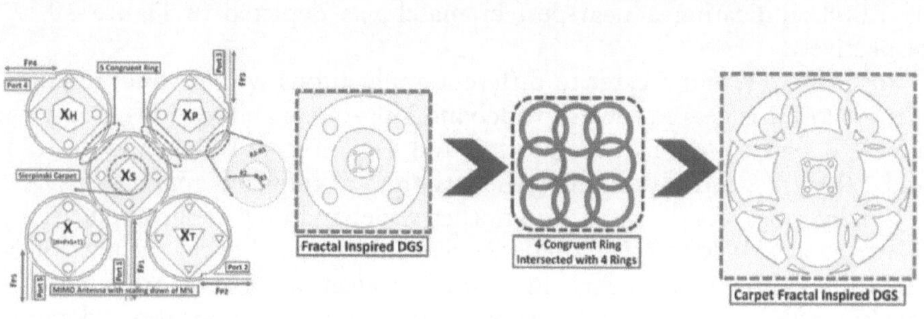

Figure 10.14 MIMO antenna with DGS concerning four congruent rings intersected with four rings (50%).

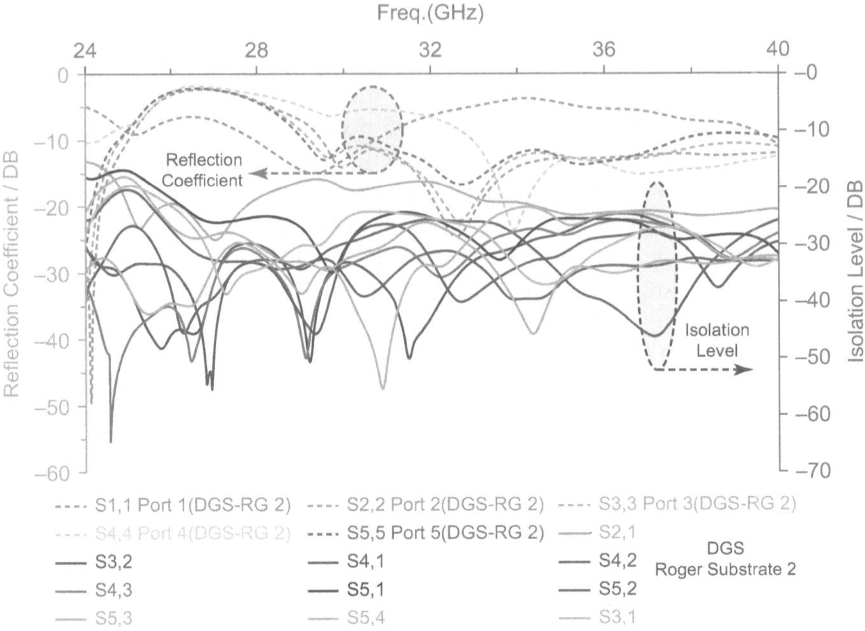

Figure 10.15 S1,1 and isolation vs. frequency concerning MIMO antenna with DGS (Con. Ring).

structure is incorporated to enhance isolation. To ensure compatibility with MIMO systems, the proposed antenna is meticulously matched at 50 ohms, as illustrated in the schematic simulated design (Figure 10.16). The matching process involves specific adjustments for each antenna port, optimizing their performance for diverse applications. For instance, Port 1 is tailored for advanced mmWave 5G applications, with a resonating frequency finely tuned to 29 GHz. The reflection coefficient at this frequency stands at an impressive 30 dB, ensuring reliable performance. The impedance matching is equally noteworthy, with a Voltage Standing Wave Ratio (VSWR) of 1.008, indicating a near-perfect match, as depicted in Figure 10.17, respectively.

Ports 2, 3, 4, and 5 cater to different applications with distinct requirements. Port 2 is geared toward wideband long-range applications, boasting a VSWR of 1.5. Meanwhile, ports 3 and 4 exhibit VSWR values of 1.25 and 1.005, respectively, aligning them with their intended applications.

Finally, port 5 is designed for another wideband long-range application, featuring a VSWR of 1.5. Using a Rogers substrate enhances the antenna's versatility, allowing it to excel in various applications. Each port is meticulously crafted to meet the demands of its designated application, demonstrating the flexibility and adaptability of the proposed MIMO antenna.

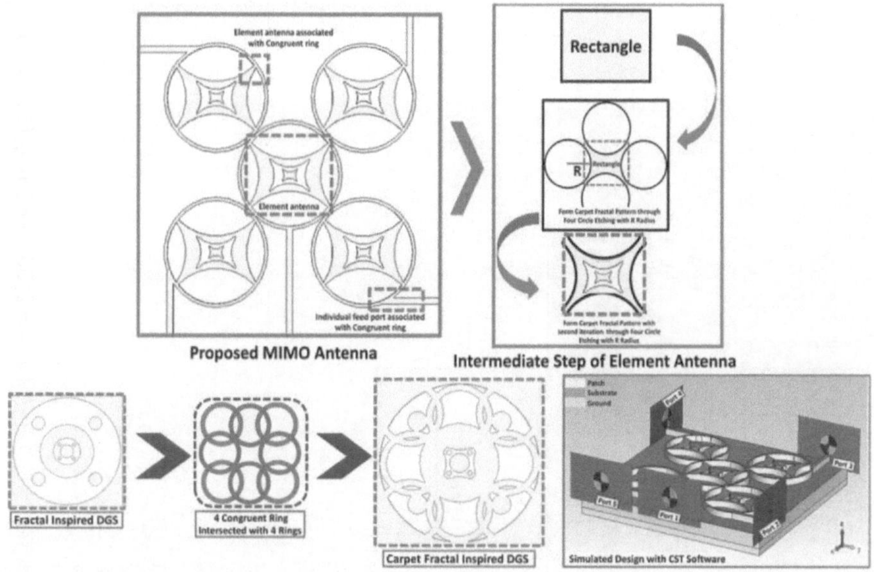

Figure 10.16 Proposed MIMO antenna with DGS concerning four congruent rings intersected with four rings (50% scaling).

The achieved VSWR values across different ports underscore the effectiveness of the matching process, ensuring optimal performance in a range of scenarios.

The comparative study presented in Tables 10.10 and 10.11 and Figure 10.18 evaluates the performance of different port designs across various parameters. The designs under consideration are labeled as port 1, port 2, port 3, port 4, and port 5.

In Table 10.10, the frequency characteristics (fr, f1, f2), bandwidth, and return loss for each port are documented. Port 1 exhibits a frequency range of 27.3 GHz to 32.8 GHz, a bandwidth of 5.5 GHz, and a return loss of 30 dB. Ports 2, 3, 4, and 5 display variations in frequency ranges, bandwidths, and return losses. Notably, port 2 demonstrates a wider bandwidth but slightly lower return loss compared to the other ports. Moving to Table 10.11, additional parameters such as peak isolation level, peak gain, peak efficiency, minimum envelope correlation coefficient (ECC), diversity gain, and 3 dB beam width are examined. The peak isolation levels across all ports range from 19 dB to 60 dB, with port 5 achieving the highest isolation. Peak gain values are near 7 dBi for all ports. The peak efficiency, ranging from –3 dB to –1.5 dB, indicates a decent level of energy utilization.

The minimum envelope correlation coefficient (ECC) is below 0.012 for all ports, suggesting a low correlation between the signal envelopes. This is advantageous for minimizing interference and enhancing the overall

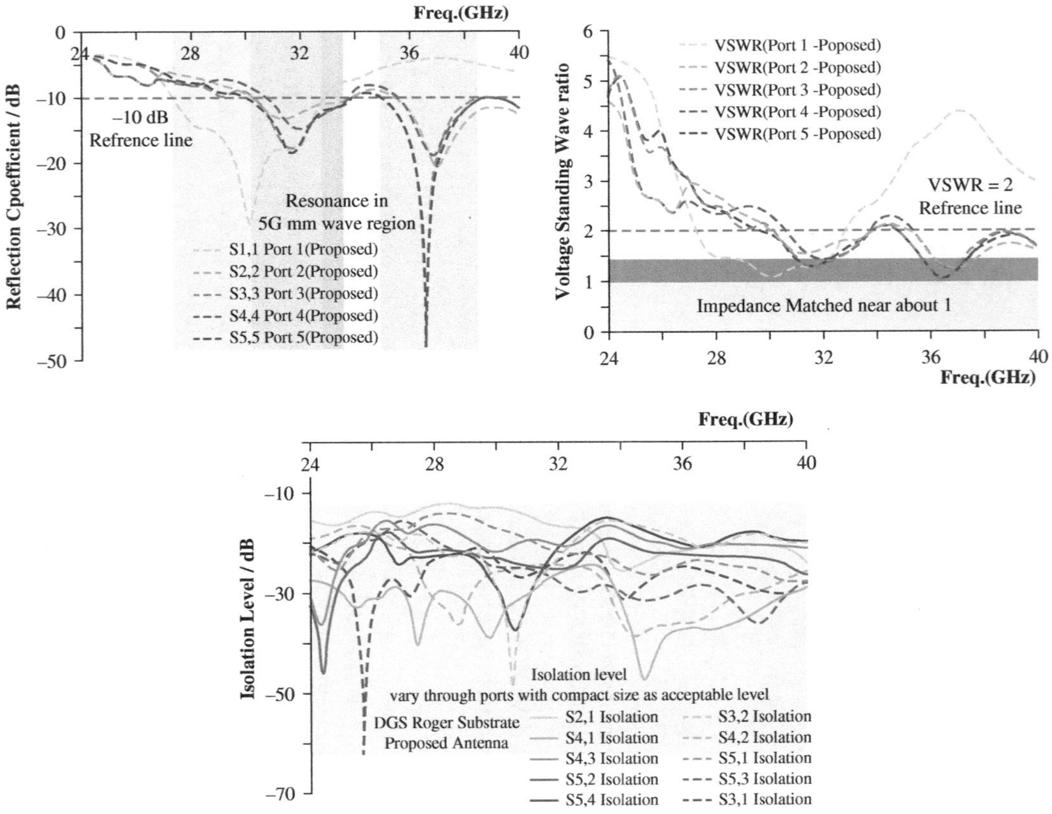

Figure 10.17 Ref. coefficient, Vswr, and isolation vs frequency graphs concerning proposed MIMO antenna with DGS concerning four congruent rings intersected with four rings.

Table 10.10 Comparative study with different parameters concerning ports 1, 2, 3, 4, and 5

Type of designs	*f_r (GHz)*	*f_l (GHz)*	*f_2 (GHz)*	*Bandwidth (GHz)*	*Return loss (dB)*
Port 1	29	27.3	32.8	5.5	30
Port 2	31.2	30.1	33.5	3.4	14
	37	35.8	40	4.2	20
Port 3	31.8	30.1	33.5	3.4	19
	36.2	35.8	40	4.2	33
Port 4	32	30.5	33.5	3	15
	37	35.8	40	4.2	19
Port 5	31.8	30.1	33.5	3.4	20
	36.2	35	40	5	50

Table 10.11 Comparative study with different parameters concerning ports 1, 2, 3, 4, and 5

Peak isolation level (dB)	*Peak gain (dBi)*	*Peak efficiency (dB)*	*Minimum envelope correlation coefficient (ECC)*	*Diversity gain (dB)*	*3dB beam width (degree)*
19>>Is<<60	Near 7	−3<<Eff<<−1.5	<<0.012	>>9.666	<<35

performance of the ports. The diversity gains, exceeding 9.666 dB for all ports, highlight their effectiveness in exploiting diversity for improved communication reliability. The 3 dB beam width, an important parameter for directional antennas, is around 35 degrees for all ports, indicating a relatively focused radiation pattern.

Furthermore, the comparative analysis reveals that each port design has its own unique set of characteristics. Port 5 stands out with the highest peak isolation level, while Port 2 boasts a wider bandwidth. All ports exhibit favorable ECC values and diversity gains, indicating their potential for reliable communication. The 3 dB beam width around 35 degrees suggests a balanced compromise between directionality and coverage for these designs.

A comparative study of the properties for ports 1, 2, 3, 4, and 5 of the proposed antenna reveals distinct characteristics. Port 1 lacks multi-notch behavior, while ports 2 through 5 exhibit this feature. High-gain directive behavior is observed in ports 1 and 3, with average performance in ports 2 and 4. Wideband behavior is present across all ports. The antenna demonstrates a compact size and size reduction in all ports. Pencil beam long-range behavior is evident in ports 2 and 5, while port 4 shows average performance. In Table 10.13, the proposed antenna's performance is compared with other published works. The proposed antenna with five ports, a size of 25 × 25 mm², and isolation <<0.012

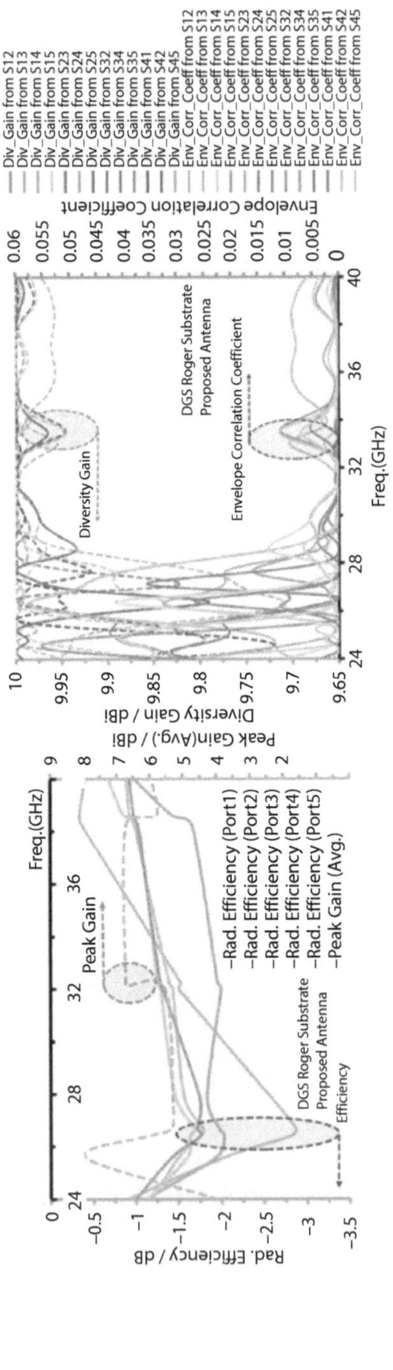

Figure 10.18 **Efficiency, peak gain and diversity gain, envelope correlation coefficient vs. frequency graphs for proposed MIMO antenna.**

Table 10.12 Comparative study with different Properties concerning ports 1, 2, 3, 4, and 5 of the proposed antenna

Properties	*Port 1*	*Port 2*	*Port 3*	*Port 4*	*Port 5*
"Multi-notch behavior"	No	Yes	Yes	Yes	Yes
"High gain directive"	Yes	Average	Yes	Average	Yes
"Wide band behavior"	Yes	Yes	Yes	Yes	Yes
"Compact size"	Yes	Yes	Yes	Yes	Yes
"Size reduction"	Yes	Yes	Yes	Yes	Yes
"Pencil beam long range" behavior	Average	Yes	Average	Average	Yes
"MM wave 5G applications"	Yes	Yes	Yes	Yes	Yes

Table 10.13 Comparative study of proposed antenna's result with other published work concerning different parameter

Reference	*Ports*	*Size of antenna (mm^2)*	*Isolation (dB)*	*Gain (dBi)*	*ECC*
[12]	6	136 × 68	10	3.9	<0.3
[13]	8	150 × 75	12.5	4.9	<0.15
[14]	4	55 × 110	28.32	8.27	<0.047
[15]	3	130 × 80	53	10.3	–
[16]	4	26.6 × 3.25	–	5.2	–
[17]	4	19.25 × 26	25	5.72	<0.04
[18]	2	55 × 110	27.3	9.49	<0.06
[19]	2	26 × 11	25	5.7	<0.01
[21]	4	30 × 30	>29	6.1	<0.16
Proposed work	5	25 × 25	>>19	7	<<0.012

dB outperforms existing designs in terms of gain and isolation. Notably, it surpasses ECC [12], ECC [13], and ECC [15] in terms of isolation, and ECC [14], ECC [18], and ECC [21] in gain. This positions the proposed antenna as a promising candidate for MM wave 5G applications, showcasing its potential for high performance and compact design in comparison to existing literature.

In Figures 10.19, 10.20, and 10.21 , the 2D power field pattern, E-field, and H-field of the proposed antenna are illustrated. Figure 10.22 provides insight into the current distribution. The radiation patterns at resonant frequencies are examined concerning ports 1, 2, 3, 4, and 5, as well as phi 0, 90 and theta 90 degrees. These figures collectively offer a comprehensive

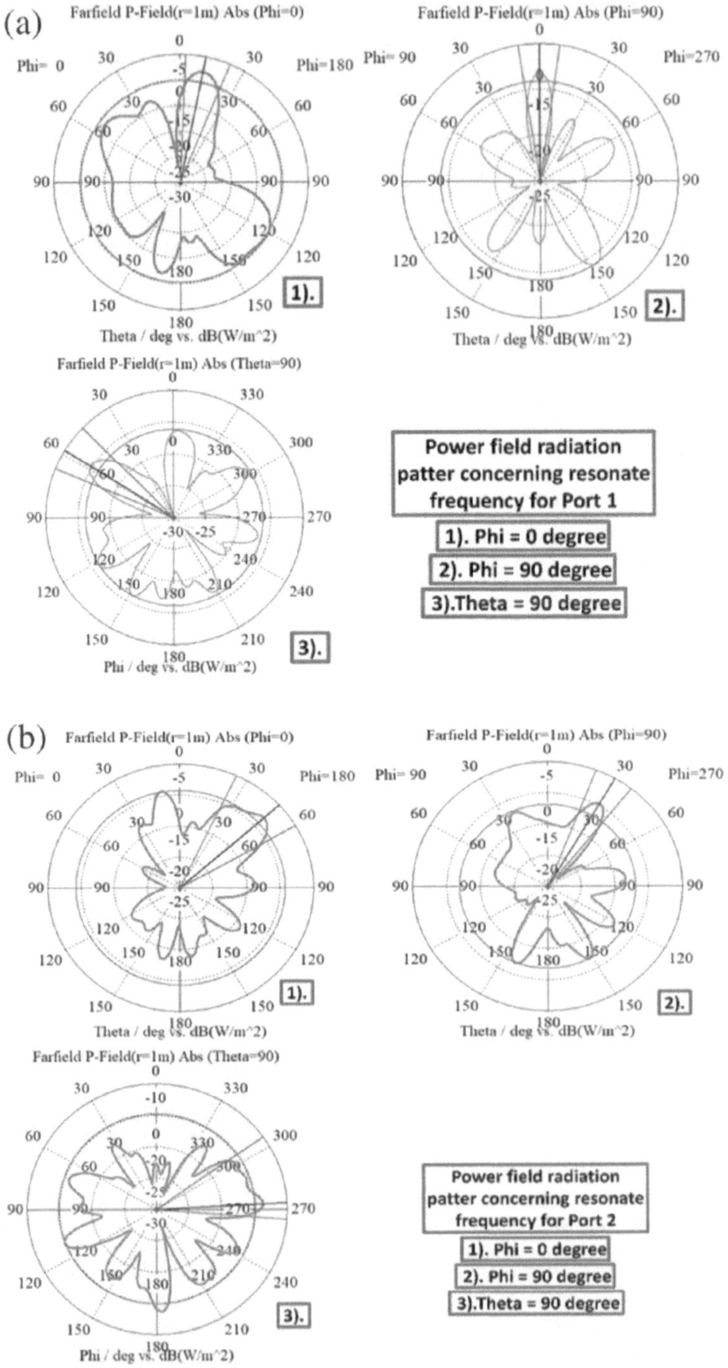

Figure 10.19 2D powerfield pattern concerning respective ports for proposed MIMO antenna.

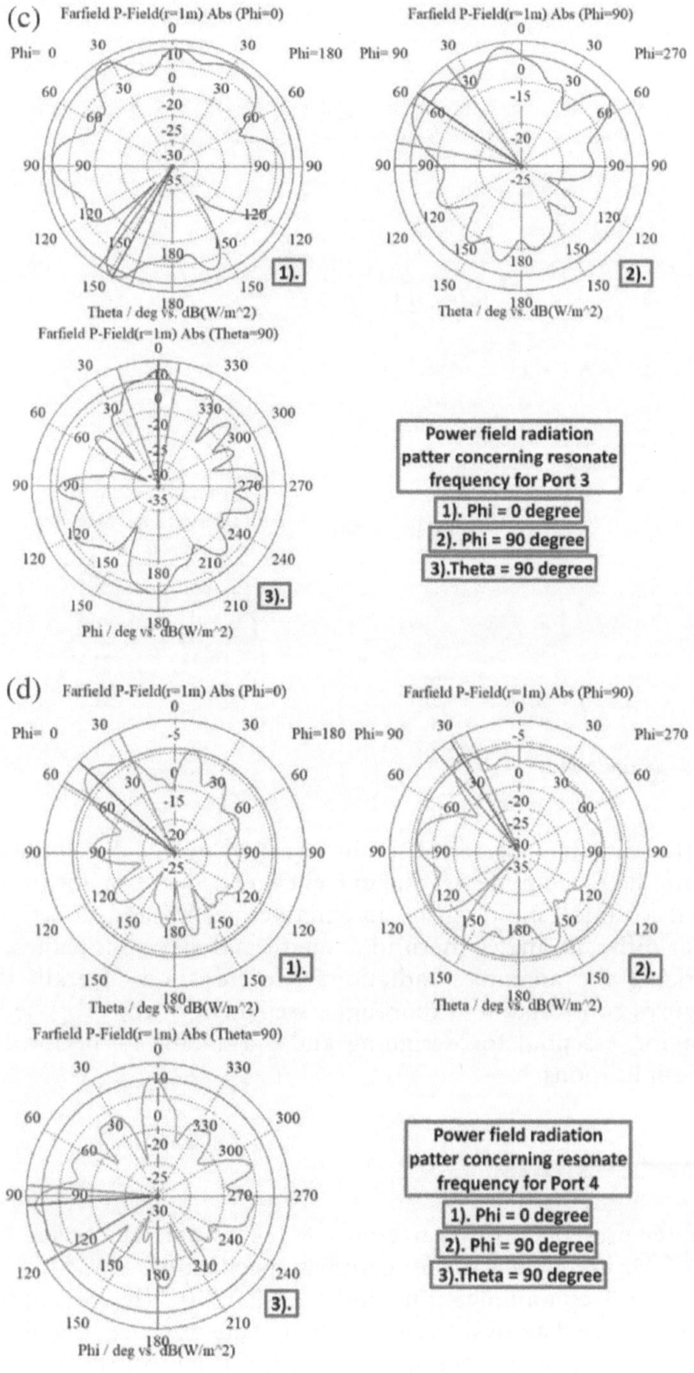

Figure 10.19 (Continued)

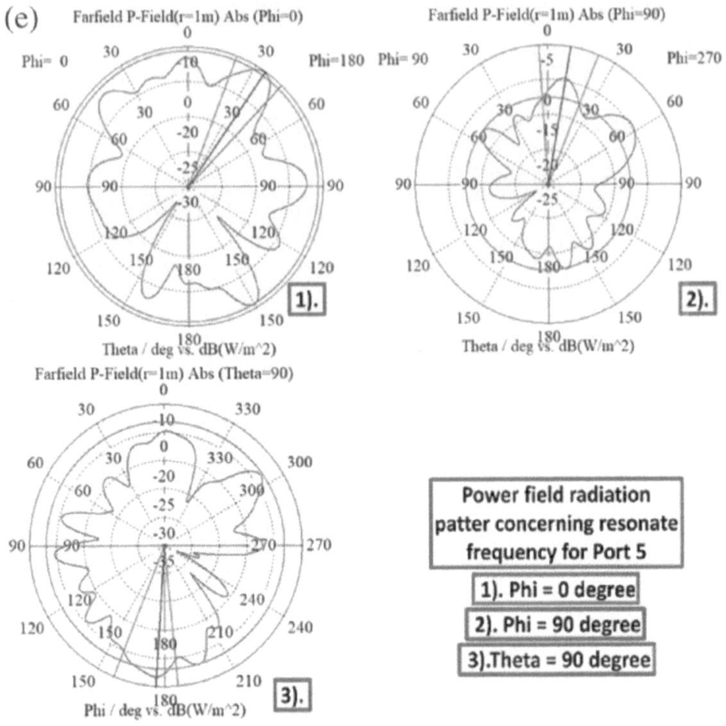

Figure 10.19 (Continued)

view of the antenna's performance in terms of power distribution, electromagnetic field components, and current flow. The resonant frequencies are crucial for understanding the antenna's efficiency and directivity. The analysis at different angles, both in azimuth and elevation planes, aids in characterizing the antenna's radiation characteristics. Overall, the presented figures contribute to a thorough assessment of the proposed antenna's behavior, essential for designing and optimizing its performance in practical applications.

10.4 CONCLUSIONS

This chapter presents a novel five-port MIMO antenna designed for the 24 to 40 GHz frequency range, catering specifically to n257, n258 satellite-to-ground communication, and n259, n260 NR 5G applications with material-based analysis. Fr4 and Roger substrate are used to analyze the approaches for the proposed antenna, and Roger substrate is suitable for desirable outcomes. The antenna draws inspiration from congruent

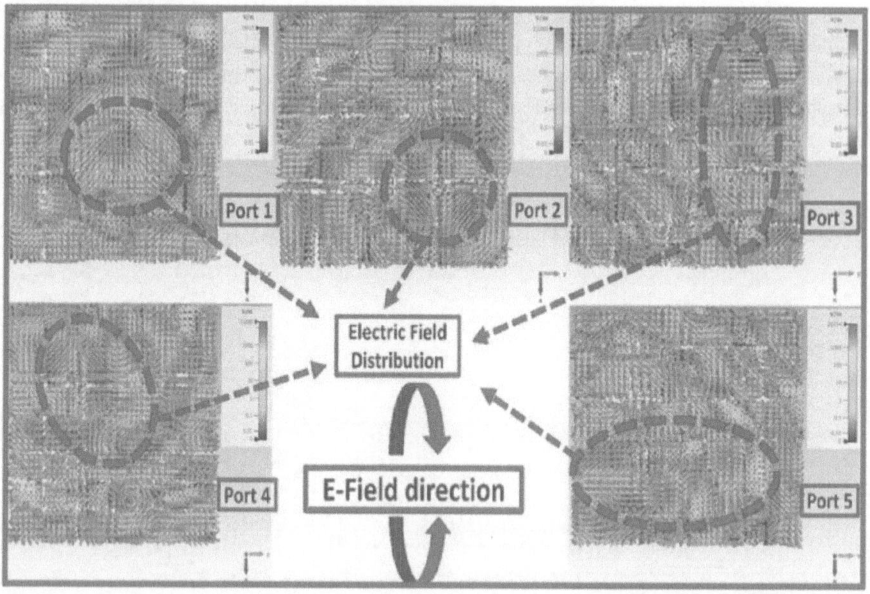

Figure 10.20 E field distribution graphs concerning resonance with respective ports for proposed MIMO antenna.

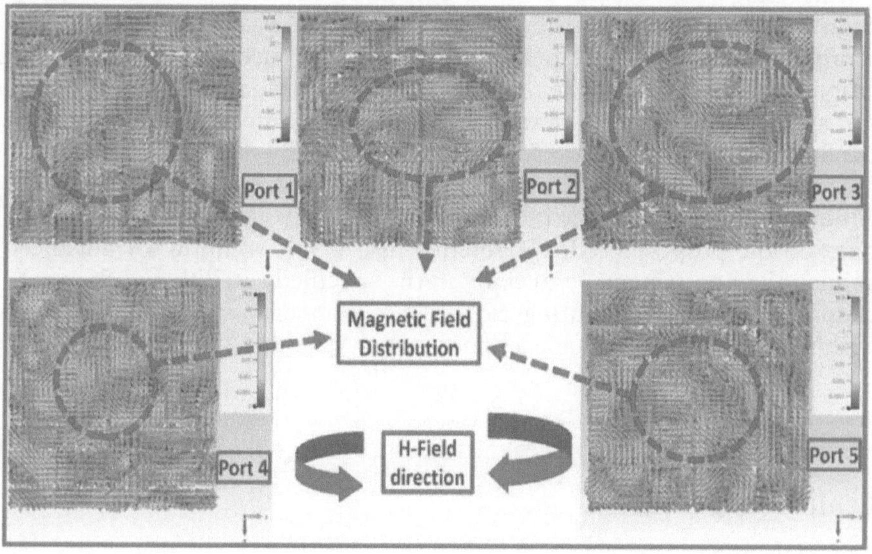

Figure 10.21 H-field distribution graphs concerning resonance with respective ports for proposed MIMO antenna.

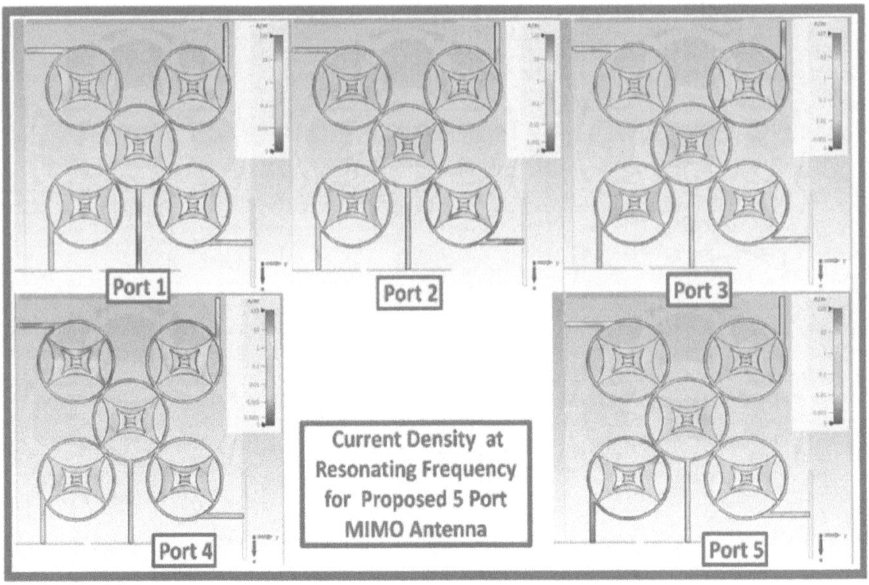

Figure 10.22 Current distribution graphs concerning resonance with respective ports for proposed MIMO antenna.

circular rings with fractal element antennas, boasting a compact size, a peak gain of 7 dBi, and a roger substrate. Notably, the MIMO configuration exhibits impressive performance metrics, with an envelope correlation coefficient (ECC) below 0.012, a diversity gain exceeding 9.666 dB, and enhanced isolation across its ports. The antenna achieves a wide bandwidth, ranging from 5.5 GHz for ports 1, 3.4, 4.2 GHz for port 2, 3.4, 4.2 GHz for port 3, 3.4.2 GHz for port 4, and 3.4, 5 GHz for port 5, ensuring robust communication capabilities. These attributes collectively position the proposed MIMO antenna as a viable solution for advanced communication systems, especially in the specified satellite and 5G applications, thereby contributing to the evolution of wireless technologies within the 24 to 40 GHz spectrum.

DECLARATIONS

Ethical Approval

Not applicable

Competing interests

There are no relevant financial or non-financial competing interests to the findings.

Funding

Our gratitude extends to SERB and the Department of Science and Technology, Government of India (Sanction Order No: EEQ/2021/000700; Dated 04 March 2022), for generously funding our research, enabling its successful execution.

Availability of data and materials

The datasets/materials produced in this study can be obtained upon reasonable requests directed to the corresponding author, ensuring transparency and accessibility for further research purposes.

REFERENCES

1. G. Bharti and J. S. Sivia, "A design of multiband nested square shaped ring fractal antenna with circular ring elements for wireless applications", *Progress in Electromagnetics Research C*, vol. 108, pp.115–125, 2021.
2. A. Kumar and A. P. S. Pharwaha, "Development of a modified Hilbert curve fractal antenna for multiband applications", *IETE Journal of Research*, vol. 68, pp.1–10, 2020.
3. S. Singhal and A. K. Singh, "Elliptical monopole based super wideband fractal antenna", *Microwave and Optical Technology Letters*, vol. 62, no. 3, pp.1324–1328, 2020.
4. G. S. Karthikeya, M. P. Abegaonkar and S. K. Koul, "Low cost high gain triple band mmwave Sierpinski antenna loaded with uniplanar EBG for 5G applications", *2017 IEEE International Conference on Antenna Innovations & Modern Technologies for Ground*, Aircraft and Satellite Applications (iAIM), pp. 1–5, Nov. 2017.
5. R. K. Garg, M. V. D. Nair, S. Singhal and R. Tomar, "A new type of compact ultra-wideband planar fractal antenna with WLAN band rejection", *Microwave and Optical Technology Letters*, vol. 62, no. 7, pp.2537–2545, 2020.
6. H. Ullah and F. A. Tahir, "A novel snowflake fractal antenna for dual-beam applications in 28 GHz band", *IEEE Access*, vol. 8, pp. 19873–19879, 2020.
7. M. Kaur and J. S. Sivia, "Giuseppe Peano and Cantor set fractals based miniaturized hybrid fractal antenna for biomedical applications using artificial neural network and firefly algorithm", *International Journal of RF and Microwave Computer-Aided Engineering*, vol. 30, no. 1, p. e22000, 2020.
8. S. Singhal, "Four arm windmill shaped superwideband terahertz MIMO fractal antenna", *Optik*, vol. 219, p. 165093, 2020.

9. M. Khalily, R. Tafazolli, P. Xiao and A. A. Kishk, "Broadband mm-wave microstrip array antenna with improved radiation characteristics for different 5G applications", *IEEE Transactions on Antennas and Propagation*, vol. 66, no. 9, pp. 4641–4647, 2018.
10. S. D. Sateaa, M. S. Hussein, Z. G. Faisal, A. M. Abood and H. D. Satea, "Design and simulation of dual-band rectangular microstrip patch array antenna for millimeter-wave", *Bulletin of Electrical Engineering and Informatics*, vol. 11, no. 1, pp.299–309, 2022.
11. S. Ghosh, G. S. Baghel and M. V. Swati, "A low-profile, high gain, dual-port, planar array antenna for mm-wave powered 5G IoT systems", *AEU-International Journal of Electronics and Communications*, vol. 109, p. 154354, 2022.
12. M. Abdullah, S. H. Kiani, L. F. Abdulrazak, A. Iqbal, M. Bashir, S. Khan and S. Kim, "High-performance multiple-input multiple-output antenna system for 5G mobile terminals", *Electronics,* vol. 8, p. 1090, 2019.
13. S. H. Kiani, A. Altaf, M. Abdullah, F. Muhammad, N. Shoaib, M. R. Anjum, R. Damaševičius and T. Blažauskas, "Eight element side edged framed MIMO antenna array for future 5G smart phones", *Micromachines*, vol. 11, p. 956, 2020.
14. H. M. Marzouk, M. I. Ahmed and A. H. A. M. Shaalan, "Novel dual-band 28/38 GHz MIMO antennas for 5G mobile applications", *Progress in Electromagnetics Research*, vol. 93, pp. 103–117, 2019.
15. Y. Liu, Y. Li, L. Ge, J. Wang and B. Ai, "A compact hepta-band mode-composite antenna for sub (6, 28, and 38) GHz applications", *IEEE Transactions on Antennas and Propagation*, vol. 68, pp. 2593–2602, 2020.
16. P. M. Sunthari and R. Veeramani, "Multiband microstrip patch antenna for 5G wireless applications using MIMO techniques", *Proceedings of the 2017 First International Conference on Recent Advances in Aerospace Engineering (ICRAAE)*, Coimbatore, India, 3–4 March 2017, pp. 1–5.
17. D. T. T. Tu, N. G. Thang, N. T. Ngoc, N. T. B. Phuong and V. Van Yem, "28/38 GHz dual-band MIMO antenna with low mutual coupling using novel round patch EBG cell for 5G applications", *Proceedings of the 2017 International Conference on Advanced Technologies for Communications (ATC)*, Quy Nhon, Vietnam, 18–20 Oct. 2017; pp. 64–69.
18. H. Marzouk, M. Ahmed and A. Shaalan, "A novel dual-band 28/38 GHz slotted microstip MIMO antenna for 5G mobile applications", *Proceedings of the 2019 IEEE International Symposium on Antennas and Propagation and USNC-URSI Radio Science Meeting*, Atlanta, GA, USA, 7–12 July 2019, pp. 607–608.
19. W. Ali, S. Das, H. Medkour and S. Lakrit, "Planar dual-band 27/39 GHz millimeter-wave MIMO antenna for 5G applications", *Microsystem Technologies*, vol. 27, pp. 1–10, 2020.
20. A. Raj, D. C. Dhubkarya, D. K. Srivastava and D. Mandal, "Design and analysis of square shape slot cut high gain Sierpinski carpetfractal antenna for wireless applications", *Microwave and Optical Technology Letters*, vol. 65, pp. 2337–2343, 2023.

21. Mian Muhammad Kamal, Shouyi Yang, Xin-cheng Ren, Ahsan Altaf, Saad Hassan Kiani, Muhammad Rizwan Anjum, Amjad Iqbal, Muhammad Asif and Sohail Imran Saeed, "Infinity shell shaped MIMO antenna array for mm-wave 5G applications", *Electronics*, vol. 10, no. 2, p. 165, 2021.
22. A. Raj and D. Mandal, "Design and experimental analysis of fractal antennae with ground-defected structure for expected 6G and 5G mm-wave communication and wireless applications", *Transactions on Emerging Telecommunications Technologies*, vol. 2023, p. e4900, 2023.
23. A. Raj and D. Mandal, "Design and fabrication of high-gain array antenna for 5G communication and wireless applications", D. K. Sarkar, P. K. Sadhu, S. Bhunia, J. Samanta and S. Paul (Eds.), *Proceedings of the 4th International Conference on Communication, Devices and Computing*, ICCDC 2023. Lecture Notes in Electrical Engineering, vol. 1046, Springer, Singapore, 2023.
24. A. Raj and D. Mandal, "Design of slot cut modified rectangular shape and square shape antenna for L band and 5G applications", *2023 First International Conference on Microwave, Antenna and Communication (MAC)*, Prayagraj, India, pp. 1–4, 2023.
25. A. Raj and D. Mandal, "Comparative analysis of the 2×2 patch antenna array with modified fractal antenna for 5G communication and wireless applications", *2022 IEEE Microwaves, Antennas, and Propagation Conference (MAPCON)*, Bangalore, India, pp. 670–675, 2022.
26. A. Raj and D. Mandal, "Design and implementation of hybrid fed array antenna for Sub-6 GHz and 5G mm-wave communication and wireless applications", *International Journal of Microwave and Wireless Technologies*, pp.1–20, 2023.
27. Arun Raj and Durbadal Mandal, "A novel substrate cylindrical cavity 4×4 fractal inspired MIMO antenna for 5G - n258", *Satellite Communication*, IEEE International Conference on Recent Advances in Systems Science and Engineering (RASSE), November 08–11, 2023, Kerala, India, pp. 1–7, 2023.
28. D. Verma, A. Raj and D. Mandal, "Design and analysis of UWB fractal array antenna for 5G and millimeter wave communication", *2023 9th International Conference on Signal Processing and Communication (ICSC)*, NOIDA, India, pp. 84–88, 2023.
29. P. Kumar, A. Raj and D. Mandal, "Design and fabrication of Sierpinski fractal-based MIMO antenna for sub 6GHz and IOT applications", *2023 2nd International Conference on Futuristic Technologies (INCOFT)*, Belagavi, Karnataka, India, pp. 1–5, 2023.
30. Divya Sharma, Rajesh Mehra and B. Raj, "Comparative analysis of photovoltaic technologies for high efficiency solar cell design", *Superlattices and Microstructures*, vol. 153, p. 106861, May 2021.
31. Pawandeep Kaur, Avtar Singh Buttar and B. Raj, "A comprehensive analysis of nanoscale transistor based biosensor: a review", *Indian Journal of Pure and Applied Physics*, vol. 59, pp. 304–318, April 2021.
32. Divya Yadav, Balwant Raj and B. Raj, "Design and simulation of low power microcontroller for IoT applications", *Journal of Sensor Letters*, vol. 18, pp. 401–409, May 2020.

33. Shailendra Singh and B. Raj, "A 2-D analytical surface potential and drain current modeling of double-gate vertical t-shaped tunnel FET", *Journal of Computational Electronics*, vol. 19, pp.1154–1163, April 2020.
34. Jeetendra Singh and B. Raj, "An accurate and generic window function for non-linear memristor model", *Journal of Computational Electronics*, vol. 18 no. 2, pp. 640–647, June 2019.
35. Manjit Kaur, Neena Gupta, Sanjeev Kumar, B. Raj and Arun Kumar Singh, "Comparative RF and crosstalk analysis of carbon based nano interconnects", *IET Circuits, Devices & Systems*, vol. 15, no. 6, pp. 493–503, Feb. 2021.
36. Nehru Kandasamy, Firdous Ahmad, D. Ajitha, B. Raj and Nagarjuna Telagam "Quantum dot cellular automata based scan flip flop and boundary scan register", *IETE Journal of Research*, vol. 66, pp. 535–548, 2020.
37. Jeetendra Singh and B. Raj, "Design and investigation of 7T2M NVSARM with enhanced stability and temperature impact on store/restore energy", *IEEE Transactions on Very Large Scale Integration Systems*, vol. 27, no. 6, pp. 1322–1328, June 2019.
38. Anil Kumar Bhardwaj, Sumeet Gupta, B. Raj and Amandeep Singh, "Impact of double gate geometry on the performance of carbon nanotube field effect transistor structures for low power digital design", *Computational and Theoretical Nanoscience*, vol. 16, pp. 1813–1820, 2019.
39. Neeraj Jain and B. Raj, "Thermal stability analysis and performance exploration of asymmetrical dual-k underlap spacer (ADKUS) SOI FinFET for security and privacy applications", *Indian Journal of Pure & Applied Physics*, vol. 57, pp. 352–360, May 2019.
40. Neeraj Jain and Balwinder Raj, "Dual-k spacer region variation at the drain side of asymmetric SOI FinFET structure: performance analysis towards the analog/RF design applications", *Journal of Nanoelectronics and Optoelectronics*, vol. 14, pp. 349–359, March 2019.
41. Jeetendra Singh, Sanjeev Sharma, B. Raj and Mamta Khosla "Analysis of barrier layer thickness on performance of In1-xGaxAs based gate stack cylindrical gate nanowire MOSFET", *Journal of Nanoelectronics and Optoelectronics*, vol. 13, pp. 1473–1477, Oct. 2018.
42. Neeraj Jain and B. Raj, "Analysis and performance exploration of high-k SOI FinFETs over the conventional low-k SOI FinFET toward analog/RF design", *Journal of Semiconductors*, vol.39, no. 12, p. 124002-1–7, Dec. 2018.
43. Candy Goyal, Jagpal Singh Ubhi and B. Raj, "A reliable leakage reduction technique for approximate full adder with reduced ground bounce noise", *Journal of Mathematical Problems in Engineering*, vol. 2018, p. 16, 15 Oct. 2018.
44. Jeetendra Singh Anuradha, B. Raj and Mamta Khosla, "Design and performance analysis of nano-scale memristor-based nonvolatile SRAM", *Journal of Sensor Letter*, vol. 16, pp. 798–805, Oct. 2018.
45. Girish Wadhwa and B. Raj, "Parametric variation analysis of charge-plasma-based dielectric modulated JLTFET for biosensor application", *IEEE Sensor Journal*, vol. 18, no. 15, pp. 6070–6077, 1 Aug. 2018.

46. Jeetendra Singh and B. Raj, "Comparative analysis of memristor models for memories design", *Journal of Semiconductors*, vol. 39, No. 7, p. 074006-1–12, July 2018.
47. Divya Yadav, Shailesh Singh Chouhan, Santosh Kumar Vishvakarma and B. Raj, "Application specific microcontroller design for IoT based WSN", *Sensor Letter*, vol. 16, pp. 374–385, May 2018.
48. Gurmohan Singh, R. K. Sarin and B. Raj, "Fault-tolerant design and analysis of quantum-dot cellular automata based circuits", *IEEE/IET Circuits, Devices & Systems*, vol. 12, pp. 638–664, 2018.
49. Jeetendra Singh and B. Raj, "Modeling of mean barrier height levying various image forces of metal insulator metal structure to enhance the performance of conductive filament based memristor model", *IEEE Nanotechnology*, vol. 17, no. 2, pp. 268–267, March 2018.
50. Aakash Jain, Sanjeev Sharma and B. Raj, "Analysis of triple metal surrounding gate (TM-SG) III–V nanowire MOSFET for photosensing application", *Opto-Electronics Journal*, vol. 26, no. 2, pp. 141–148, May 2018.
51. Aakash Jain, Sanjeev Sharma and B. Raj, "Design and analysis of high sensitivity photosensor using cylindrical surrounding gate MOSFET for low power sensor applications", *Engineering Science and Technology, an International Journal*, vol. 19, no. 4, pp. 1864–1870, Dec. 2016.
52. G. Saiphani Kumar, Amandeep Singh and B. Raj, "Design and analysis of gate all around CNTFET based SRAM cell design", *Journal of Computational Electronics*, vol. 17, no. 1, pp. 138–145, March 2018.
53. Gurinder pal Singh, B. S. Sohi and Balwinder Raj, "Material properties analysis of graphene base transistor (GBT) for VLSI analog circuits", *Indian Journal of Pure & Applied Physics*, vol. 55, pp. 896–902, Dec. 2017.
54. Sunil Kumar and B. Raj, "Estimation of stability and performance metric for inward access transistor based 6T SRAM cell design using n-type/p-type DMDG-GDOV TFET", *IEEE VLSI Circuits and Systems Letter*, vol. 3, no. 2, pp. 25–39, June 2017
55. Shashikant Sharma, Anjan Kumar, Manisha Pattanaik and B. Raj, "Forward body biased multimode multi-threshold CMOS technique for ground bounce noise reduction in static CMOS adders", *International Journal of Information and Electronics Engineering*, vol. 3, no. 3, pp. 567–572, 2013.
56. Hamendra Singh, Pankaj Kumar and Balwinder Raj, "Performance analysis of majority gate SET based 1-bit full adder", *International Journal of Computer and Communication Engineering*,vol. 2, no. 4, pp. 1–8, 2013.
57. Anil Kumar Bhardwaj, Sumeet Gupta and B. Raj, "Investigation of parameters for Schottky barrier (SB) height for Schottky barrier based carbon nanotube field effect transistor device", *Journal of Nanoelectronics and Optoelectronics*, vol. 15, pp. 783–791, July 2020.
58. Priya Bansal and B. Raj, "Memristor: a versatile nonlinear model for dopant drift and boundary issues," *Journal of Computational and Theoretical Nanoscience*, vol. 14, no. 5, pp. 2319–2325, May 2017.
59. Neeraj Jain and B. Raj, "An analog and digital design perspective comprehensive approach on Fin-FET (Fin-Field Effect transistor) technology – a review", *Reviews in Advanced Sciences and Engineering*, vol. 5, pp. 1–14, 2016.

60. Balwinder Raj, A. K. Saxena and S. Dasgupta, "Analytical modeling for the estimation of leakage current and subthreshold swing factor of nanoscale double gate FinFET device", *Microelectronics International*, vol. 26, pp. 53–63, 2009.
61. Shailendra Singh Soniya, Girish Wadhwa and Balwinder Raj, "An analytical modeling for dual source vertical tunnel field effect transistor", *International Journal of Recent Technology and Engineering*, vol. 8 no. 2, pp. 603–608, July 2019.
62. Shailendra Singh and B. Raj, "Design and analysis of heterojunction vertical T-shaped tunnel field effect transistor", *Journal of Electronics Material*, vol. 48, no. 10, pp. 6253–6260, Oct. 2019.
63. Candy Goyal, Jagpal Singh Ubhi and B. Raj, "A low leakage CNTFET based inexact full adder for low power image processing applications", *International Journal of Circuit Theory and Applications*, vol. 47, no. 9, pp. 1446–1458, Sep. 2019.
64. B. Raj, A. K. Saxena and S. Dasgupta, "A compact drain current and threshold voltage quantum mechanical analytical modeling for FinFETs", *Journal of Nanoelectronics and Optoelectronics*, vol. 3, no. 2, pp. 163–170, 2008.
65. Girish Wadhwa and B. Raj, "An analytical modeling of charge plasma based tunnel field effect transistor with impacts of gate underlap region", *Superlattices and Microstructures*, vol. 142, pp. 106512, June 2020.
66. Shailendra Singh and B. Raj, "Modeling and simulation analysis of SiGe heterojunction double GateVertical t-shaped tunnel FET", *Superlattices and Microstructures*, vol. 142, p. 106496, June 2020.
67. Neeraj Jain and Balwinder Raj, "Device and circuit co-design perspective comprehensive approach on FinFET technology – a review", *Journal of Electron Devices*, vol. 23, no. 1, pp. 1890–1901, 2016.
68. Sunil Kumar and B. Raj, "Analysis of I_{ON} and ambipolar current for dual-material gate-drain overlapped DG-TFET", *Journal of Nanoelectronics and Optoelectronics*, vol. 11, pp. 323–333, June 2016.
69. Naveed Anjum, Tarun Bali and B. Raj, "Design and simulation of handwritten multiscript character recognition", *International Journal of Advanced Research in Computer and Communication Engineering*, vol. 2, no. 7, pp. 2544–2549, July 2013.
70. Karmjit Singh and B. Raj, "Performance and analysis of temperature dependent multi-walled carbon nanotubes as global interconnects at different technology nodes", *Journal of Computational Electronics*, vol. 14 no. 2, pp. 469–476, June 2015.
71. Sunil Kumar and B. Raj, "Compact channel potential analytical modeling of DG-TFET based on evanescent–mode approach," *Journal of Computational Electronics*, vol. 14, no. 2, pp. 820–827, July 2015.
72. Karmjit Singh and B. Raj, "Temperature dependent modeling and performance evaluation of multi-walled CNT and single-walled CNT as global interconnects", *Journal of Electronic Materials*, vol. 44, no. 12, pp. 4825–4835, Dec. 2015.
73. V. K. Sharma, M. Pattanaik and B. Raj, "INDEP approach for leakage reduction in nanoscale CMOS circuits", *International Journal of Electronics*, vol. 102, no. 2, pp. 200–215, 2014.

74. Karmjit Singh and B. Raj, "Influence of temperature on MWCNT bundle, SWCNT bundle and copper interconnects for nanoscaled technology nodes", *Journal of Materials Science: Materials in Electronics*, vol. 26, no. 8, pp. 6134–6142, 2015.
75. Naveed Anjum, Tarun Bali and B. Raj, "Design and simulation of handwritten Gurumukhi and Devanagri numerical recognition", *International Journal of Computer Applications*, vol. 73, no. 12, pp. 16–21, 2013.
76. S. Khandelwal, V. Gupta, B. Raj and R. D. Gupta, "Process variability aware low leakage reliable nano scale DG-FinFET SRAM cell design technique", *Journal of Nanoelectronics and Optoelectronics*, vol. 10, no. 6, pp. 810–817, Dec. 2015.
77. V. K. Sharma, M. Pattanaik and B. Raj, "ONOFIC approach: low power high speed nanoscale VLSI circuits design", *International Journal of Electronics*, vol. 101, no. 1, pp. 61–73, 2014.
78. S. Khandelwal, Balwinder Raj and R. D. Gupta, "FinFET based 6T SRAM cell design: analysis of performance metric, process variation and temperature effect", *Journal of Computational and Theoretical Nanoscience*, vol. 12, pp. 2500–2506, 2015.
79. Sumit Singh, Shekhar Yadav, Jagdeep Rahul, Anurag Srivastava and B. Raj, "Impact of HfO_2 in graded channel dual insulator double gate MOSFET", *Journal of Computational and Theoretical Nanoscience*, vol. 12, no. 6, pp. 950–953, April 2015.
80. Vijay Kumar Sharma, Manisha Pattanaik and B. Raj, "PVT variations aware low leakage INDEP approach for nanoscale CMOS circuits", *Microelectronics Reliability*, vol. 54, pp. 90–99, 2014.
81. B. Raj, A. K. Saxena and S. Dasgupta, "Quantum mechanical analytical modeling of nanoscale DG FinFET: evaluation of potential, threshold voltage and source/drain resistance", *Journal of Material Science in Semiconductor Processing*, vol. 16, no. 4, pp. 1131–1137, 2013.
82. Maisagalla Gopal, Siva Sankar D Prasad and Balwinder Raj, "8T SRAM cell design for dynamic and leakage power reduction", *International Journal of Computer Applications*, vol. 71, no. 9, pp. 43–48, June 2013.
83. Manisha Pattanaik, B. Raj, Shashikant Sharma and Anjan Kumar, "Diode based trimode multi-threshold CMOS technique for ground bounce noise reduction in static CMOS adders", *Advanced Materials Research*, vol. 548, pp. 885–889, 2012.
84. Balwinder Raj, A. K. Saxena and S. Dasgupta, "Nanoscale FinFET based SRAM cell design: analysis of performance metric, process variation, underlapped FinFET and temperature effect", *IEEE Circuits and System Magazine*, vol. 11, no. 2, pp. 38–50, 2011.
85. V. K. Sharma, M. Pattanaik and Balwinder Raj, "Leakage current ONOFIC approach for deep submicron VLSI circuit design", *International Journal of Electrical, Computer, Electronics and Communication Engineering*, vol. 7, no. 4, pp. 239–244, 2013.
86. Parminder Kaur, Sandeep Singh Gill and B. Raj, "Comparative analysis of OFETs materials and devices for sensor applications", *Journal of Silicon*, vol. 14, pp. 4463–4471, 2022.

87. Sanjeev Kumar Sharma, Parveen Kumar, Balwant Raj and B. Raj, "$In_{1-x}Ga_xAs$ double metal gate-stacking cylindrical nanowire MOSFET for highly sensitive photo detector", *Journal of Silicon*, vol. 14, pp. 3535–3541, 2022.
88. B. Raj, A. K. Saxena and S. Dasgupta, "Analytical modeling of quasi planar nanoscale double gate FinFET with source/drain resistance and field dependent carrier mobility: a quantum mechanical study", *Journal of Computer*, vol. 4, no. 9, pp.1–8, 2009.
89. S. Bhushan, S. Khandelwal and B. Raj, "Analyzing different mode FinFET based memory cell at different power supply for leakage reduction", *Seventh International Conference on Bio-Inspired Computing: Theories and Application* (BIC-TA 2012), Advances in Intelligent Systems and Computing, vol. 202, pp. 89–100, 2013.
90. Jeetendra Singh and B. Raj, "Temperature dependent analytical modeling and simulations of nanoscale memristor", *Engineering Science and Technology, an International Journal*, vol. 21, pp. 862–868, Oct. 2018.
91. Shradhya Singh, Shashi Bala, Balwant Raj and B. Raj, "Improved sensitivity of dielectric modulated junctionless transistor for nanoscale biosensor design", *Sensor Letter*, vol.18, pp.328–333, April 2020.
92. Vivek Kumar, Santosh Kumar Vishvakarma and B. Raj, "Design and performance analysis of ASIC for IoT applications", *Sensor Letter*, vol. 18, pp. 31–38, Jan. 2020.
93. Akanksha Jaiswal, R. K. Sarin, B. Raj and Shikha Sukhija, "A novel circular slotted microstrip-fed patch antenna with three triangle shape defected ground structure for multiband applications", *Advanced Electromagnetic*, vol. 7, no. 3, pp. 56–63, Aug. 2018.
94. Girish Wadhwa and B. Raj, "Label free detection of biomolecules using charge-plasma-based gate underlap dielectric modulated junctionless TFET", *Journal of Electronic Materials*, vol. 47, no. 8, pp. 4683–4693, Aug. 2018.
95. Gurmohan Singh, R. K. Sarin and B. Raj, "Design and performance analysis of a new efficient coplanar quantum-dot cellular automata adder", *Indian Journal of Pure & Applied Physics*, vol. 55, pp. 97–103, Feb. 2017.
96. Parminder Kaur, Vikas Pandey and B. Raj, "Comparative study of efficient design, control and monitoring of solar power using IoT", *Sensor Letter*, vol. 18, pp. 419–426, May 2020.
97. Anil Kumar Bhardwaj, Sumeet Gupta and B. Raj, "Development & analysis of compact model for double gate Schottky barrier CNTFET", *Journal of Nanoelectronics and Optoelectronics*, vol. 15, pp. 1199–1208, Aug. 2020.
98. Girish Wadhwa, Priyanka Kamboj, Jeetendra Singh and B. Raj, "Design and investigation of junctionless DGTFET for biological molecule recognition", *Transactions on Electrical and Electronic Materials*, vol. 22, pp. 282–289, 2021.
99. Sachin Kumar Verma, Shailendra Singh, Girish Wadhwa and B. Raj, "Detection of biomolecules using charge-plasma based gate underlap dielectric modulated dopingless TFET", *Transactions on Electrical and Electronic Materials*, vol. 21, pp. 528–535, June 2020.

100. Neeraj Jain and B. Raj, "Impact of underlap spacer region variation on electrostatic and analog/RF performance of symmetrical high-k SOI FinFET at 20 nm channel length", *Journal of Semiconductors*, vol. 38, no. 12, pp. 122002, Dec. 2017.
101. Shailendra Singh and B. Raj, "Analytical modeling and simulation analysis of T-shaped III–V heterojunction vertical T-FET", *Superlattices and Microstructures*, vol. 147, pp. 106717, Nov. 2020.
102. Gurmohan Singh, R. K. Sarin and B. Raj, "Design and analysis of area efficient QCA based reversible logic gates", *Journal of Microprocessors and Microsystems*, vol. 52, pp. 59–68, May 2017.
103. Shailendra Singh and B. Raj, "Parametric variation analysis on hetero-junction vertical t-shape TFET for suppressing ambipolar conduction", *Indian Journal of Pure and Applied Physics*, vol. 58, pp. 478–485, June 2020.
104. Shailendra Singh Soniya, Girish Wadhwa and Balwinder Raj, "Design and analysis of dual source vertical tunnel field effect transistor for high performance", *Transactions on Electrical and Electronics Materials,* vol. 21, pp. 74–82, Oct. 2019.
105. Manjit Kaur, Neena Gupta, Sanjeev Kumar, B. Raj and Arun Kumar Singh, "RF performance analysis of intercalated graphene nanoribbon based global level interconnects", *Journal of Computational Electronics,* vol. 19, pp.1002–1013, June 2020.
106. Girish Wadhwa and B. Raj, "Design and performance analysis of junctionless TFET biosensor for high sensitivity", *IEEE Nanotechnology*, vol. 18, pp. 567–574, 2019.
107. Jeetendra Singh and B. Raj, "Enhanced nonlinear memristor model encapsulating stochastic dopant drift", *Journal of Nanoelectronics and Optoelectronics*, vol. 14, pp. 958–963, 2019.
108. Anjana Bhardwaj, Pradeep Kumar, B. Raj and Sunny Anand, "Design and performance optimization of doping-less vertical nanowire TFET using gate stack technique", *Journal of Electronic Materials*, vol. 41, no. 7, pp. 4005–4013, 2022.
109. Jeetendra Singh and B. Raj, "Tunnel current model of asymmetric MIM structure levying various image forces to analyze the characteristics of filamentary memristor", *Applied Physics A*, vol. 125, no. 3, pp. 203.1–203.11, Feb. 2019.
110. Candy Goyal, Jagpal Singh Ubhi and B. Raj, "Low leakage zero ground noise nanoscale full adder using source biasing technique", *Journal of Nanoelectronics and Optoelectronics*, vol. 14, pp. 360–370, March 2019.
111. Gurmohan Singh, R. K. Sarin and B. Raj, "A novel robust exclusive-OR function implementation in QCA nanotechnology with energy dissipation analysis", *Journal of Computational Electronics*, vol. 15, no. 2, pp. 455–465, June 2016.
112. Girish Wadhwa, Priyanka Kamboj and Balwinder Raj, "Design optimisation of junctionless TFET biosensor for high sensitivity", *Advances in Natural Sciences: Nanoscience and Nanotechnology*, vol. 10, no. 7, p. 45001, 2019.
113. Priya Bansal and B. Raj, "Memristor modeling and analysis for linear dopant drift kinetics", *Journal of Nanoengineering and Nanomanufacturing*, vol. 6, pp. 1–7, 2016.

114. D. Vaithiyanathan Ashima and B. Raj, "Performance analysis of charge plasma induced graded channel Si nanotube", *Journal of Engineering Research*, EMSME Special no. pp. 146–154, Aug. 2021.
115. Abhishek Singh Tomar, Vijay Kumar Magraiya and B. Raj, "Scaling of access and data transistor for high performance DRAM cell design", *Quantum Matter*, vol. 2, pp. 412–416, Oct. 2013.
116. Neeraj Jain and B. Raj, "Parasitic capacitance and resistance model development and optimization of raised source/drain SOI FinFET structure for analog circuit applications", *Journal of Nanoelectronics and Optoelectronins*, vol. 13, pp. 531–539, April 2018.
117. Shradhya Singh, S. K. Vishvakarma and B. Raj, "Analytical modeling of split-gate junction-less transistor for a biosensor application", *Sensing and Bio-Sensing*, vol. 18, pp. 31–36, April 2018.
118. Maisagalla Gopal and Balwinder Raj, "Low power 8T SRAM cell design for high stability video applications", *ITSI Transaction on Electrical and Electronics Engineering*, vol. 1, no. 5, pp. 91–97, 2013.
119. Balwinder Raj, Jatin Mitra, Deepak Kumar Bihani, V. Rangharajan, A. K. Saxena and S. Dasgupta, "Analysis of noise margin, power and process variation for 32 nm FinFET based 6T SRAM cell", *Journal of Computer*, vol. 5, no. 6, pp.1–8, 2010.

Chapter 11

2D nanomaterials based THz electromagnetic radiators

Shobhit Saxena

11.1 INTRODUCTION

The purpose of this chapter is to show the importance of using nanomaterials in improving the performance of electromagnetic radiators operating in the Terahertz (THz) range [1]. The submillimeter band (THz band) covers the frequency range from 0.1 THz to 10 THz [2]. Terahertz antennas can be effectively utilized in wireless communication systems where high data rates are required [3]. Besides that, they also find applications in medical imaging, security screening, astronomy, sensors, remote sensing, etc. It is very challenging to design THz antennas as the dimensions are very small (due to the small wavelength). Fabricating such small structures requires advanced fabrication and manufacturing technologies. Most of the substrates used at lower frequency bands fail to perform well in the THz band [4]. Thus the selection of a substrate that can efficiently work in the THz band is also a very important design criterion. Achieving narrowband/wideband performance in the THz frequency region is also a challenging task as there is a requirement for precise tuning [5].

A nanomaterial is about the size 1 – 100 nm. For a material to be counted as a nanomaterial, at least one dimension of the material must be in the nanoscale (i.e., 1–100 nm) [6]. Based on dimensions, nanomaterials can be classified into 0D, 1D, 2D, and 3D as shown in Figure 11.1. A 2D nanomaterial is a material whose two dimensions remain large and only one dimension is reduced to the nanoscale level. Such structures are like very thin sheets generally known as

Graphene is a very well-known example of a nanosheet [7]. It is the thinnest two-dimensional nanomaterial in the world. Graphene is a single thin layer of graphite, which is a soft black material used in pencil lead [8]. It is an allotrope of carbon in the form of a single layer of atoms in a two-dimensional hexagonal lattice. Graphene has some interesting properties; it is harder than diamond yet more elastic than rubber and lighter than aluminum [9]. Graphene also has a very high thermal conductivity ratio, even better than silver, copper, and diamond. In reality, graphene conducts heat two times better than diamond. Materials that conduct heat very well also

DOI: 10.1201/9781003612797-11

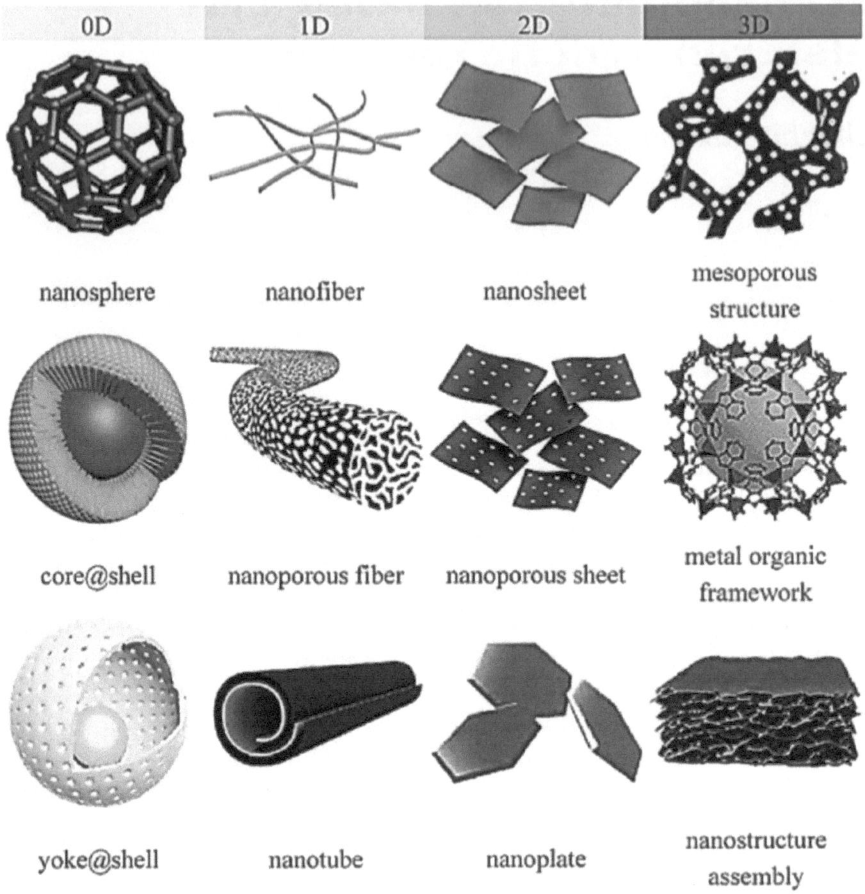

Figure 11.1 Various types of nanomaterials/nanostructures [6].

conduct electricity very well, as both processes transfer energy using electrons. The flat hexagonal lattice of graphene offers relatively less resistance to electrons. Hence, graphene's electrical conductivity is better than copper's. Graphene also has high electron mobility, which is 100 times faster than silicon, the conventional material used in the electronics industry for decades [10].

To show the importance of 2D nanomaterial (graphene), four antennas namely Antenna-A, Antenna-B, Antenna-C, and Antenna-D are presented in this chapter. Antenna-A is a simple patch antenna having only copper in the patch and ground. In Antenna-B, a very narrow and thin graphene film is deposited on a patch to see the effect on the impedance matching of the antenna. Next, Antenna-C and Antenna-D were designed with efficient heat transfer capabilities in mind, as the antennas have to be mounted on a chip [11, 12]. All electromagnetic and thermal simulations were carried out in CST Microwave Studio.

11.2 DESIGNING OF ANTENNA-A AND ANTENNA-B

A simple microstrip line-fed patch antenna (Antenna-A) shown in Figure 11.2 was designed. Various high-frequency substrates from Rogers (like RT/Duroid 5870, RT/Duroid 6006, RO 4350B, etc.) were considered in the designing of the above antenna. Table 11.1 summarizes the effect of changing substrate on the performance of the Antenna-A. Figure 11.3 shows the $|S_{11}|$ performance of Antenna-A.

Next, to show the importance of nanomaterials in improving antenna performance, a graphene film of dimensions 65.9 × 5.37 μm² and a thickness of 0.0025 μm is deposited on the patch of Antenna-A (Figure 11.4). Desirable impedance matching can be obtained by carefully optimizing the dimensions of this graphene film. It is also observed in Antenna-B that the placement of the graphene film on different locations of the patch yields

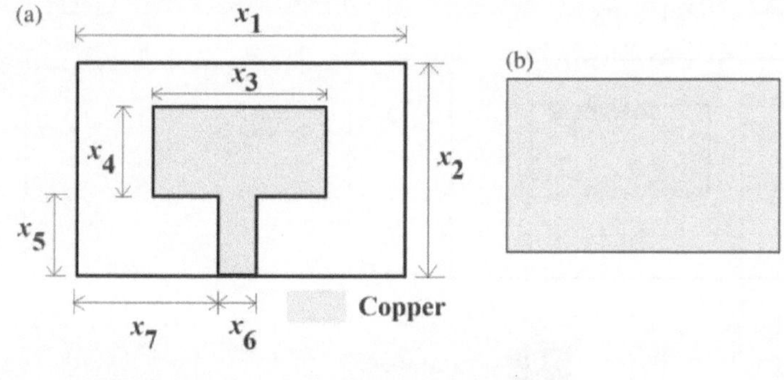

Figure 11.2 Antenna-A: (a) Top view and (b) bottom view.

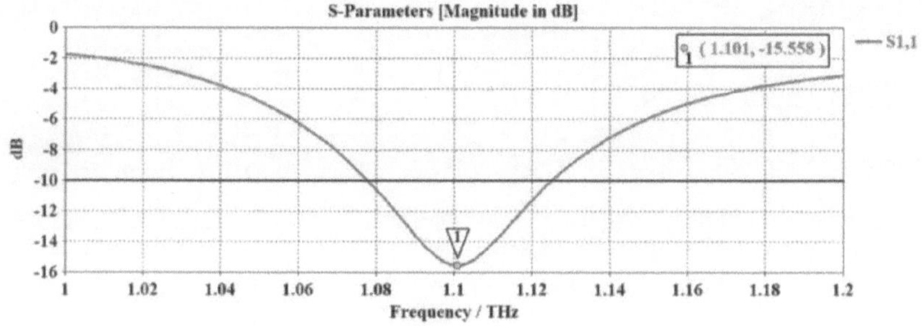

Figure 11.3 $|S_{11}|$ performance of Antenna-A.

different results. Numerous simulations were carried out in this work to get the optimum results [13–17].

It can be easily seen from Figure 11.3, Figure 11.5, Table 11.1, and Table 11.3 that the use of graphene film on the patch in Antenna-B resulted in improved $|S_{11}|$ performance. Thus, better impedance matching can be achieved in antennas by incorporating 2D nanomaterials in the design [18–23].

Figure 11.6 shows the effect of chemical potential on the $|S_{11}|$ performance of Antenna-B. It can be easily seen from Figure 11.6. that optimum $|S_{11}|$ performance is observed at a chemical potential of 0.1 V in Antenna-B. It can also be seen from Figures 11.7 and 11.8 that the use of graphene film affects the electric field (E-field) and surface current distribution [24–31].

Figure 11.9 shows the far-field radiation pattern of Antenna-A at 1.101 THz. It can be easily seen from the figure that Antenna-A achieves a gain of 7.723 dBi and a directivity of 8.404 dBi at 1.101 THz. The far-field radiation pattern of Antenna-B is shown in Figure 11.10. It can be seen from Figure 11.10 that a gain of 6.306 dBi and a directivity of 8.440 dBi have been achieved in Antenna-B at 1.1089 THz. It must be noted from Figure 11.3, Figure 11.5, Figure 11.9, and Figure 11.10 that the improvement

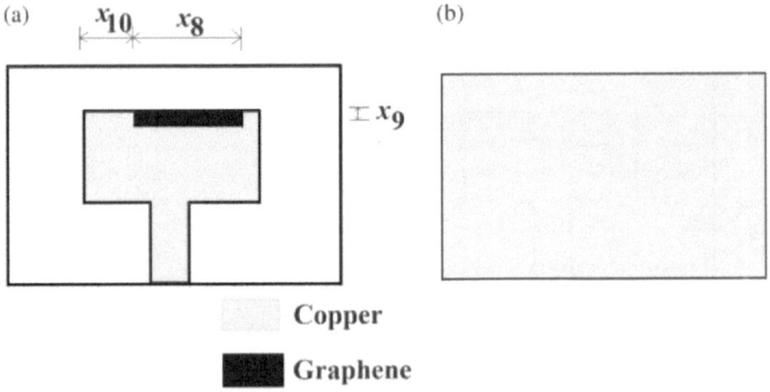

Figure 11.4 Antenna-B: (a) Top view and (b) bottom view.

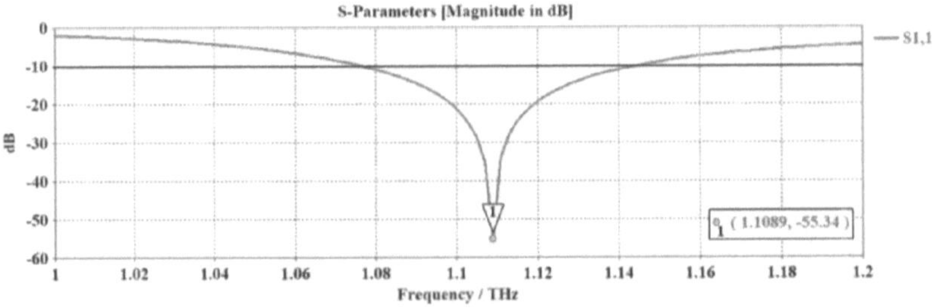

Figure 11.5 $|S_{11}|$ performance of Antenna-B.

Table 11.1 Performance of Antenna-A on various high-frequency substrates

Substrate	ε_r	*tan* δ	*Frequency band (THz)*	$\lvert S_{11}\rvert$ *(dB)/frequency (THz)*
RT/Duroid 5870	2.33	0.0012	1.0526–1.0964	−14.905/1.0745
RT/Duroid 5880	2.20	0.0009	1.0782–1.1248	−15.558/1.10
RO 4350 B	3.66	0.0037	0.9572-–0.98	−11.233/0.9679
RT/Duroid 6002	2.94	0.0012	0.9568–0.98	−11.232/0.9678
RT/Duroid 6202	2.90	0.0015	0.9568–0.9814	−11.405/0.9683
RO 3003	3.00	0.0013	0.95–0.9695	−10.92/0.9587

Table 11.2 Antenna-A and Antenna-B design parameters (in μm)

Parameter	*Value*	*Parameter*	*Value*
x_1	320.688	x_6	10
x_2	185.652	x_7	155.344
x_3	106.896	x_8	40
x_4	78.744	x_9	20
x_5	53.448	x_{10}	38

Table 11.3 Performance of Antenna-B on various high-frequency substrates

Substrate	ε_r	*tan* δ	G_{Length} *(μm)*	*Frequency band (THz)*	$\lvert S_{11}\rvert$ *(dB)/frequency (THz)*
RT/Duroid 5870	2.33	0.0012	41.896	1.0503–1.1185	–50.445/1.0824
RT/Duroid 5880	2.20	0.0009	37.896	1.0762–1.1448	–55.34/1.1089
RO 4350 B	3.66	0.0037	57.896	0.9498–1.0151	–38.981/0.9797
RT/Duroid 6002	2.94	0.0012	60.896	0.95–1.0173	–47.428/0.9816
RT/Duroid 6202	2.90	0.0015	59.896	0.9503–1.0171	–48.597/0.9815
RO 3003	3.00	0.0013	60.896	0.9414–1.0074	–46.43/0.9721

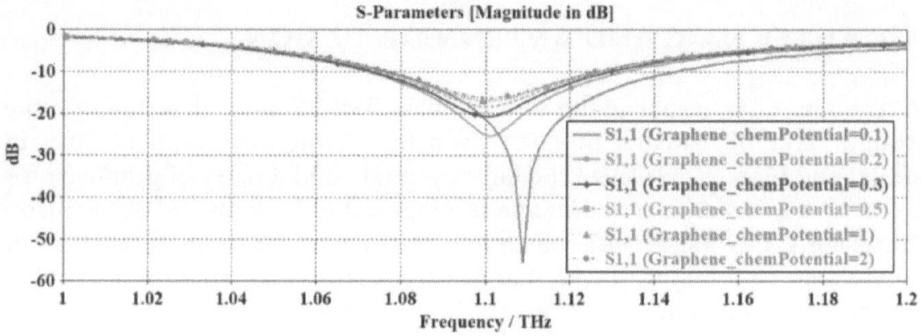

Figure 11.6 Effect of graphene chemical potential on the performance of Antenna-B.

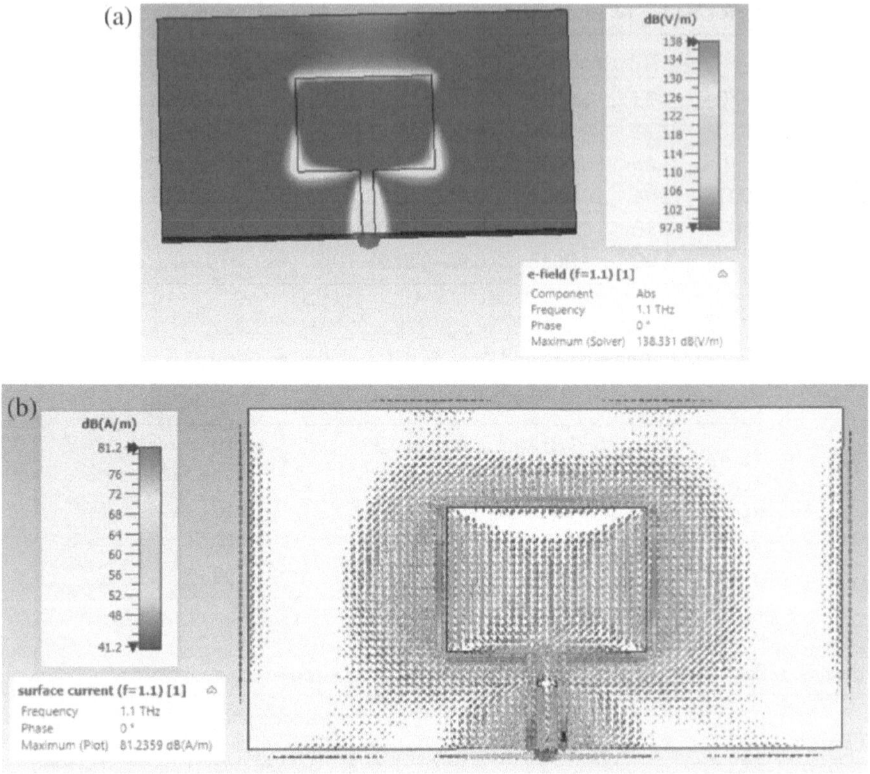

Figure 11.7 Antenna-A (a) E-field at 1.1 THz and (b) surface current distribution at 1.1 THz.

of $|S_{11}|$ in Antenna-B indicates reduced reflections and better transfer of power between the transmission line and antenna. However, the gain of the antenna is dependent on several other factors, and introducing graphene film doesn't mean that there will always be an improvement in gain [32–37].

11.3 ANTENNA-C AND ANTENNA-D DESIGN

In Antenna-C, a graphene film of thickness 0.0025 μm is deposited on the ground plane of Antenna-B. The reason for adding this graphene layer in the ground plane is to utilize the high thermal conductivity of graphene for better heat transfer, as the antenna is supposed to be mounted on a silicon chip. Figure 11.11 shows the patch and the ground plane of Antenna-C. It can be seen from Figure 11.12 that there is a slight degradation in the $|S_{11}|$ of Antenna-C compared to Antenna-B, but incorporating graphene in the ground plane can result in better thermal management [38–43].

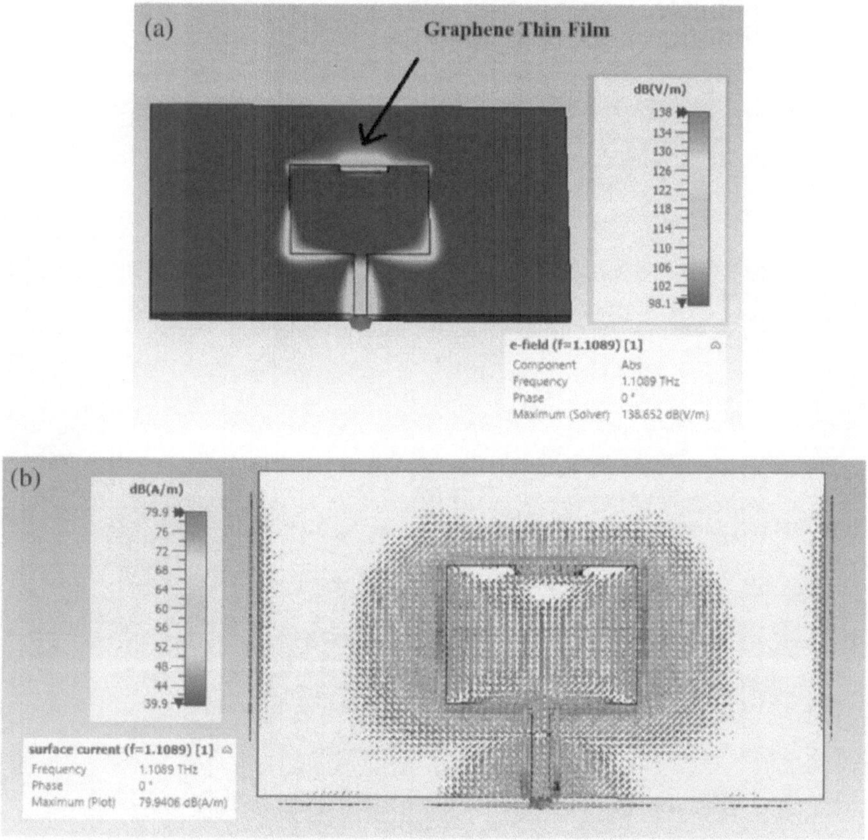

Figure 11.8 Antenna-B: (a) E-field at 1.1089 THz; (b) surface current distribution at 1.1089 THz.

In Antenna-D, to further improve the heat dissipation, a hole/airway has been created in the antenna patch, substrate, and ground (the graphene layer in the ground plane remains untouched) (Figure 11.13). The purpose of creating this hole is to improve airflow and thermal convection. Thus, it can help in preventing overheating and performance degradation of other electronic components placed on a silicon chip. It is very important to consider the size and location of the hole as it can affect the radiation characteristics and performance of the antenna. The optimum value of the outer radius "r" of the cylinder used for creating this hole is found to be 10 μm. It can be seen from Figure 11.14 that there is only a slight change in the $|S_{11}|$ performance of Antenna-D compared to Antenna-C.

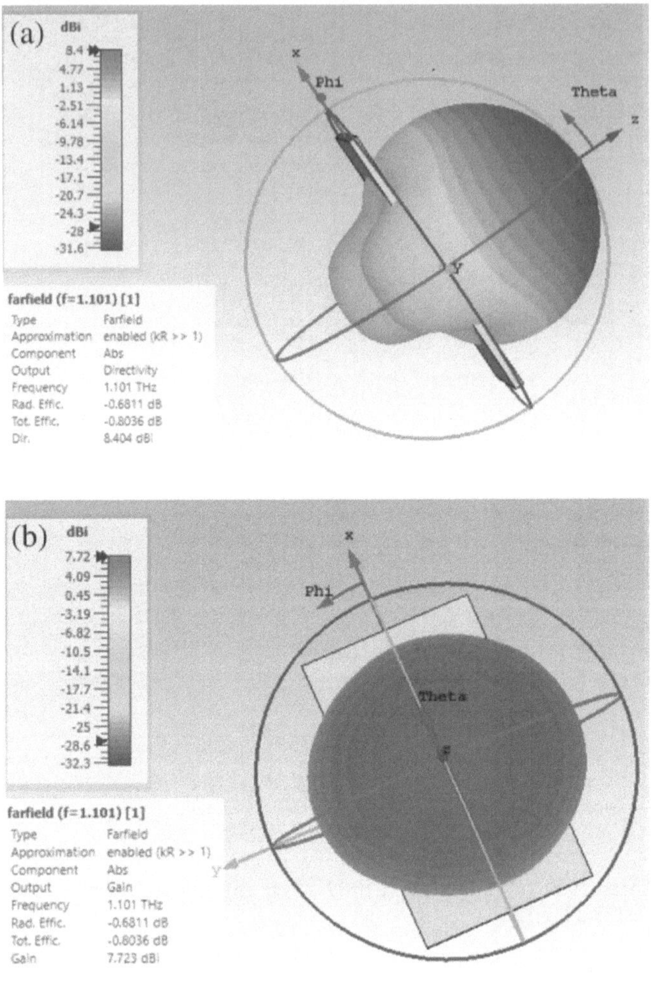

Figure 11.9 Far-field radiation pattern of Antenna-A at 1.101 THz: (a) View-1 (directivity) and (b) View-2 (gain).

The E-field and surface current distribution of Antenna-C and Antenna-D are shown in Figures 11.15 and 11.16, respectively. Figure 11.17 shows the far field radiation pattern of Antenna-C at 1.1089 THz. It can be easily seen from Figure 11.17 that Antenna-C achieves a gain of 6.305 dBi and a directivity of 8.439 dBi at 1.1089 THz. The far-field radiation pattern of Antenna-D is shown in Figure 11.18. It can be seen from Figure 11.18 that a gain of 6.120 dBi and a directivity of 8.417 dBi have been achieved in Antenna-D at 1.08 THz.

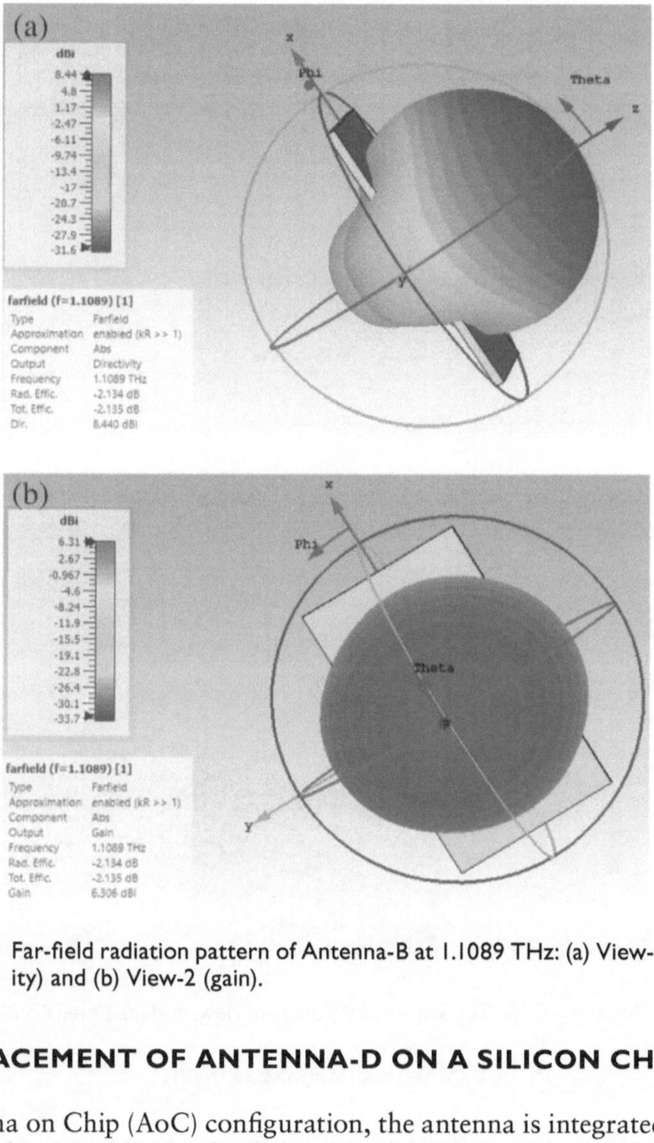

Figure 11.10 Far-field radiation pattern of Antenna-B at 1.1089 THz: (a) View-1 (directivity) and (b) View-2 (gain).

11.4 PLACEMENT OF ANTENNA-D ON A SILICON CHIP

In Antenna on Chip (AoC) configuration, the antenna is integrated directly onto the chip [13, 14]. As electronic devices are getting miniaturized, this technique is quite useful in the manufacturing of compact wireless communication systems [44–48].

In this work, Antenna-D is integrated onto a chip with dimensions of dimension 2000 µm × 2000 µm. The chip comprises a silicon (Si) substrate of thickness 200 µm and an insulating silicon dioxide (SiO_2) layer of thickness 0.1 µm as shown in Figure 11.19. It can be seen from Figure 11.20 that a satisfactory $|S_{11}|$ performance has been achieved when Antenna-D

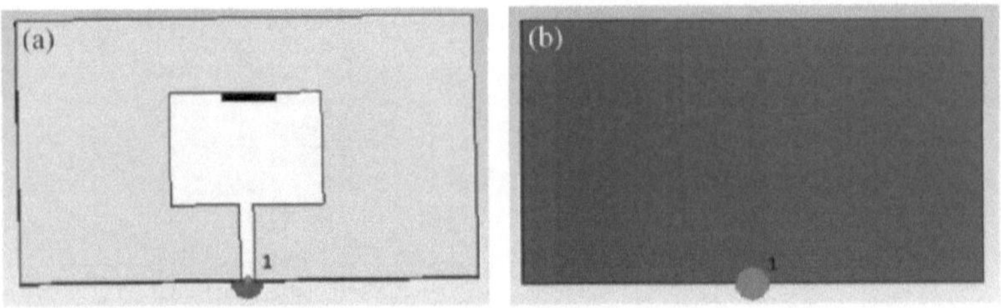

Figure 11.11 Antenna-C: (a) Top view and (b) bottom view.

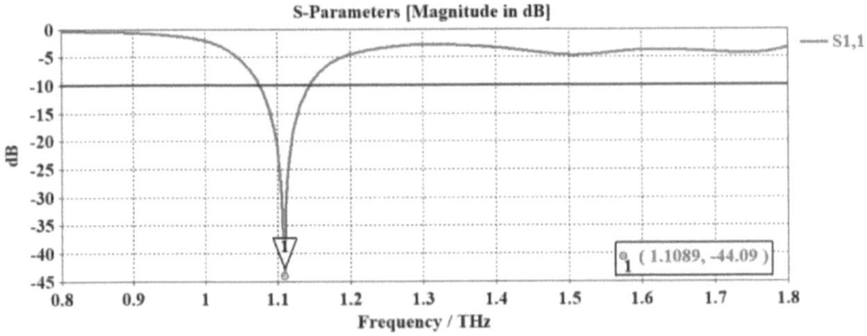

Figure 11.12 $|S_{11}|$ Performance of Antenna-C.

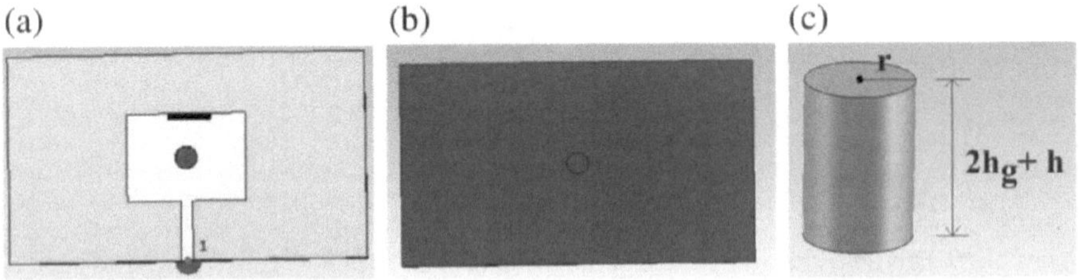

Figure 11.13 Antenna-D: (a) Top view and (b) bottom view, and (c) dimensions of cylinder.

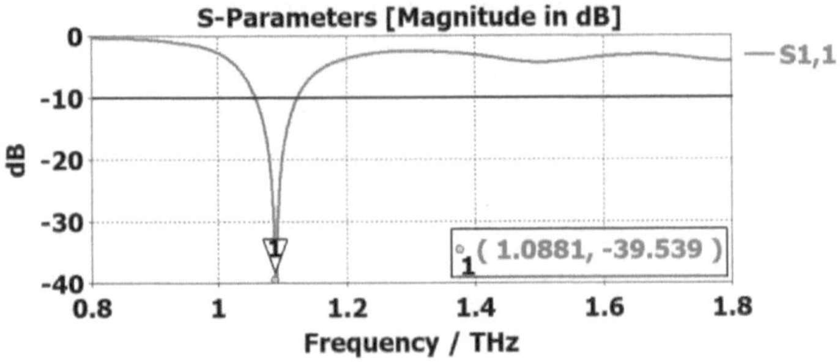

Figure 11.14 $|S_{11}|$ Performance of Antenna-D.

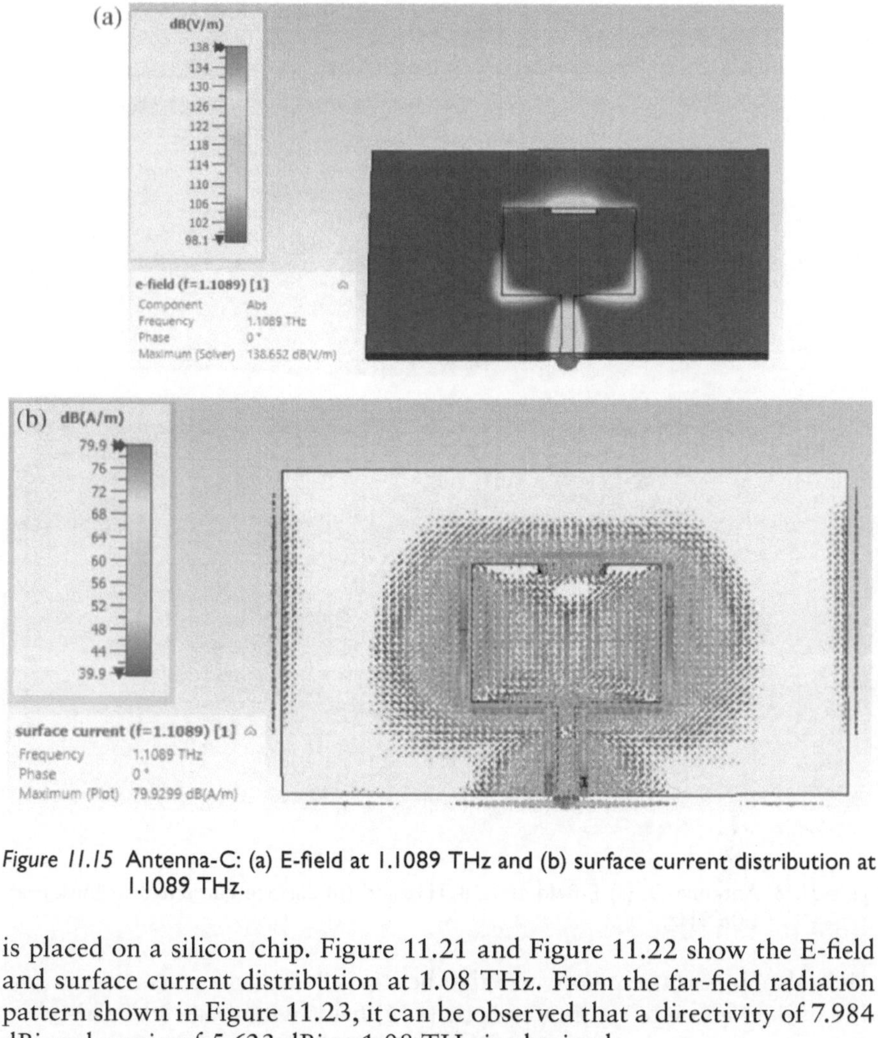

Figure 11.15 Antenna-C: (a) E-field at 1.1089 THz and (b) surface current distribution at 1.1089 THz.

is placed on a silicon chip. Figure 11.21 and Figure 11.22 show the E-field and surface current distribution at 1.08 THz. From the far-field radiation pattern shown in Figure 11.23, it can be observed that a directivity of 7.984 dBi and a gain of 5.623 dBi at 1.08 THz is obtained.

11.5 THERMAL ANALYSIS OF ANTENNA-D PLACED ON A SILICON CHIP

Heat flow density is the rate of heat flow transfer per unit area. Excessive generation of heat can adversely affect the performance of device [15]. Thus, it is always better to have a lower value of heat flow density in AoC. It can be easily seen from Figure 11.24 that the presented AoC configuration has a low value of heat flow density [49–56].

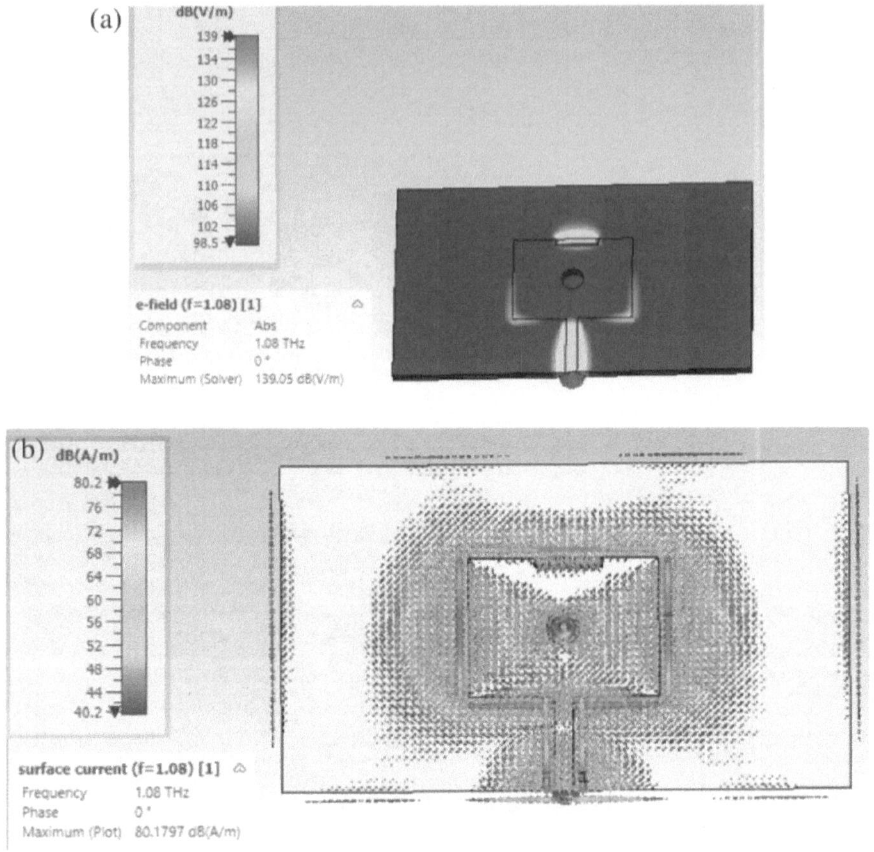

Figure 11.16 Antenna-D: (a) E-field at 1.08 THz and (b) surface current distribution at 1.08 THz.

It is also important to note here that heat is induced whenever there is any kind of interaction of electromagnetic (EM) fields with materials. This phenomenon is more relevant in the THz band as the temperature is increased due to the absorption of EM fields by materials [57–71]. Thus, it is very important to utilize heat dissipation techniques to prevent overheating. The temperature (EM-thermal coupling) of Antenna-D placed on a silicon chip is shown in Figure 11.25.

11.6 CONCLUSION AND FUTURE SCOPE

In this chapter, THz antenna configurations, namely Antenna-A, Antenna-B, Antenna-C and Antenna-D are presented. The importance of using 2D nanomaterial (graphene) in Antenna-B is also highlighted [72–84]. It has

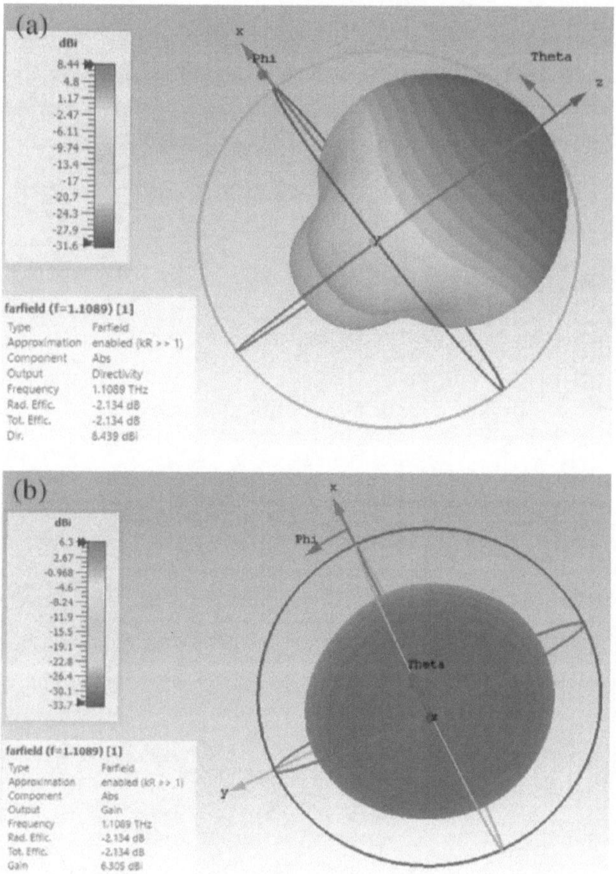

Figure 11.17 Far-field radiation pattern of Antenna-C at 1.1089 THz: (a) View-1 (directivity) and (b) View-2 (gain).

been shown that perfect impedance matching is obtained in Antenna-B compared to Antenna-A. For thermal management purposes, two new configurations, Antenna-C and Antenna-D are presented. Antenna-C makes use of graphene film in the ground plane for better thermal conductivity, while Antenna-D makes use of a hole for better airflow and heat transfer. Satisfactory electromagnetic and thermal performance is observed when Antenna-D is mounted on a silicon chip [85–93].

There is a lot of scope remaining for improving the EM characteristics as well as the thermal performance of the device. Other nanomaterials can be utilized to design THz radiators with better radiation characteristics. The thermal management performance of the antenna can be further improved by utilizing some novel heat dissipation techniques without affecting the EM properties of the device [94–105].

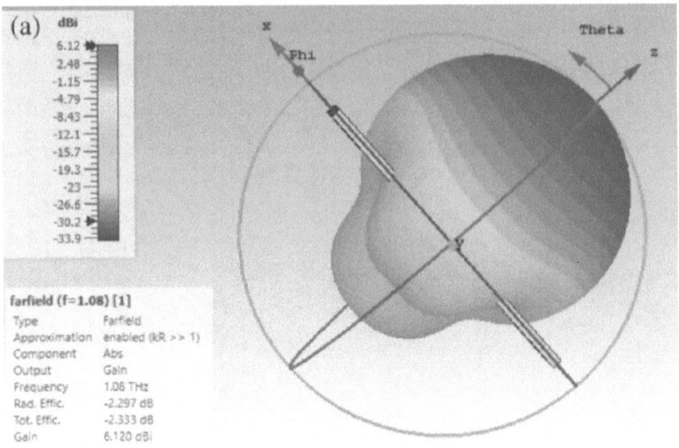

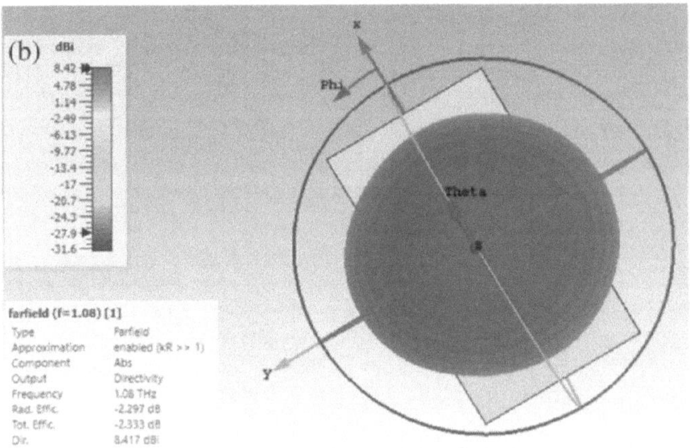

Figure 11.18 Far-field radiation pattern of Antenna-D at 1.08 THz: (a) View-1 (gain) and (b) View-2 (directivity).

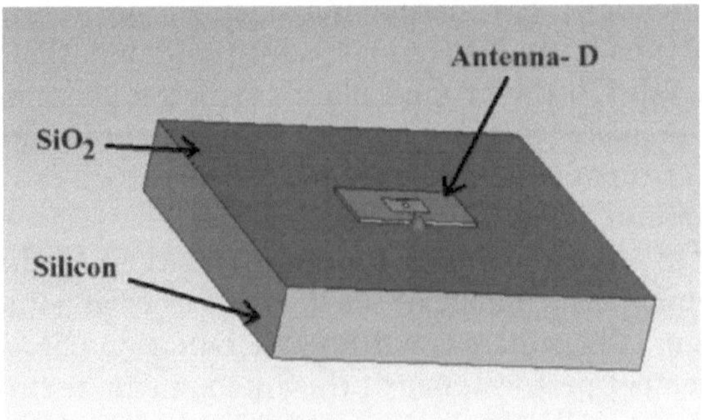

Figure 11.19 Antenna-D placed on a silicon chip.

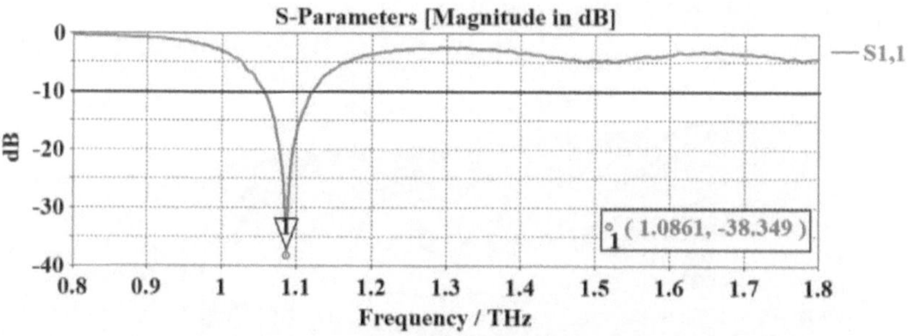

Figure 11.20 $|S_{11}|$ Performance of Antenna-D placed on a silicon chip.

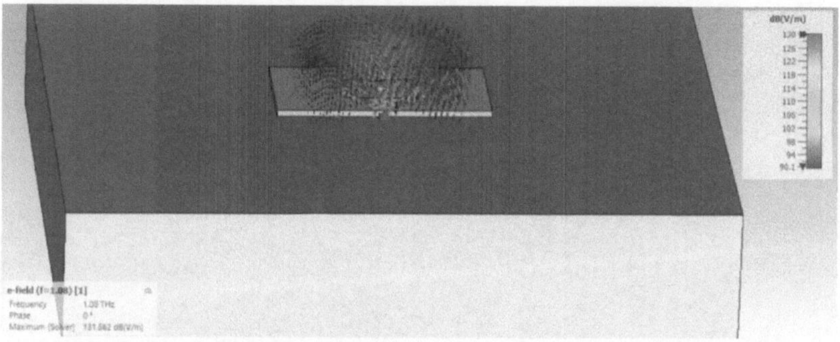

Figure 11.21 E-field distribution of Antenna-D placed on a silicon chip at 1.08 THz.

Figure 11.22 Surface current distribution of Antenna-D placed on a silicon chip at 1.08 THz.

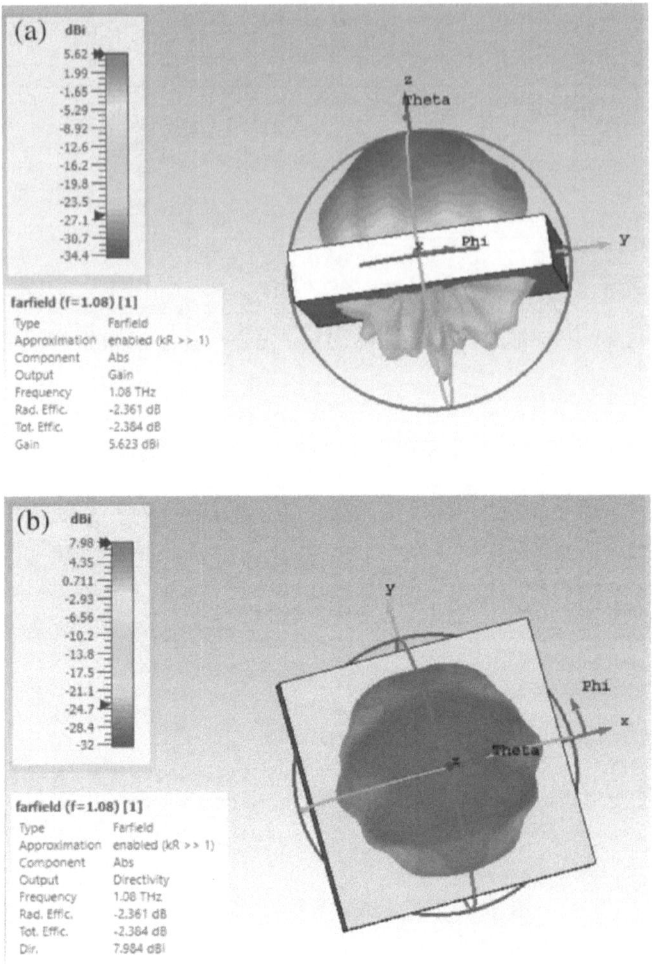

Figure 11.23 Far-field radiation pattern of Antenna-D placed on a silicon chip at 1.08 THz: (a) View-1 (gain) and (b) View-2 (directivity).

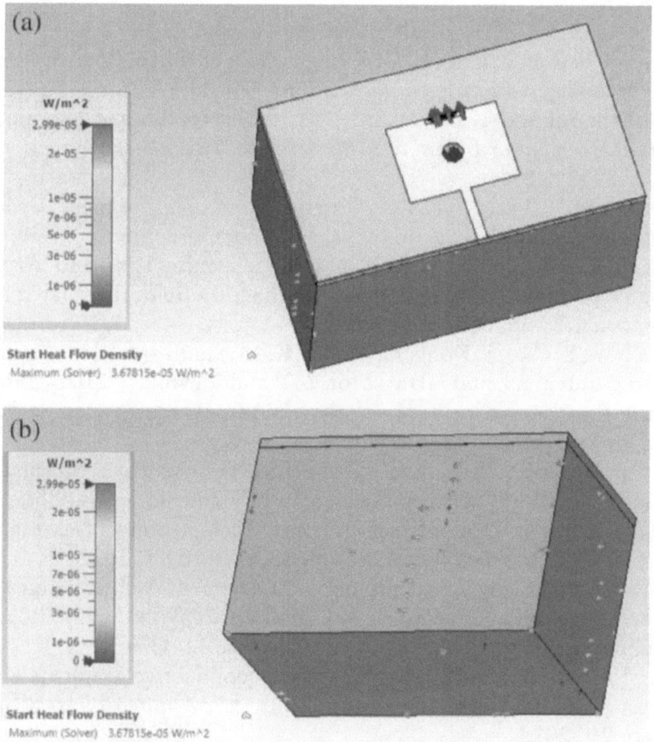

Figure 11.24 Heat flow density in Antenna-D placed on a silicon chip: (a) top view and (b) side view.

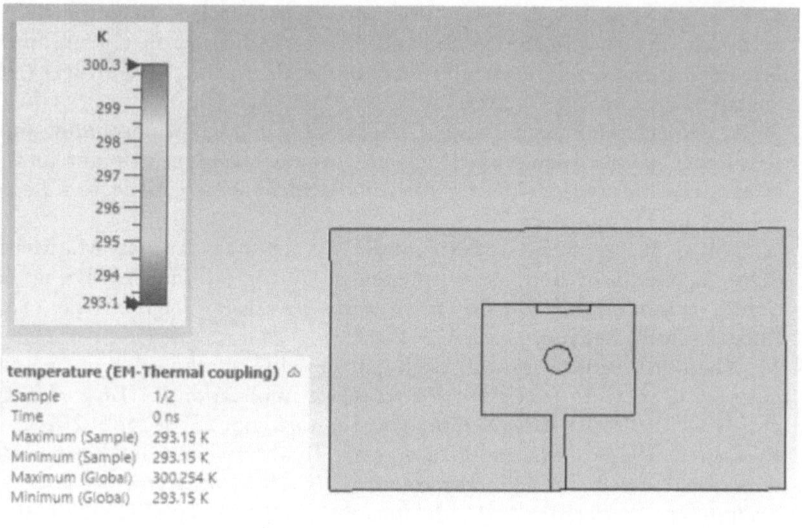

Figure 11.25 Temperature (EM-thermal coupling) of Antenna-D placed on a silicon chip.

REFERENCES

1. Y. He, Y. Chen, L. Zhang, S.-W. Wong and Z. N. Chen, "An overview of terahertz antennas," *China Communications*, vol. 17, no. 7, pp. 124–165, 2020.
2. T. K. Ostmann and T. Nagatsuma, "A review on terahertz communications research", *Journal of Infrared, Millimeter, and Terahertz Waves*, vol. 32, no. 2, pp. 143–171, 2011.
3. X. Chen and X. Liu, "Terahertz antennas and measurement", *Handbook of Antenna Technologies,* Springer Nature, Singapore, pp. 620–659, 2016.
4. Qudsia Rubani, Sindhu Hak Gupta, Shuchismita Pani and Arun Kumar, "Design and analysis of a terahertz antenna for wireless body area networks", *Optik*, vol. 179, pp. 684–690, 2019.
5. C. H. Chan, L. Gao, S. Kong, G.-B. Wu, K. M. Shum and K. F. Chan, "On-chip terahertz antennas and arrays for 6G and beyond", *2023 International Workshop on Antenna Technology (iWAT)*, Aalborg, Denmark, pp. 1–3, 2023.
6. Manisha Byakodi, Narlawar Sagar Shrikrishna, Riya Sharma, Shekhar Bhansali, Yogendra Mishra, Ajeet Kaushik and Sonu Gandhi, "Emerging 0D, 1D, 2D, and 3D nanostructures for efficient point-of-care biosensing", *Biosensors and Bioelectronics:* X, vol. 12, p. 100284, 2022.
7. K. Novoselov, V. Falko, L. Colombo, P. Gellert, M. Schwab and K. Kim, "A roadmap for graphene", *Nature*, vol. 490, no. 7419, pp. 192–200, 2012.
8. X. Huang, T. Leng, X. Zhang, J. C. Chen, K. H. Chang, A. K. Geim, K. S. Novoselov and Z. Hu, "Binder-free highly conductive graphene laminate for low cost printed radio frequency applications", *Applied Physics Letters*, vol. 106, no. 20, 2015.
9. M. Akbari, M. W. A. Khan, M. Hasani, T. Bj¨orninen, L. Syd¨anheimo and L. Ukkonen, "Fabrication and characterization of graphene antenna for low-cost and environmentally friendly RFID tags", *IEEE Antennas and Wireless Propagation Letters*, vol. 15, no. c, pp. 1569–1572, 2016.
10. S. K. Tripathi and A. Kumar, "Performance comparison of graphene terahertz antenna and copper terahertz antenna", *Advances in Computational Intelligence and Communication Technology: Proceedings of Springer CICT,* Singapore, pp. 489–497, 2019.
11. W. A. Ahmad, M. Kucharski, A. Di Serio, H. J. Ng, C. Waldschmidt and D. Kissinger, "Planar highly efficient high-gain 165 GHz on-chip antennas for integrated radar sensors", *IEEE Antennas and Wireless Propagation Letters,* vol. 18, no. 11, pp. 2429–2433, 2019.
12. H. Dinis, M. Zamith, J. Fernandes, J. Magalhaes and P. M. Mendes, "On-chip, efficient and small antenna array for millimeter-wave applications", *International Workshop on Antenna Technology (iWAT)*, Seoul, Korea (South), 2015, pp. 227–228, 2015.
13. M. Alibakhshikenari, et al., "High-gain on-chip antenna design on silicon layer with aperture excitation for terahertz applications", *IEEE Antennas and Wireless Propagation Letters,* vol. 19, no. 9, pp. 1576–1580, 2020.
14. K. Tasaki, R. Takigawa and H. Kanaya, "Terahertz band on-chip one-sided directional wide band slot array antenna", *IEEE 23rd Electronics Packaging Technology Conference (EPTC)*, Singapore, Singapore, pp. 326–329, 2021.

15. W. Zhang, Z. Zhang and L. Zhu, "Thermal effects on high-power microwave antennas: a review", *Progress in Electromagnetics Research*, vol. 154, pp. 1–13, 2015.
16. Manjit Kaur, Neena Gupta, Sanjeev Kumar, B. Raj and Arun Kumar Singh, "Comparative RF and crosstalk analysis of carbon based nano interconnects", *IET Circuits, Devices & Systems*, vol. 15, no. 6, pp. 493–503, Feb. 2021.
17. Nehru Kandasamy, Firdous Ahmad, D. Ajitha, B. Raj and Nagarjuna Telagam, "Quantum dot cellular automata based scan flip flop and boundary scan register", *IETE Journal of Research*, vol. 66, pp. 535–548, 2020.
18. Jeetendra Singh and B. Raj, "Design and investigation of 7T2M NVSARM with enhanced stability and temperature impact on store/restore energy", *IEEE Transactions on Very Large Scale Integration Systems*, vol. 27, no. 6, pp. 1322–1328, June 2019.
19. Anil Kumar Bhardwaj, Sumeet Gupta, B. Raj and Amandeep Singh, "Impact of double gate geometry on the performance of carbon nanotube field effect transistor structures for low power digital design", *Computational and Theoretical Nanoscience*, vol. 16, pp. 1813–1820, 2019.
20. Neeraj Jain and B. Raj, "Thermal stability analysis and performance exploration of asymmetrical dual-k underlap spacer (ADKUS) SOI FinFET for security and privacy applications", *Indian Journal of Pure & Applied Physics*, vol. 57, pp. 352–360, May 2019.
21. Neeraj Jain and Balwinder Raj, "Dual-k spacer region variation at the drain side of asymmetric SOI FinFET structure: performance analysis towards the analog/RF design applications", *Journal of Nanoelectronics and Optoelectronics*, vol. 14, pp. 349–359, March 2019.
22. Jeetendra Singh, Sanjeev Sharma, B. Raj and Mamta Khosla "Analysis of barrier layer thickness on performance of In1-xGaxAs based gate stack cylindrical gate nanowire MOSFET", *Journal of Nanoelectronics and Optoelectronics*, vol. 13, pp. 1473–1477, Oct. 2018.
23. Neeraj Jain and B. Raj, "Analysis and performance exploration of high-k SOI FinFETs over the conventional low-k SOI FinFET toward analog/RF design", *Journal of Semiconductors*, vol. 39, no. 12, p. 124002-1–7, Dec. 2018.
24. Candy Goyal, Jagpal Singh Ubhi and B. Raj, "A reliable leakage reduction technique for approximate full adder with reduced ground bounce noise", *Journal of Mathematical Problems in Engineering*, vol. 2018, p. 16, 15 Oct. 2018.
25. Jeetendra Singh Anuradha, B. Raj and Mamta Khosla, "Design and performance analysis of nano-scale memristor-based nonvolatile SRAM", *Journal of Sensor Letter*, vol. 16, pp. 798–805, Oct. 2018.
26. Girish Wadhwa and B. Raj, "Parametric variation analysis of charge-plasma-based dielectric modulated JLTFET for biosensor application", *IEEE Sensor Journal*, vol. 18, no. 15, pp. 6070–6077, 1 Aug. 2018.
27. Jeetendra Singh and B. Raj, "Comparative analysis of memristor models for memories design", *Journal of Semiconductors*, vol. 39, no. 7, p. 074006-1–12, July 2018.

28. Divya Yadav, Shailesh Singh Chouhan, Santosh Kumar Vishvakarma and B. Raj, "Application specific microcontroller design for IoT based WSN", *Sensor Letter*, vol. 16, pp. 374–385, May 2018.
29. Gurmohan Singh, R. K. Sarin and B. Raj, "Fault-tolerant design and analysis of quantum-dot cellular automata based circuits", *IEEE/IET Circuits, Devices & Systems*, vol. 12, pp. 638–64, 2018.
30. Jeetendra Singh and B. Raj, "Modeling of mean barrier height levying various image forces of metal insulator metal structure to enhance the performance of conductive filament based memristor model", *IEEE Nanotechnology*, vol. 17, no. 2, pp. 268–267, March 2018.
31. Aakash Jain, Sanjeev Sharma and B. Raj, "Analysis of triple metal surrounding gate (TM-SG) III–V nanowire MOSFET for photosensing application", *Opto-Electronics Journal*, vol. 26, no. 2, pp. 141–148, May 2018.
32. Aakash Jain, Sanjeev Sharma and B. Raj, "Design and analysis of high sensitivity photosensor using cylindrical surrounding gate MOSFET for low power sensor applications", *Engineering Science and Technology, an International Journal*, vol. 19, no. 4, pp. 1864–1870, Dec. 2016.
33. G. Saiphani Kumar, Amandeep Singh and B. Raj, "Design and analysis of gate all around CNTFET based SRAM cell design", *Journal of Computational Electronics*, vol. 17, no. 1, pp. 138–145, March 2018.
34. Gurinder pal Singh, B. S. Sohi and Balwinder Raj, "Material properties analysis of graphene base transistor (GBT) for VLSI analog circuits", *Indian Journal of Pure & Applied Physics*, vol. 55, pp. 896–902, Dec. 2017.
35. Sunil Kumar and B. Raj, "Estimation of stability and performance metric for inward access transistor based 6T SRAM cell design using n-type/p-type DMDG-GDOV TFET", *IEEE VLSI Circuits and Systems Letter*, vol. 3, no. 2, pp. 25–39, June 2017.
36. Shashikant Sharma, Anjan Kumar, Manisha Pattanaik and B. Raj, "Forward body biased multimode multi-threshold CMOS technique for ground bounce noise reduction in static CMOS adders", *International Journal of Information and Electronics Engineering*, vol. 3, no. 3, pp. 567–572, 2013.
37. Hamendra Singh, Pankaj Kumar and Balwinder Raj, "Performance analysis of majority gate SET based 1-bit full adder", *International Journal of Computer and Communication Engineering*, vol. 2, no. 4, pp. 1–8, 2013.
38. Anil Kumar Bhardwaj, Sumeet Gupta and B. Raj, "Investigation of parameters for Schottky Barrier (SB) height for Schottky Barrier based carbon nanotube field effect transistor device", *Journal of Nanoelectronics and Optoelectronics*, vol. 15, pp. 783–791, July 2020.
39. Priya Bansal and B. Raj, "Memristor: a versatile nonlinear model for dopant drift and boundary issues", *Journal of Computational and Theoretical Nanoscience*, vol. 14, no. 5, pp. 2319–2325, May 2017.
40. Neeraj Jain and B. Raj, "An analog and digital design perspective comprehensive approach on Fin-FET (Fin-Field Effect transistor) technology – a review", *Reviews in Advanced Sciences and Engineering*, vol. 5, pp. 1–14, 2016.
41. Balwinder Raj, A. K. Saxena and S. Dasgupta, "Analytical modeling for the estimation of leakage current and subthreshold swing factor of nanoscale double gate FinFET device", *Microelectronics International*, vol. 26, pp. 53–63, 2009.

42. Shailendra Singh Soniya, Girish Wadhwa and Balwinder Raj, "An analytical modeling for dual source vertical tunnel field effect transistor", *International Journal of Recent Technology and Engineering*, vol. 8 no. 2, pp. 603–608, July 2019.
43. Shailendra Singh and B. Raj, "Design and analysis of heterojunction vertical T-shaped tunnel field effect transistor", *Journal of Electronics Material*, vol. 48, no. 10, pp. 6253–6260, Oct. 2019.
44. Candy Goyal, Jagpal Singh Ubhi and B. Raj, "A low leakage CNTFET based inexact full adder for low power image processing applications", *International Journal of Circuit Theory and Applications*, vol. 47, no. 9, pp. 1446–1458, Sep. 2019.
45. B. Raj, A. K. Saxena and S. Dasgupta, "A compact drain current and threshold voltage quantum mechanical analytical modeling for FinFETs", *Journal of Nanoelectronics and Optoelectronics*, vol. 3, no. 2, pp. 163–170, 2008.
46. Girish Wadhwa and B. Raj, "An analytical modeling of charge plasma based tunnel field effect transistor with impacts of gate underlap region", *Superlattices and Microstructures*, vol. 142, p. 106512, June 2020.
47. Shailendra Singh and B. Raj, "Modeling and simulation analysis of SiGe heterojunction double GateVertical t-shaped tunnel FET", *Superlattices and Microstructures*, vol. 142, p. 106496, June 2020.
48. Neeraj Jain and Balwinder Raj, "Device and circuit co-design perspective comprehensive approach on FinFET technology – a review", *Journal of Electron Devices*, vol. 23, no. 1, pp. 1890–1901, 2016.
49. Sunil Kumar and B. Raj, "Analysis of I_{ON} and ambipolar current for dual-material gate-drain overlapped DG-TFET", *Journal of Nanoelectronics and Optoelectronics*, vol. 11, pp. 323–333, June 2016.
50. Naveed Anjum, Tarun Bali and B. Raj, "Design and simulation of handwritten multiscript character recognition", *International Journal of Advanced Research in Computer and Communication Engineering*, vol. 2, no. 7, pp. 2544–2549, July 2013.
51. Karmjit Singh and B. Raj, "Performance and analysis of temperature dependent multi-walled carbon nanotubes as global interconnects at different technology nodes", *Journal of Computational Electronics*, vol. 14 no. 2, pp. 469–476, June 2015.
52. Sunil Kumar and B. Raj, "Compact channel potential analytical modeling of DG-TFET based on evanescent–mode approach", *Journal of Computational Electronics*, vol. 14, no. 2, pp. 820–827, July 2015.
53. Karmjit Singh and B. Raj, "Temperature dependent modeling and performance evaluation of multi-walled CNT and single-walled CNT as global interconnects", *Journal of Electronic Materials*, vol. 44, no. 12, pp. 4825–4835, Dec. 2015.
54. V. K. Sharma, M. Pattanaik and B. Raj, "INDEP approach for leakage reduction in nanoscale CMOS circuits", *International Journal of Electronics*, vol. 102, no. 2, pp. 200–215, 2014.
55. Karmjit Singh and B. Raj, "Influence of temperature on MWCNT bundle, SWCNT bundle and copper interconnects for nanoscaled technology nodes", *Journal of Materials Science: Materials in Electronics*, vol. 26, no. 8, pp. 6134–6142, 2015.

56. Naveed Anjum, Tarun Bali and B. Raj, "Design and simulation of handwritten Gurumukhi and Devanagri numerical recognition", *International Journal of Computer Applications*, vol. 73, no. 12, pp. 16–21, 2013.
57. S. Khandelwal, V. Gupta, B. Raj and R. D. Gupta, "Process variability aware low leakage reliable nano scale DG-FinFET SRAM cell design technique", *Journal of Nanoelectronics and Optoelectronics*, vol. 10, no. 6, pp. 810–817, Dec. 2015.
58. V. K. Sharma, M. Pattanaik and B. Raj, "ONOFIC approach: low power high speed nanoscale VLSI circuits design", *International Journal of Electronics*, vol. 101, no. 1, pp. 61–73, 2014.
59. S. Khandelwal, Balwinder Raj and R. D. Gupta, "FinFET based 6T SRAM cell design: analysis of performance metric, process variation and temperature effect", *Journal of Computational and Theoretical Nanoscience*, vol. 12, pp. 2500–2506, 2015.
60. Sumit Singh, Shekhar Yadav, Jagdeep Rahul, Anurag Srivastava and B. Raj, "Impact of HfO_2 in graded channel dual insulator double gate MOSFET", *Journal of Computational and Theoretical Nanoscience*, vol. 12, no. 6, pp. 950–953, April 2015.
61. Vijay Kumar Sharma, Manisha Pattanaik and B. Raj, "PVT variations aware low leakage INDEP approach for nanoscale CMOS circuits", *Microelectronics Reliability*, vol. 54, pp. 90–99, 2014.
62. B. Raj, A. K. Saxena and S. Dasgupta, "Quantum mechanical analytical modeling of nanoscale DG FinFET: evaluation of potential, threshold voltage and source/drain resistance", *Journal of Material Science in Semiconductor Processing*, vol. 16, no. 4, pp. 1131–1137, 2013.
63. Maisagalla Gopal, Siva Sankar D. Prasad and Balwinder Raj, "8T SRAM cell design for dynamic and leakage power reduction", *International Journal of Computer Applications*, vol. 71, no. 9, pp. 43–48, June 2013.
64. Manisha Pattanaik, B. Raj, Shashikant Sharma and Anjan Kumar, "Diode based trimode multi-threshold CMOS technique for ground bounce noise reduction in static CMOS adders", *Advanced Materials Research*, vol. 548, pp. 885–889, 2012.
65. Balwinder Raj, A. K. Saxena and S. Dasgupta, "Nanoscale FinFET based SRAM cell design: analysis of performance metric, process variation, underlapped FinFET and temperature effect", *IEEE Circuits and System Magazine*, vol. 11, no. 2, pp. 38–50, 2011.
66. V. K. Sharma, M. Pattanaik and Balwinder Raj, "Leakage current ONOFIC approach for deep submicron VLSI circuit design", *International Journal of Electrical, Computer, Electronics and Communication Engineering*, vol. 7, no. 4, pp. 239–244, 2013.
67. Parminder Kaur, Sandeep Singh Gill and B. Raj, "Comparative analysis of OFETs materials and devices for sensor applications", *Journal of Silicon*, vol. 14, pp. 4463–4471, 2022.
68. Sanjeev Kumar Sharma, Parveen Kumar, Balwant Raj and B. Raj, "$In_{1-x}Ga_xAs$ double metal gate-stacking cylindrical nanowire MOSFET for highly sensitive photo detector", *Journal of Silicon*, vol. 14, pp. 3535–3541, 2022.

69. B. Raj, A. K. Saxena and S. Dasgupta, "Analytical modeling of quasi planar nanoscale double gate FinFET with source/drain resistance and field dependent carrier mobility: a quantum mechanical study", *Journal of Computer*, vol. 4, no. 9, pp. 1–8, 2009.
70. S. Bhushan, S. Khandelwal and B. Raj, "Analyzing different mode FinFET based memory cell at different power supply for leakage reduction", *Seventh International Conference on Bio-Inspired Computing: Theories and Application*, vol. 202, pp. 89–100, 2013.
71. Jeetendra Singh and B. Raj, "Temperature dependent analytical modeling and simulations of nanoscale memristor", *Engineering Science and Technology, an International Journal*, vol. 21, pp. 862–868, Oct. 2018.
72. Shradhya Singh, Shashi Bala, Balwant Raj and B. Raj, "Improved sensitivity of dielectric modulated junctionless transistor for nanoscale biosensor design", *Sensor Letter*, vol. 18, pp. 328–333, April 2020.
73. Vivek Kumar, Santosh Kumar Vishvakarma and B. Raj, "Design and performance analysis of ASIC for IoT applications", *Sensor Letter*, vol. 18, pp. 31–38, Jan. 2020.
74. Akanksha Jaiswal, R. K. Sarin, B. Raj and Shikha Sukhija, "A novel circular slotted microstrip-fed patch antenna with three triangle shape defected ground structure for multiband applications", *Advanced Electromagnetic*, vol. 7, no. 3, pp. 56–63, Aug. 2018.
75. Girish Wadhwa and B. Raj, "Label free detection of biomolecules using charge-plasma-based gate underlap dielectric modulated junctionless TFET", *Journal of Electronic Materials*, vol. 47, no. 8, pp. 4683–4693, Aug. 2018.
76. Gurmohan Singh, R. K. Sarin and B. Raj, "Design and performance analysis of a new efficient coplanar quantum-dot cellular automata adder", *Indian Journal of Pure & Applied Physics*, vol. 55, pp. 97–103, Feb. 2017.
77. Parminder Kaur, Vikas Pandey and B. Raj, "Comparative study of efficient design, control and monitoring of solar power using IoT", *Sensor Letter*, vol. 18, pp. 419–426, May 2020.
78. Anil Kumar Bhardwaj, Sumeet Gupta and B. Raj, "Development & analysis of compact model for double gate Schottky barrier CNTFET", *Journal of Nanoelectronics and Optoelectronics*, vol. 15, pp. 1199–1208, Aug. 2020.
79. Girish Wadhwa, Priyanka Kamboj, Jeetendra Singh and B. Raj, "Design and investigation of junctionless DGTFET for biological molecule recognition", *Transactions on Electrical and Electronic Materials*, vol. 22, pp. 282–289, 2021.
80. Sachin Kumar Verma, Shailendra Singh, Girish Wadhwa and B. Raj, "Detection of biomolecules using charge-plasma based gate underlap dielectric modulated dopingless TFET", *Transactions on Electrical and Electronic Materials*, vol. 21, pp. 528–535, June 2020.
81. Neeraj Jain and B. Raj, "Impact of underlap spacer region variation on electrostatic and analog/RF performance of symmetrical high-k SOI FinFET at 20 nm channel length", *Journal of Semiconductors*, vol. 38, no. 12, p. 122002, Dec. 2017.
82. Shailendra Singh and B. Raj, "Analytical modeling and simulation analysis of T-shaped III–V heterojunction vertical T-FET", *Superlattices and Microstructures*, vol. 147, p. 106717, Nov. 2020.

83. Gurmohan Singh, R. K. Sarin and B. Raj, "Design and analysis of area efficient QCA based reversible logic gates", *Journal of Microprocessors and Microsystems*, vol. 52, pp. 59–68, May 2017.
84. Shailendra Singh and B. Raj, "Parametric variation analysis on hetero-junction vertical t-shape TFET for suppressing ambipolar conduction", *Indian Journal of Pure and Applied Physics*, vol. 58, pp. 478–485, June 2020.
85. Shailendra Singh Soniya, Girish Wadhwa and Balwinder Raj, "Design and analysis of dual source vertical tunnel field effect transistor for high performance", *Transactions on Electrical and Electronics Materials,* vol. 21, pp. 74–82, Oct. 2019.
86. Manjit Kaur, Neena Gupta, Sanjeev Kumar, B. Raj and Arun Kumar Singh, "RF performance analysis of intercalated graphene nanoribbon based global level interconnects", *Journal of Computational Electronics,* vol. 19, pp. 1002–1013, June 2020
87. Girish Wadhwa and B. Raj, "Design and performance analysis of junctionless TFET biosensor for high sensitivity", *IEEE Nanotechnology*, vol. 18, pp. 567–574, 2019.
88. Jeetendra Singh and B. Raj, "Enhanced nonlinear memristor model encapsulating stochastic dopant drift", *Journal of Nanoelectronics and Optoelectronics*, vol. 14, pp. 958–963, 2019.
89. Anjana Bhardwaj, Pradeep Kumar, B. Raj and Sunny Anand, "Design and performance optimization of doping-less vertical nanowire TFET using gate stack technique", *Journal of Electronic Materials*, vol. 41, no. 7, pp. 4005–4013, 2022.
90. Jeetendra Singh and B. Raj, "Tunnel current model of asymmetric MIM structure levying various image forces to analyze the characteristics of filamentary memristor", *Applied Physics A*, vol. 125, no. 3, pp. 203.1–203.11, Feb. 2019.
91. Candy Goyal, Jagpal Singh Ubhi and B. Raj, "Low leakage zero ground noise nanoscale full adder using source biasing technique", *Journal of Nanoelectronics and Optoelectronics*, vol. 14, pp. 360–370, March 2019.
92. Gurmohan Singh, R. K. Sarin and B. Raj, "A novel robust exclusive-OR function implementation in QCA nanotechnology with energy dissipation analysis", *Journal of Computational Electronics*, vol. 15, no. 2, pp. 455–465, June 2016.
93. Girish Wadhwa, Priyanka Kamboj and Balwinder Raj" Design optimisation of junctionless TFET biosensor for high sensitivity", *Advances in Natural Sciences: Nanoscience and Nanotechnology*, vol. 10, no. 7, pp. 045001, 2019.
94. Priya Bansal and B. Raj, "Memristor modeling and analysis for linear dopant drift kinetics", *Journal of Nanoengineering and Nanomanufacturing*, vol. 6, pp. 1–7, 2016.
95. D. Vaithiyanathan Ashima and B. Raj, "Performance analysis of charge plasma induced graded channel Si nanotube", *Journal of Engineering Research*, EMSME Special issue, pp. 146–154, Aug. 2021.
96. Abhishek Singh Tomar, Vijay Kumar Magraiya and B. Raj, "Scaling of access and data transistor for high performance DRAM cell design", *Quantum Matter*, vol. 2, pp. 412–416, Oct. 2013.

97. Neeraj Jain and B. Raj, "Parasitic capacitance and resistance model development and optimization of raised source/drain SOI FinFET structure for analog circuit applications", *Journal of Nanoelectronics and Optoelectronics*, vol. 13, pp. 531–539, April 2018.
98. Shradhya Singh, S. K. Vishvakarma and B. Raj, "Analytical modeling of split-gate junction-less transistor for a biosensor application", *Sensing and Bio-Sensing*, vol. 18, pp. 31–36, April 2018.
99. Maisagalla Gopal and Balwinder Raj, "Low power 8T SRAM cell design for high stability video applications", *ITSI Transaction on Electrical and Electronics Engineering*, vol. 1, no. 5, pp. 91–97, 2013.
100. Balwinder Raj, Jatin Mitra, Deepak Kumar Bihani, V. Rangharajan, A. K. Saxena and S. Dasgupta, "Analysis of noise margin, power and process variation for 32 nm FinFET based 6T SRAM cell", *Journal of Computer*, vol. 5, no. 6, pp. 1–8, 2010.
101. Divya Sharma, Rajesh Mehra and B. Raj, "Comparative analysis of photovoltaic technologies for high efficiency solar cell design", *Superlattices and Microstructures*, vol. 153, p. 106861, May 2021.
102. Pawandeep Kaur, Avtar Singh Buttar and B. Raj, "A comprehensive analysis of nanoscale transistor based biosensor: a review", *Indian Journal of Pure and Applied Physics*, vol. 59, pp. 304–318, April 2021.
103. Divya Yadav, Balwant Raj and B. Raj, "Design and simulation of low power microcontroller for IoT applications", *Journal of Sensor Letters*, vol. 18, pp. 401–409, May 2020.
104. Shailendra Singh and B. Raj, "A 2-D analytical surface potential and drain current modeling of double-gate vertical t-shaped tunnel FET", *Journal of Computational Electronics*, vol. 19, pp. 1154–1163, April 2020.
105. Jeetendra Singh and B. Raj, "An accurate and generic window function for non-linear memristor model", *Journal of Computational Electronics*, vol. 18 no. 2, pp. 640–647, June 2019.

Chapter 12

Nanomaterials for semiconductor devices

Brajbhushan Tripathi, Ankit kumar Dipjyoti Borah, Mandeep Singh, Tarun chaudhary, and Balwinder Raj

12.1 INTRODUCTION

Introduction to nanomaterial-based semiconductor devices: Semiconductor devices play a crucial role in contemporary electronics, facilitating a broad spectrum of applications, from transistors and diodes to solar cells and sensors. With the evolution of nanotechnology, the utilization and incorporation of nanomaterials have introduced novel possibilities to enhance the performance and functionalities of semiconductor devices [1–4].

Nanomaterials, distinguished by their nanoscale dimensions ranging typically from 1 to 100 nm, manifest distinct and often superior properties in comparison to their bulk counterparts. In the domain of semiconductors, nanomaterials exhibit significant promise in overcoming constraints associated with traditional technologies.

A primary advantage of nanomaterial-based semiconductor devices lies in their capacity to harness quantum effects prevalent at the nanoscale. Quantum dots, nanowires, and other nanostructures display quantum confinement, enabling meticulous control over electronic and optical characteristics. This precise control facilitates the creation and manufacturing of devices boasting unparalleled efficiency and functionality [5–9].

Several nanomaterials find applications in semiconductor devices, including:

Quantum dots: Nanoscale semiconductor particles showcasing quantum confinement effects, allowing tunable optical and electronic properties. Quantum dots are applied in LEDs, solar cells, and biological imaging.

Nanowires: One-dimensional structures with nanometer-scale diameters offering distinctive electronic and mechanical properties. Nanowires are employed in transistors, sensors, and nanoelectromechanical systems (NEMS).

Graphene: A monolayer of carbon atoms arranged in a hexagonal lattice, graphene possesses remarkable electrical conductivity, mechanical strength, and thermal properties. It is utilized in transistors, flexible electronics, and energy storage devices [10–13].

DOI: 10.1201/9781003612797-12

Nanotubes: Cylindrical structures composed of carbon atoms, nanotubes exhibit notable strength and electrical conductivity. They find applications in transistors, sensors, and as reinforcing elements in composite materials.

Nanocomposites: Materials consisting of a matrix embedded with nanoscale reinforcements, enhancing overall material properties. Nanocomposites are employed in various semiconductor devices for improved performance and durability.

Nanomaterial-based semiconductor devices offer advantages such as heightened sensitivity, reduced power consumption, and novel functionalities. The integration of these materials into semiconductor technology holds the potential to revolutionize electronics, ushering in more efficient and versatile devices across diverse applications. As nanotechnology research advances, the ongoing development of increasingly sophisticated nanomaterial-based semiconductor devices holds tremendous promise for the future of electronics and technology [14–19].

12.2 NANOWIRE-BASED DEVICES

12.2.1 Nanowires

A nanowire is a type of nanostructure with a diameter of approximately one nanometer. Nanowires typically exhibit variable lengths and a thickness or diameter of tens of nanometers or less. Quantum mechanical effects become significant at the nanoscale, leading to nanowires often being referred to as quantum wires. Various types of nanowires have been developed, including metallic ones (such as Ni, Pt, and Au), semiconducting ones (like Si, InP, and GaN), and insulating ones (such as SiO_2 and TiO_2).

The physical properties of nanowires are influenced by the diameter-dependent band gap, carrier density of the material state, carrier shape, and other factors. Nanowires are considered one-dimensional (1D) materials due to their frequently high aspect ratios (length-to-width ratio), often reaching 1000 or more. The conductivity characteristics of nanowires differ from those of bulk materials or larger, three-dimensional structures made from the same materials. The quantum confinement of electrons in nanowires laterally results in energy levels distinct from the conventional continuum of energy levels or bands observed in bulk materials [20–25].

Semiconductor and superlattice nanowires can be utilized in the production of basic electrical devices such as transistors, junction diodes, field-effect transistors (FETs), and logic gates. Nanowires are under development for various applications, including bioelectrical–mechanical and biomedical sensor applications. Metallic nanowires can be employed in constructing thermoelectric cooling systems. Well-organized arrays of magnetic nanowires hold significant potential for recording media applications, while semiconductor nanowire junctions can be utilized in various optoelectronic applications [26–30].

12.2.2 Nanowire-based gate-all-around

Understanding the mechanics of nanoscale devices requires an understanding of quantum phenomena that are introduced when the devices' size is reduced. Quantum corrections become indispensable as the channel cross-section approaches the wavelength of the free electron, making an accurate estimation of the device's electrostatic properties necessary. Traditional local continuum models lose some of their accuracy in describing nanoscale behavior as lateral nanowire transistors (LNW) continue to shrink from 7 nm to 5 nm or smaller as a result of ongoing technology scaling. Thus, it is believed that the development of the semiconductor industry depends on sophisticated physics-based models, especially those that integrate drift-diffusion and density gradient in technology computer-aided design (TCAD) [31–36].

Through scaling, integrated circuit performance has increased dramatically over the last 60 years. Improved total carrier mobility, especially below the 90 nm technology node, has been demonstrated by strain engineering. With the introduction of high-k/metal gate stacks at the 45 nm technology node, problems like gate leakage and static power dissipation have been addressed. New materials and topologies have been introduced as a result of the difficulties caused by short-channel effects, which have restricted the downscaling of MOS transistors. Particularly for FinFETs, triple-gate designs have become competitive alternatives for reducing the impact of short-channel effects, particularly below the 14 nm technology threshold. Gate-all-around (GAA) nanowire transistors are currently thought to be the most promising architectures for rapid scaling, even as we venture below the 7 nm technological node [1].

One of the challenges facing CMOS technology advancement is the development of robust electrical transport models. To comprehend semiconductor device limits, nanoscale CMOS device simulation is essential. Accurate representation of carrier transport is offered by expensive transport solvers such as Monte Carlo. Using three-dimensional quantum mechanical simulations, we investigate the potential benefits of augmenting gate-all-around (GAA) nanowire architectures with additional wires. The Si-NW FET has almost flawless sub-threshold characteristics and high transconductance. The electrical loss resulting from shorter gates is compensated for by smaller nanowire diameter. While full quantum models provide a detailed physical explanation based on the tight-binding Hamiltonian, semi-classical models describe nanowire inversion layers in sub-7 nm nanowire simulations. In low-field and quasi-ballistic regimes with conventional TCAD tools, quantum correction techniques such as the density gradient approach are essential [1].

12.2.3 Radial junction nanowire devices for photovoltaics and photodetection

Nanowire devices have been proposed as an efficient substitute for their planar counterparts in a variety of optoelectronic applications. Sadly, obstacles to doping growth, characterization, and p–n junction formation along axial or radial axes in nanowire devices have slowed down progress considerably. Struggling to create a radial p–n junction exacerbates the problems. Few studies have been conducted on high-performance radial junction nanowire optoelectronic devices, despite expectations of similar performance to axial counterparts [2].

Since the development of nanotechnology, nanowires have been suggested as an economical and successful substitute for their thin-film equivalents in a range of optoelectronic uses, such as photovoltaics, photodetectors, and LEDs/lasers. Particularly fast growth rates, easy strain relaxation enabling heteroepitaxy, high absorption and emission per unit volume, and enhanced defect tolerance are just a few of the special benefits that nanowires have to offer. For optoelectronics, these properties make nanowires especially interesting since they allow the creation of device designs that would be difficult or even impossible with planar devices. Nanowire devices can be classified as axial or radial junction devices based on which axis is dominant for charge carrier separation. The bulk of planar/nanowire devices are axial junction nanowire devices; radial junction nanowire devices have been proposed as an effective substitute since 2005. It is possible to independently optimize optical and electronic behaviors in radial junctions due to the separation of the charge carrier separation axis and the optical axis [2].

Mainly because of the difficulties in forming a controlled p–n junction and attaining accurate doping, controlling the behavior of nanowire devices made by bottom-up methods has proven to be difficult. High-performance radial junction nanowire devices have proven challenging to realize due to the large surface area of nanowires and surface recombination issues caused by the application of standard passivation layers, especially in the case of core–shell nanowires. Antenna junction devices perform better than first anticipated, according to empirical findings [37–41].

Studies conducted recently suggest that radial junction nanowire devices are better than their axial counterparts, especially when the absorber lifespan is constrained. Radial junction devices were developed, departing from the conventional bottom-up approach, using a top-down approach involving etched p-type InP nanowires to address challenges. The p-type InP nanowires were coated with a conformal layer of n-type aluminum zinc oxide (AZO)/ZnO in order to create the p–n junction. High-performance solar cells with a record-breaking 31.3 mA/cm^2 short-circuit current were produced as a result of the transformation of an absorber layer with a bulk lifetime below 100 ps. Furthermore, using this radial junction architecture, photodetectors with the ability to detect single-photon levels across a wide spectrum at room temperature were made possible [2].

The lift-off process for chromium nanodisks, the plasma etching of InP nanowires, and extensive optimization of large-area electron beam lithography were necessary to produce radial junction nanowire solar cells. The uniform and conformal deposition of AZO/ZnO over the top-down etched nanowire array was crucial for attaining high efficiency. Radial junction nanowire solar cells demonstrated an efficiency of more than 16% in measurements, along with an outstanding active area short-circuit current of 31.3 mA/cm^2 and an active area efficiency of more than 17% per unit. In addition to solar cells, the article described the creation of radial junction nanowire photodetectors. An extraordinarily large built-in electric field across the radial junction was achieved by depositing n-type AZO/ZnO with a broad bandgap over heavily doped p-type InP. With a sensitivity of just one photon, the photodetector demonstrated remarkable self-powered detection and an impressive internal quantum efficiency of 59.4%. The device displayed an extended dynamic range that allowed photodetection down to a few picowatts, and it demonstrated a nearly linear improvement in sensitivity [42–48].

12.2.4 Nanowire array solar cells

Various materials can be included in a single semiconductor matrix using modern nanowire (NW) manufacturing processes, making it easier to assemble nanowires with axial or radial p–n homo-(hetero) junctions. Carrier collection mostly happens perpendicular to the direction of light absorption in radial p–n junctions. This special arrangement provides almost full absorption of incident light in the axial direction and reduced carrier collection lengths in the radial direction. One notable benefit of nanowire solar cells is their unique radial p–n junction characteristic. Any rod-like structure can be used to form radial p–n junctions. To reduce recombination losses in the structure, rods with radii smaller than the carrier diffusion length must be used [49–53]. To further reduce carrier recombination losses, the distances between these rods in the matrix shouldn't be greater than the diffusion length values. Structures with thicknesses matching [1/absorption coefficient] ought to be taken into consideration in order to guarantee sufficient absorption of incident light. Consequently, when used in photovoltaics, nanowire (NW) arrays provide special benefits. These characteristics propel continuous advancements toward third-generation photovoltaics through upgrades and breakthroughs in solar cell technology [3].

12.3 CARBON NANOTUBE-BASED DEVICES

12.3.1 Carbon nanotube

Carbon nanotubes (CNTs) are spherical nanostructures made of rolled carbon sheets with different graphene layers that can have open or closed ends. Length, diameter, and wall count are some of the variables that affect the composition and characteristics of CNTs. Longer carbon nanotubes (CNTs) have higher electrical and thermal conductivity, but Young's modulus and mechanical strength don't change with length (Figure 12.1).

Single-walled carbon nanotubes (SWCNTs), double-walled carbon nanotubes (DWCNTs), and multi-walled carbon nanotubes (MWCNTs) are the three primary categories of carbon nanotubes. The diameters of these groups range from 0.8 to 2 nm, 2.75 to 4.78 nm, and 5 to 50 nm, respectively. CNTs can have metallic or semi-conducting electrical properties, and their conductivity is dependent on the arrangement of structural elements [54–57]. Depending on their diameter and index, CNTs can exhibit metallic, semi-metallic, or semiconducting behavior in their energy gaps. Because of their unique qualities, CNTs have attracted interest in the field of electrical engineering, particularly in the area of electrical conductivity. High conductivity, low weight, strength, and current-carrying capacity are some of their characteristics. Although discussions about carbon nanotubes predominate in the field of nanotechnology, scientists are also investigating other carbon-based nanomaterials with unique properties for a range of applications [4].

Because of their complex structure, carbon nanocomposites exhibit anisotropic behavior in both nanostructure and replication degree, which affects their electrical properties. High 2D electrical conductivity is a result of the intrinsic anisotropy of all sp^2 carbon materials, which is caused by the relocation of the last non-hybridized valence electrons [58–62].

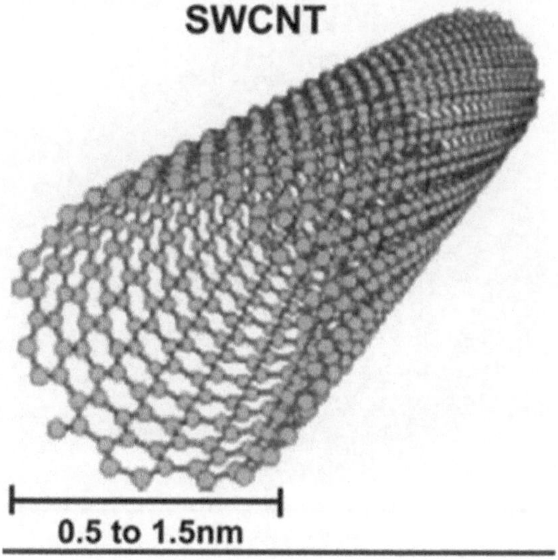

Figure 12.1 Single-walled carbon nanotube [12].

12.3.2 Factors influencing CNT's electrical conductivity

CNTs have outstanding mechanical qualities, low density, and electrical conductivity. CNTs must be incorporated into metallic structures in order to fully utilize these properties. Researchers are using metals like copper and aluminum to combine with CNTs to improve their electrical properties. Because carbon nanotubes (CNTs) are more mechanically strong and electrically conductible than traditional metals, they can be used in applications like touchscreens, gas sensors, and electromagnetic interference (EMI) shielding. Several techniques are used to increase CNT conductivity, such as:

1. Combining CNT with metals such as aluminum and copper.
2. Using techniques to reduce weight, improve thermal conduction, and increase reliability in comparison to copper-based wiring.
3. Applying techniques akin to those for metalized carbon fiber to metalize externally continuous lengths of CNT fibers.
4. Extended and enhanced SWCNTs to support wire conductors with lower inter-tube resistances.
5. To reduce inter-tube resistances, molten metal can be injected between CNT tubes.
6. Connecting metallic conductors to CNTs (Figure 12.2).

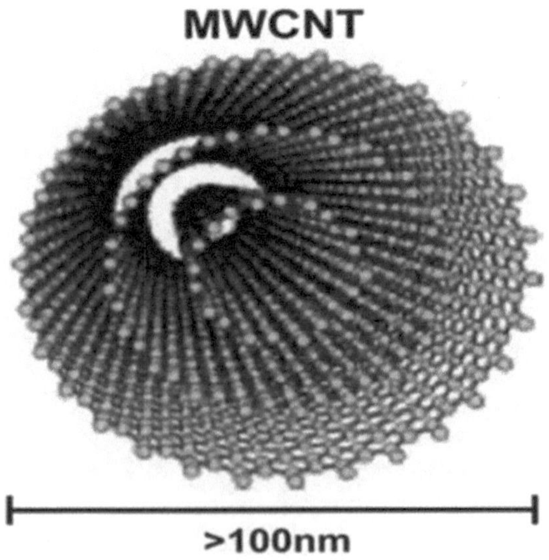

Figure 12.2 Multi-walled carbon nanotube [12].

12.3.3 Gate-all-around carbon nanotube-based field-effect transistor

The ultra-slim structure of carbon nanotubes (CNTs) allows for aggressive channel length scaling, which guarantees accurate gate control. Electrostatic gate control is thought to work best with field-effect transistors (FETs) configured in a gate-all-around (GAA) structure. Consequently, for better device design, it makes sense to integrate CNTs into a GAA device geometry. Notwithstanding the fact that earlier research has shown enhanced electrostatics on a CNT FET with an omega-shaped dielectric coating, difficulties with obtaining a thin and consistent dielectric coating are brought on by the chemical inertness of CNT surfaces. A functional layer, like deoxyribonucleic acid (DNA), is suggested for surface modification in order to get around this problem. Furthermore, the CNT has to be freestanding prior to dielectric and gate metal deposition in the absence of substrate assistance, which adds another level of complexity [5].

This chapter describes the wrap-around process and presents the first demonstration of a GAA-CNFET. The nanotube is functionalized with NO_2 prior to atomic layer deposition (ALD), which makes it easier to deposit a 7 nm film of Al_2O_3 uniformly around the tube. After that, the dielectric post-ALD deposition is wrapped with a 20 nm WN gate [63–67]. Figure 12.3 shows a schematic of the GAA-CNFET, where WN and Al_2O_3 are removed by wet chemical etching, with the exception of the gate area.

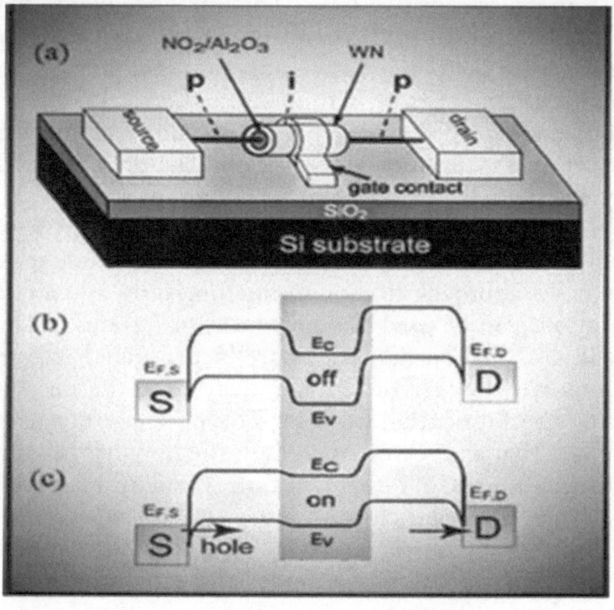

Figure 12.3 (a) The GAA, (b) energy band diagram of OFF-state, and (c) energy band diagram of ON-state [5].

Chemical or electrostatic doping is possible, thanks to the tube's open sections at the CN's source/drain (S/D) connections. Band diagrams showing the OFF and ON states of a p/i/p transistor are shown in Figure 12.3(b) and (c), which demonstrate how a p/i/p or n/i/n profile can be created depending on doping in the outer segments [5].

This study explores the impact of the NO_2 functionalization group on the growth of ALD Al_2O_3 on freestanding CNs. A test capacitor with noticeable frequency dependence and hysteresis in CV measurements was made on a Si substrate with the same functionalization and a 10 nm ALD Al_2O_3 process. This suggests the presence of oxide-trapped and fixed charges. When GAA-CNFETs are annealed after ALD at temperatures higher than 500° C, the devices exhibit reduced hysteresis and become less noisy. This improves the oxide stack's quality without changing the CV characteristics' slope. To sum up, we have discussed the manufacturing procedure and assessments for the first externally assembled GAA-CNFET [5].

12.3.4 Carbon nanotube-based single-layer piezoresistive polyimide pressure sensor

Conventional flexible sensors usually have layers of polymer deposited on top of a substrate, which presents issues with adhesion and delamination. This work presents a novel flexible pressure sensor that incorporates 2 wt percent carbon nanotubes (CNTs) into the polymer, using the polyimide substrate as a pressure-sensitive layer. This method, which makes use of polyimide's ideal temperature, chemical composition, and mechanical characteristics, eliminates the possibility of adhesion issues or delamination, in contrast to conventional sensors with multiple stacked polymer layers [68–71]. Softer polymer-based sensors, such as polydimethylsiloxane (PDMS), are more susceptible to plastic deformation in the pressure sensor. This is mitigated by the high elastic modulus of polyimide (9.8 GPa). Significant pressure response drifts in PDMS were noted by the researchers, highlighting the benefits of flexible sensors based on polyimide. These sensors present a potentially novel approach to monitoring long-term aging effects, where pressure variations can be used to detect leaks or failures, such as those in industrial sealings. Carbon nanotubes are incorporated into the polymer directly to improve conductivity, doing away with the need for solvents and expediting the fabrication process. The principle of measurement is based on the fact that as pressure increases, the polymer shrinks, resulting in a decrease in average CNT–CNT distances and an increase in material conductivity. These material changes are measured with gold interdigital electrodes [72–77].

The manufacturing procedure includes the utilization of a polyimide precursor containing 20 wt percent polyamic acid, the integration of carbon nanotubes through a mechanical mixing technique, and the spin-coating of the films. The produced sensors, having a thickness of 4.7

μm, are designed with gold interdigital electrodes and incorporated into hydrogenated nitrile rubber components, mimicking rubber sealings. The experimental configuration involves the attachment of copper wires to the sensors, their connection to a digital multimeter, and the application of pressure using a 3D-printed stamp. The outcomes illustrate the sensors' consistent performance, stability, and cyclic response to diverse pressure conditions, showcasing potential applications in industrial environments. Further investigations include the plasma etching of polymer–CNT layers, analysis of pressure response curves, testing for cyclic stability, and conducting experiments on thermal aging [78–84]. The research concludes by demonstrating the functionality of a single-layer pressure sensor, eliminating the necessity for additional substrate layers. The sensor exhibits stability and resilience, enduring pressures within the tested maximum range of 55N. The relationship between pressure and sensor response proves to be linear from the initial pressure cycle. Future research will be dedicated to enhancing sensitivity and expanding the detectable pressure range. Several avenues for achieving these objectives include the reduction of carbon nanotube (CNT) weight percentage to approach the percolation threshold. The aim is to significantly amplify the change in resistance in response to pressure, particularly by utilizing piezoresistive materials near the percolation threshold. Exploration into the effects of modifying sensor geometry, thickness, linewidths, and shapes will be undertaken as potential optimization strategies. The ultimate goal is to integrate the finalized sensors into industrial sealings, facilitating continuous pressure monitoring for practical applications [85–92]. This forward-looking research endeavors to contribute valuable advancements to the field of pressure-sensing technology [6].

12.3.5 IR detector based on carbon nanotube

The common electron beam (E-beam) lithography technique presents significant challenges when creating arrays of individual carbon nanotube (CNT)-based infrared detectors. Using an atomic force microscope (AFM)-based nanomanipulation system, our study solves this problem by precisely positioning a single semiconducting CNT between two electrodes to form a Schottky diode. The creation of arrays of individual CNT-based infrared detectors is made easier by this regulated and trustworthy assembly process. A Schottky photodiode is created at the CNT–metal contact by forming a Schottky barrier. The internal or external electric field within this barrier separates photogenerated electron–hole pairs, resulting in the generation of a photocurrent. When a semiconductive CNT is used, Schottky contacts usually form with both electrodes, resulting in diodes that are opposed to each other and produce photocurrents that cancel each other out. To enhance the separation of these photocurrents, we design a structure with different electrodes using various contact metals [7].

The CNT detector's results using a fixed electrode arrangement show that when the infrared spot's center moves between electrodes, the photocurrent polarity changes from positive to negative. The photovoltaic effect, not the heating effect, is the likely source of the photocurrent based on this observation. Moreover, the CNT detector using a different electrode structure yields results that show a signal-to-dark current ratio that is three orders of magnitude higher than that of a uniform detector. This enhances the detector's overall performance and highlights the usefulness of the specially designed heterogeneous structure. Figure 12.1 shows the setup of every single infrared (IR) detector based on carbon nanotubes (CNTs). AFM lithography in conjunction with conventional photolithography is used to create these detectors, which are made up of one CNT and two electrodes [93–97]. An IR detector installed in a detector chamber and connected to a measurement box, where the current signal is measured using a Keithley Pico ammeter, is one of the experimental configurations used in the text to investigate the setup and testing of CNT-based IR detectors. To evaluate the IR response of the CNT detectors, an IR laser is used. The report describes how the detectors were made. One multi-walled nanotube (MWNT) with ohmic contacts is one of the detectors; it shows I–V curves and changes in conductance under infrared light. The study also looks into Schottky diodes that have -single-walled nanotubes (SWNT), which show better performance [7].

An asymmetric metal–CNT contact is created by combining various materials as contact electrodes, resulting in a varied asymmetric structure designed to improve detector efficiency. A CNT infrared detector with asymmetric Ag–Pd electrodes shows how this novel design increases the total photocurrent. The study focuses on single CNT-based photodiodes and highlights how asymmetric electrodes affect performance. Test findings demonstrate the effect of the laser spot on photocurrent, and a three-orders-of-magnitude increase in the signal-to-dark current ratio is obtained using a variety of metals as contact electrodes [7].

12.4 QUANTUM DOT-BASED DEVICES

Quantum dots, nanoscale semiconductor particles, find applications in various semiconductor devices, leveraging their unique optical and electrical properties. Common types include CdSe, CdTe, InP, PbS, PbSe, ZnO, and $CuInS_2$, each serving different purposes in LEDs, solar cells, photodetectors, and sensors. Material selection is based on device requirements, considering factors such as stability, toxicity, and emission wavelength. Ongoing research aims to develop new quantum dot types for expanded applications and improved performance [98–103].

The distinctive electrical and optical properties of quantum dots stem from their nanoscale dimensions and quantum mechanical nature.

Quantum confinement results in quantized charge carriers, yielding discrete energy levels electrically. This discrete energy structure enables precise control over electronic transitions. Optically, quantum dots exhibit size-dependent emission and absorption spectra known as quantum size effects. The emitted and absorbed light colors depend on the dot's size, determined by its bandgap. Quantum dots, due to their tunable optical behavior, find utility in applications like biological imaging and light-emitting devices. Additionally, they enhance carrier lifetimes and electron-hole separation, contributing to improved solar cell and photodetector performance.

Quantum dots' versatility and appeal for diverse technological applications arise from the combination of their unique electrical and optical properties. The progress in semiconductor lasers over the past three decades has seen significant improvements in device performance through specialized designs. Notably, the use of semiconductor heterostructures allowed separate optimization of optical and carrier confinement. The incorporation of quantum films, or quantum wells, further decreased threshold current density by raising the density of states and discretizing permitted energy states. The conceptualization of quantum dots emerged from theoretical research aimed at maximizing quantum effects in semiconductor lasers [104–107].

Quantum-dot-like structures demonstrated a substantial reduction in threshold current density compared to quantum film-based lasers. Achieving high-quality quantum-dot structures at the device level took over ten years, with a breakthrough occurring through self-assembly techniques during epitaxial growth. Continuous development has led to quantum-dot lasers with device data comparable to or surpassing traditional quantum-film lasers. Quantum dots, with three-dimensional carrier confinement on a nanometer scale, exhibit unique properties. This behavior results in a discrete energy spectrum controlled by geometric size, akin to weakly or non-interacting single atoms. The high density of states in quantum dots at laser energies significantly enhances the probability of stimulated emission in laser structures. Quantum dots offer advantages such as increased gain, reduced carrier diffusion length, lower active laser volume, lower band edge absorption, and diminished leakage currents, making them suitable for specific applications [8].

12.4.1 InSb/GaAs quantum dot solar cell

Quantum dots have been the focus of extensive theoretical and experimental investigations, particularly in relation to their optical and electrical characteristics. Their significance in optoelectronic devices, such as quantum dot solar cells (QDSCs) and QD lasers, stems from their adjustable bandgap and carrier confinement capabilities. Materials with narrow bandgaps, such as InAs (Eg = 0.42 eV) and InSb (Eg = eV), are deemed promising in advancing solar cells and infrared detectors for the next generation. Researchers are actively collaborating to enhance the efficiency of GaAs solar cells, leveraging their advantageous physical and optical properties. The optimization of GaAs solar cells with quantum dots as a semiconductor host is recognized as a superior approach. While most quantum dot solar cell research has concentrated on InAs/GaAs QDs, limited information exists on InSb/GaAs due to growth challenges associated with antimony (Sb) compounds. Addressing this gap, this work presents the modeling and simulation of InSb/GaAs quantum dot solar cells, exploring the impact of quantum dots on solar cell characteristics, and characterizing the optical properties using photoluminescence (PL) spectra.

All simulations were conducted under AM1.5 solar spectrum conditions at a temperature of 300 K. The research initiated with simulations of InSb/GaAs QDSCs, analyzing the impact of varying quantum dot layer counts (0, 10, 20, and 30) on external quantum efficiency (EQE) and current–voltage characteristics. Photoluminescence (PL) spectra were then computed to assess the optical characteristics of the QDSCs. Furthermore, a comprehensive analysis of the effects of variations in quantum dot thickness on the observed curves was presented. Conclusion: Our research involves the modeling and simulation of an InSb/GaAs quantum dot solar cell, focusing on enhancing the Quantum Dot Solar Cell (QDSC) by examining the effects of various parameters on both optical and electrical properties, such as the quantity and thickness of embedded quantum dots. Notably, an efficiency of 16.19% was achieved with 30 layers of InSb/GaAs quantum dots, representing a substantial relative improvement of about 29.31% over the baseline solar cell (Figure 12.4).

Furthermore, utilizing the same quantity of quantum dots, we observed a significant 11% increase in external quantum efficiency (EQE) for longer wavelength photons in the range of 900–1400 nm. The short circuit current for a 10-layer TnSb/GaAs QDSC increased from 20.42 to 22.33 mA/cm^2 while maintaining a quantum dot thickness of 10 nm. Moreover, the EQE gain increased proportionately for the 10 TnSb/GaAs quantum dot layers, rising from 5% to 10%. It is noteworthy that the photoluminescence (PL) spectra of the QD structure at a wavelength of 1.2 μm increased from 0.6 to 1.0 due to the constant thickness of the quantum dots [9].

12.4.2 Nanocomposite sensitized solar cells using Ag–CdSe quantum dots

This chapter provides an in-depth analysis of a solar cell that uses an Ag-CdSe quantum dot (QD) nanocomposite and a glass/FTO/TiO_2 photoanode structure, together with an iodide and triiodide electrolyte and a Pt electrode. The study explores the effects of Ag nanoparticles and CdSe QD filling factors on the light absorption of solar cells in detail. Furthermore, a number of factors pertaining to light absorption are investigated in detail, including the thicknesses of the TiO_2 layer, iodide/triiodide layer, and nanocomposite layer. The findings show that absorbance is impacted by

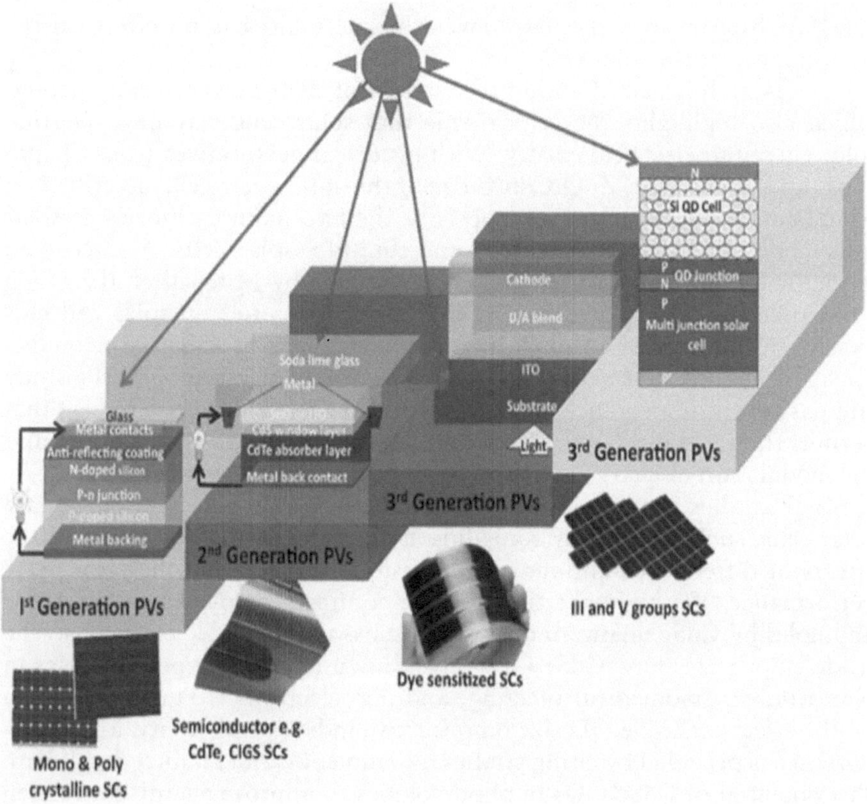

Figure 12.4 Schematic of representative structures of quantum dot solar cell [16].

Table 12.1 Characteristic parameters of the standard solar cell and QDSC with a different number of QD layers [14]

Number of QD layers	*Jsc (mA/cm²)*	*Voc (V)*	*FF (%)*	*η (%)*
0	24.69	1.351	77.75	25.96
10	29.11	1.353	76.97	30.33
20	31.63	1.354	75.44	32.33
30	32.60	1.355	76.87	33.96

Ag nanoparticles at approximately 700 nm, declines at approximately 800 nm, and stays constant between 400 nm and 600 nm as the silver fraction increases. The performance of solar cells is only slightly affected by the addition of CdSe. Increasing the composite thickness to 200 nm causes the solar cell's absorbance (A) to reach its maximum values and extend toward longer wavelengths. Di = 200 nm is where the highest values of the iodide/

triiodide layer thickness (di) are observed. This layer thickness also influences A. Remarkably, the TiO_2 layer thickness (dt) has no effect on the performance of the solar cell.

The research is placed within the context of solar power in this chapter, which also highlights the critical role that solar cells play in converting solar energy to electrical energy. With historical perspectives from Chapin et al. and updates on ZnO/CdS/$CuInSe_2$ thin-film solar cells described by AbuShama et al., it offers a synopsis of the two main categories of solar cells: crystalline silicon solar cells and thin-film solar cells. Also covered is the investigation of n-CdS/p-CdTe solar cells by Mitchell et al., which emphasizes the effect of different CdS film resistivities on solar cell efficiency. In the process of exploring the advantages of using semiconductor quantum dots (QDs) in solar cell light-harvesting assemblies, the study emphasizes the importance of size quantization as a crucial component that permits the generation of multiple charge carriers from a single photon and the modulation of charge transfer [10].

Many studies that examine the manufacturing processes of CdTe/CdS solar cells, such as those by Kapadnis et al. and Zhou et al., highlight the effects of different conditions and deposition methods on the solar cells' performance. We highlight the effective confinement of excited electrons and holes by CdSe quantum dots and discuss their tunable emissions. The study emphasizes how CdSe QDs have drawn interest for possible uses in photovoltaics, biomedical imaging, and optoelectronics. Further research on the effects of CdSe QD size on refractive index, permittivity, and transmittance is provided by citing studies by Tran et al. and Hamizi and Johan. The potential of CdSe QDs in photovoltaics to improve energy conversion efficiency is acknowledged in the study. Lastly, an exploration of silver nanoparticles (Ag-NPs) is given, highlighting how they differ from their larger counterparts in terms of their physical, chemical, and biological properties.

In order to improve solar cell performance, Ag-doped CdS quantum dots (QDs) are incorporated; the chapter emphasizes the significant increase in absorbance. It is also mentioned how the introduction of Ag-doped CdSe nanocrystal quantum dots improved the performance of CdSe QDs in photovoltaic devices by adding additional charge carriers. The nanocomposite mixture of CdSe QD and Ag nanoparticles is thoroughly examined in this work, taking into account variations in the mixture's fraction of Ag nanoparticles as well as varying layer thicknesses. The structure of the proposed solar cell is described in detail. It uses Pt as the electrode, iodide/triiodide as the electrolyte, and glass/FTO/TiO_2/Ag-CdSe nanocomposite as the photoanode. The study's findings and related discussions are presented in the conclusion of the chapter [10].

Our new solar cell design has a Pt electrode, an iodide/triiodide electrolyte, and a glass/FTO/TiO_2 photoanode with a layer of Ag–CdSe quantum dots (QD) integrated into a nanocomposite combination. The objective of

Table 12.2 Comparison between semiconductor photodetector and -quantum dot photodetector [13]

Material and ref	*np–p (%)*	*λ NIR (nm)*	*λ Vis (nm)*	*L (cd/m²)*	*On/off ratio*	*PNIR*
Semi-PD and OLED [23]	0.7	1500	520	1235		1.1
Full organic (RGB) [26]	0.11	810	512	150		100
QDPD and OLED [43]	1.3	1300	520	6	<1001	0.1
Organic PD and Pe-LED [44]	2	850	520	200	5	5
Full organic [47]	0.07	850	green		15	100
Full QDs (This work)	6.3	970	624	155	10^4	0.014

our investigation was to assess the impact of differing CdSe QD and Ag nanoparticle filling factors, along with variations in layer thickness, on the solar cell response. Most remarkably, we found that the absorbance of the solar cell increases at 700 nm and decreases at 800 nm as the filling factor of Ag nanoparticles increases. Changing the filling factor of CdSe, on the other hand, has very little effect. Reaching 200 nm for the composite thickness causes the absorbance value (A) to peak and spread out to longer wavelengths. The iodide/triiodide layer (di) thickness varies, and the maximum values of A are obtained at di = 200 nm. Interestingly, there is no discernible difference in the performance of the solar cell depending on the thickness (dt) of the TiO_2 layer. Our solar cell, therefore, includes elements that can be adjusted to get the maximum absorbance [10].

12.4.3 Nitrogen-doped graphene quantum dots for femtosecond ultrafast pulsed fiber lasers

Nitrogen-doped graphene quantum dots, or N-GQDs, are a novel class of quantum dots with distinct electronic properties. While their electron transport and biochemical characteristics have been thoroughly studied, more research is required to fully understand their optical characteristics – in particular, their nonlinear optical response – and possible uses in ultrafast photonics and photoelectronics. The nonlinear optical response and ultrafast carrier dynamics of N-GQDs in a water-based solution are systematically investigated in this work.

Specifically, N-GQDs show an ultrafast carrier cooling delay in the visible range of about 143 fs. Their nonlinear optical response analysis indicates that as the intensity of the incident pulse increases, they undergo a transition from saturable absorption to reverse saturable absorption. N-GQDs are useful as saturable absorbers for producing ultrashort pulses in mode-locked Er-doped fiber lasers because of this property. These findings open up new avenues for investigation into the optical characteristics of N-GQDs and suggest possible uses in ultrafast photonics. Increasingly in demand as carbon nanomaterials, graphene quantum dots (GQDs), which are categorized as nanomaterials, are zero-dimensional graphene structures with nanometric spatial scales. Due to their special physical characteristics, GQDs are used in a number of industries, including water treatment, energy storage, and bioimaging sensing. Doping elements such as fluorine, boron, selenium, and nitrogen can be added to a material to modulate its band gap, which affects optical, electronic, and local chemical properties.

The ability of nitrogen-doped graphene quantum dots to form diverse groups by bonding with carbon atoms and their atomic radius resemblance to carbon has attracted a lot of attention. Different quantum dot materials have different electronic and optical response properties, such as BP quantum dots, MXenes quantum dots, and excessive metallic chalcogenide quantum dots. BP quantum dots, for example, exhibit tunable wide-band nonlinear optical characteristics and size-dependent electronic–optical properties. In a similar vein, MXenes quantum dots that have a high electron density close to the Fermi level have interesting optical responses and good metal conductivity. These quantum dot materials' unique electronic characteristics and optical responses have been used in a variety of optoelectronic devices. A more complex situation is introduced by the energy level structure of nitrogen-doped graphene quantum dots, which is characterized by a large pi-conjugated system and an abundance of functional groups. Nitrogen-doped graphene quantum dots have been extensively explored in fields like photoelectric sensing, photocatalysis, and fluorescence quenching, but their nonlinear optical response, ultrafast carrier dynamics, and photonic applications are still largely unknown.

N-GQDs' nonlinear optical response and ultrafast carrier dynamics are investigated in this study, which highlights the possibility of using them in ultrafast photonics. Femtosecond transient absorption spectra are investigated, revealing similar ultrafast carrier cooling delay times in agreement with the graphene time scale. The nonlinear optical response study demonstrates that N-GQDs can produce high-energy dissipative soliton pulses by going from saturable absorption to reverse saturable absorption as the intensity of the incident pulse increases. Picosecond pulses produced by an Er-doped fiber laser using N-GQDs as a saturable absorber in aqueous solution are presented, illustrating the growing potential of N-GQDs for photonic applications. This work represents a major step forward in our understanding and use of nitrogen-doped graphene quantum dots in ultrafast photonics.

Table 12.3 Comparison of output performance of mode-locking fiber lasers based on various QD materials [17]

Materials	*Central wavelength (nm)*	*3db bandwidth (nm)*	*Repetition rate (MHz)*	*Pulse duration (ps)*	*Pump power (mW)*	*Output power (mW)*
BP QDs	1567.5	4.5	15.25	1.67	–	–
PbS QDs	1563	4.78	13.9	0.559	1000	23.5
GaTe QDs	1530.9	18.1	8.79	0.115	345	3.84
$NbSe_2$ QDs	1556	2.45	7.7	0.756	200	8.3
$CsPbBr_3$ QDs	1600	4.5	8.53	14.4	130	5.16
MXene QDs	1536.3	14.8	8.7	0.17	360	7.33
N-GQDs	1559	1.92	3.964	1.36	135	3.66

To summarize, the aim of our study was to examine the optical characteristics of a solution containing nitrogen-doped graphene quantum dots (N-GQDs) in water, with a particular focus on comprehending the dynamics of carriers and nonlinear optical features. We showcased the usefulness of N-GQDs in the field of ultrafast photonics. We investigated the carrier dynamics of N-GQDs using time-resolved absorption (TA) spectroscopy and found an extraordinarily short dynamic time of 143 fs. Remarkably, as the incident pulse power increased, the saturable absorption behavior of N-GQDs changed from saturable absorption to reverse saturable absorption. In addition, we demonstrated the N-GQDs solution's usefulness as a saturable absorber (SA) for a mode-locked fiber laser, which produced picosecond pulses. This experimental validation underscores the exceptional properties of N-GQDs as a material for ultrafast photonics, highlighting its significant potential for applications in photoelectronics and ultrafast photonics [11].

12.5 CONCLUSION

In conclusion, this exhaustive review thoroughly examines the intricate domain of nanomaterials and their profound influence on semiconductor devices. The examination of nanowires, carbon nanotubes (CNTs), and quantum dots encompasses a comprehensive exploration of their unique characteristics and diverse applications across a spectrum of semiconductor devices, including gate-all-around field-effect transistors (GAAFETs), solar panels, lasers, LEDs, transistors, and MOSFETs.

Nanowires, recognized for their distinct structural and electrical properties, demonstrate substantial potential in enhancing transistor performance, particularly in the context of GAAFETs. Carbon nanotubes, acknowledged for their exceptional electrical conductivity, emerge as promising solutions to surmount traditional transistor limitations, contributing to advancements in various electronic devices.

The review conducts a thorough analysis of the incorporation of quantum dots into semiconductor devices, spotlighting their noteworthy contributions to a diverse array of applications. From improving solar panels to facilitating the development of efficient lasers and LEDs, quantum dots play a pivotal role in harnessing their size-dependent properties to enhance device performance.

The integration of nanomaterials into semiconductor devices provides numerous advantages, including a low threshold current density, temperature-insensitive performance, high-speed characteristics, and resistance to material defects. Collectively, these features lay the groundwork for the evolution of the next generation of semiconductor technologies.

In summary, our exploration of nanowires, carbon nanotubes, and quantum dots in this review underscores their pivotal roles in shaping the landscape of semiconductor devices. As researchers delve deeper into uncovering the full potential of these nanomaterials, the future holds promising opportunities for advancing electronic and optoelectronic technologies. The seamless integration of nanomaterials with semiconductor devices is poised to drive innovation, efficiency, and performance to unprecedented levels in the realm of nanotechnology.

REFERENCES

1. S. Dey, T. P. Dash, S. Das, E. Mohapatra, J. Jena and C. K. Maiti 6, "Gate-all-around Si-nanowire transistors: simulation at nanoscale", *2018 IEEE Electron Device Kolkata Conference (EDKCON)*, Kolkata, India, 24–25 Nov. 2018.
2. Vidur Raj, Yi Zhu, Kaushal Vora, Hark Hoe Tan and Chennupati Jagadish, "Pushing limits of photovoltaics and photodetection using radial junction nanowire devices", *2022 IEEE International Conference on Emerging Electronics (ICEE)*, 978-1-6654-9185-3/22/2022 IEEE.
3. S. Petrosyan, V. Khachatryan and A. Yesayan, "Nanowire array solar cells", *2022 IEEE Latin American Electron Devices Conference (LAEDC)*, Cancun, Mexico, 2022, pp. 1–4.
4. F. A. Kareem and P. A. Michael, "The properties of carbon nano tubes as electrical conductors a review", *2023 International Conference on Innovations in Engineering and Technology (ICIET)*, Muvattupuzha, India, 2023, pp. 1–4.
5. Z. Chen, D. Farmer, S. Xu, R. Gordon, P. Avouris and J. Appenzeller, "Externally assembled gate-all-around carbon nanotube field-effect transistor", *IEEE Electron Device Letters*, vol. 29, no. 2, pp. 183–185, Feb. 2008.
6. Jiangbo Zhang, Ning Xi, K. W. C. Lai, Hongzhi Chen, Yilun Luo and Guangyong Li, "Single carbon nanotube based photodiodes for infrared detection", *2007 7th IEEE Conference on Nanotechnology (IEEE NANO)*, Hong Kong, China, 2007, pp. 1156–1160.
7. T. De Rijk, M. Cen-Puc, J. K. Piening and W. Lang, "Single layer piezoresistive polyimide pressure sensor based on carbon nanotubes", *2022 IEEE Sensors*, Dallas, TX, USA, 2022, pp. 1–4.
8. J. P. Reithmaier and A. Forchel, "Semiconductor quantum dots", *IEEE Circuits and Devices Magazine*, vol. 19, no. 6, pp. 24–29, Nov. 2003.
9. D. L. Huffaker, G. Park, Z. Zhou, O. B. Shchekin and D. G. Deppe, "1.3 μm room-temperature GaAs-based quantum dot laser", *Applied Physics Letters*, vol. 73, no. 18, pp. 2564–2566, 1998.
10. A. Aissat, F. Benyettou, A. Aissat and J. P. Vilcot, "InSb/GaAs quantum dot solar cell", *2016 International Renewable and Sustainable Energy Conference (IRSEC)*, Marrakech, Morocco, 2016, pp. 1–4.
11. I. C. Novoa-De Leo´n, J. Johny, S. Va´zquez-Rodr´ıguez, N. Garc´ıa-Go´mez, S. Carranza-Bernal, I. Mendivil, S. Shaji and S. Sepu´lveda-Guzma´n, "Tuning the luminescence of nitrogen-doped graphene quantum dots synthesized by pulsed laser ablation in liquid and their use as a selective photoluminescence on–off–on probe for ascorbic acid detection", *Carbon*, vol. 150, pp. 455–464, 2019.

12. St. Bartmoss Clair, "Review of the current state of nanotechnology and making future predictions with a discussion of forecasting methods", *Nanotechnology and Predicting the Future*, pp. 1–8, 20 Dec. 2016.
13. H. Tang, et al., "Up-Conversion device based on quantum dots with high-conversion efficiency over 6%", *IEEE Access*, vol. 8, pp. 71041–71049, 2020.
14. M. Boubakeur, A. Aissat and J. P. Vilcot, "Study of GaAs1-xNx/GaAs quantum dot structure for solar cell applications", *2019 7th International Renewable and Sustainable Energy Conference (IRSEC)*, Agadir, Morocco, 2019, pp. 1–5.
15. R. J. El-Khozondar, H. J. El-Khozondar, Y. F. Nassar, E. H. E. Bayoumi and H. Awad, "Efficiency improvement of Ag-doped CdSe quantum dot sensitized solar cells", *2023 8th International Engineering Conference on Renewable Energy & Sustainability (ieCRES)*, Gaza, Palestine, State of, 2023, pp. 1–5.
16. Megha A. Deshmukh, Sang–Joon Park, Bhavna S. Hedau and Tae-Jun Ha, "Recent progress in solar cells based on carbon nanomaterials", *Solar Energy*, vol. 220, pp. 953–990, 2021.
17. Yichao Zhang, Guanyun Ding, Dongyun Kai, Xuhuan Xie, Fangteng Zhang, Youjun Zeng, Guanyu Liu, Haoyi Wu and Zhaogang Nie, "Nitrogen-doped graphene quantum dots for femtosecond ultrafast pulsed fiber lasers", *Optical Fiber Technology*, vol. 80, p. 103433, 2023.
18. Neeraj Jain and Balwinder Raj, "Dual-k spacer region variation at the drain side of asymmetric SOI FinFET structure: performance analysis towards the analog/RF design applications", *Journal of Nanoelectronics and Optoelectronics*, vol. 14, pp. 349–359, March 2019.
19. Jeetendra Singh, Sanjeev Sharma, B. Raj and Mamta Khosla, "Analysis of barrier layer thickness on performance of In1-xGaxAs based gate stack cylindrical gate nanowire MOSFET", *Journal of Nanoelectronics and Optoelectronics*, vol. 13, pp. 1473–1477, Oct. 2018.
20. Neeraj Jain and B. Raj, "Analysis and performance exploration of high-k SOI FinFETs over the conventional low-k SOI FinFET toward analog/RF design", *Journal of Semiconductors*, vol. 39, no. 12, pp. 124002-1–7, Dec. 2018.
21. Candy Goyal, Jagpal Singh Ubhi and B. Raj, "A reliable leakage reduction technique for approximate full adder with reduced ground bounce noise", *Journal of Mathematical Problems in Engineering*, vol. 2018, p. 16, 15 Oct. 2018.
22. Anuradha, Jeetendra Singh, B. Raj and Mamta Khosla, "Design and performance analysis of nano-scale memristor-based nonvolatile SRAM", *Journal of Sensor Letter*, vol. 16, pp. 798–805, Oct. 2018.
23. Girish Wadhwa and B. Raj, "Parametric variation analysis of charge-plasma-based dielectric modulated JLTFET for biosensor application", *IEEE Sensor Journal*, vol. 18, no. 15, pp. 6070–6077, 1 Aug. 2018.
24. Jeetendra Singh and B. Raj, "Comparative analysis of memristor models for memories design", *Journal of Semiconductors*, vol. 39, no. 7, pp. 074006-1–12, July 2018.
25. Divya Yadav, Shailesh Singh Chouhan, Santosh Kumar Vishvakarma and B. Raj, "Application specific microcontroller design for IoT based WSN", *Sensor Letter*, vol. 16, pp. 374–385, May 2018.

26. Gurmohan Singh, R. K. Sarin and B. Raj, "Fault-tolerant design and analysis of quantum-dot cellular automata based circuits", *IEEE/IET Circuits, Devices & Systems*, vol. 12, pp. 638–64, 2018.
27. Jeetendra Singh and B. Raj, "Modeling of mean barrier height levying various image forces of metal insulator metal structure to enhance the performance of conductive filament based memristor model", *IEEE Nanotechnology*, vol. 17, no. 2, pp. 268–267, March 2018.
28. Aakash Jain, Sanjeev Sharma and B. Raj, "Analysis of triple metal surrounding gate (TM-SG) III–V nanowire MOSFET for photosensing application", *Opto-Electronics Journal*, vol. 26, no. 2, pp. 141–148, May 2018.
29. Aakash Jain, Sanjeev Sharma and B. Raj, "Design and analysis of high sensitivity photosensor using cylindrical surrounding gate MOSFET for low power sensor applications", *Engineering Science and Technology, an International Journal*, vol. 19, no. 4, pp. 1864–1870, Dec. 2016.
30. G. Saiphani Kumar, Amandeep Singh andd B. Raj, "Design and analysis of gate all around CNTFET based SRAM cell design", *Journal of Computational Electronics*, vol. 17, no. 1, pp. 138–145, March 2018.
31. Gurinder pal Singh, B. S. Sohi and Balwinder Raj, "Material properties analysis of graphene base transistor (GBT) for VLSI analog circuits", *Indian Journal of Pure & Applied Physics*, vol. 55, pp. 896–902, Dec. 2017.
32. Sunil Kumar and B. Raj, "Estimation of stability and performance metric for inward access transistor based 6T SRAM cell design using n-type/p-type DMDG-GDOV TFET", *IEEE VLSI Circuits and Systems Letter*, vol. 3, no. 2, pp. 25–39, June 2017.
33. Shashikant Sharma, Anjan Kumar, Manisha Pattanaik and B. Raj, "Forward body biased multimode multi-threshold CMOS technique for ground bounce noise reduction in static CMOS adders", *International Journal of Information and Electronics Engineering*, vol. 3, no. 3, pp. 567–572, 2013.
34. Hamendra Singh, Pankaj Kumar and Balwinder Raj, "Performance analysis of majority gate set based 1-bit full adder", *International Journal of Computer and Communication Engineering*, vol. 2, no. 4, pp. 1–8, 2013.
35. Anil Kumar Bhardwaj, Sumeet Gupta and B. Raj, "Investigation of parameters for Schottky barrier (SB) height for Schottky Barrier based carbon nanotube field effect transistor device", *Journal of Nanoelectronics and Optoelectronics*, vol. 15, pp. 783–791, July 2020.
36. Priya Bansal and B. Raj, "Memristor: a versatile nonlinear model for dopant drift and boundary issues", *Journal of Computational and Theoretical Nanoscience*, vol. 14, no. 5, pp. 2319–2325, May 2017.
37. Neeraj Jain and B. Raj, "An analog and digital design perspective comprehensive approach on Fin-FET (Fin-Field Effect transistor) technology – a review", *Reviews in Advanced Sciences and Engineering*, vol. 5, pp. 1–14, 2016.
38. Balwinder Raj, A. K. Saxena and S. Dasgupta, "Analytical modeling for the estimation of leakage current and subthreshold swing factor of nanoscale double gate FinFET device", *Microelectronics International*, vol. 26, pp. 53–63, 2009.
39. Soniya, Shailendra Singh, Girish Wadhwa and Balwinder Raj, "An analytical modeling for dual source vertical tunnel field effect transistor", *International Journal of Recent Technology and Engineering*, vol. 8 no. 2, pp. 603–608, July 2019.

40. Shailendra Singh and B. Raj, "Design and analysis of heterojunction vertical T-shaped tunnel field effect transistor", *Journal of Electronics Material*, vol. 48, no. 10, pp. 6253–6260, Oct. 2019.
41. Candy Goyal, Jagpal Singh Ubhi and B. Raj, "A low leakage CNTFET based inexact full adder for low power image processing applications", *International Journal of Circuit Theory and Applications*, vol. 47, no. 9, pp. 1446–1458, Sep. 2019.
42. B. Raj, A. K. Saxena and S. Dasgupta, "A compact drain current and threshold voltage quantum mechanical analytical modeling for FinFETs", *Journal of Nanoelectronics and Optoelectronics*, vol. 3, no. 2, pp. 163–170, 2008.
43. Girish Wadhwa and B. Raj, "An analytical modeling of charge plasma based tunnel field effect transistor with impacts of gate underlap region", *Superlattices and Microstructures*, vol. 142, p. 106512, June 2020.
44. Shailendra Singh and B. Raj, "Modeling and Simulation analysis of SiGe heterojunction double GateVertical t-shaped tunnel FET", *Superlattices and Microstructures*, vol. 142, pp. 106496, June 2020.
45. Neeraj Jain and Balwinder Raj, "Device and circuit co-design perspective comprehensive approach on FinFET technology – a review", *Journal of Electron Devices*, vol. 23, no. 1, pp. 1890–1901, 2016.
46. Sunil Kumar and B. Raj, "Analysis of I_{ON} and ambipolar current for dual-material gate-drain overlapped DG-TFET", *Journal of Nanoelectronics and Optoelectronics*, vol. 11, pp. 323–333, June 2016.
47. Naveed Anjum, Tarun Bali and B. Raj, "Design and simulation of handwritten multiscript character recognition", *International Journal of Advanced Research in Computer and Communication Engineering*, vol. 2, no. 7, pp. 2544–2549, July 2013.
48. Karmjit Singh and B. Raj, "Performance and analysis of temperature dependent multi-walled carbon nanotubes as global interconnects at different technology nodes", *Journal of Computational Electronics*, vol. 14 no. 2, pp. 469–476, June 2015.
49. Sunil Kumar and B. Raj, "Compact channel potential analytical modeling of DG-TFET based on evanescent–mode Approach", *Journal of Computational Electronics*, vol. 14, no. 2, pp. 820–827, July 2015.
50. Karmjit Singh and B. Raj, "Temperature dependent modeling and performance evaluation of multi-walled CNT and single-walled CNT as global interconnects", *Journal of Electronic Materials*, vol. 44, no. 12, pp. 4825–4835, Dec. 2015.
51. V. K. Sharma, M. Pattanaik and B.Raj, "INDEP approach for leakage reduction in nanoscale CMOS circuits", *International Journal of Electronics*, vol. 102, no. 2, pp. 200–215, 2014.
52. Karmjit Singh and B. Raj, "Influence of temperature on MWCNT bundle, SWCNT bundle and copper interconnects for nanoscaled technology nodes", *Journal of Materials Science: Materials in Electronics*, vol. 26, no. 8, pp. 6134–6142, 2015.
53. Naveed Anjum, Tarun Bali and B. Raj, "Design and simulation of handwritten Gurumukhi and Devanagri numerical recognition", *International Journal of Computer Applications*, vol. 73, no. 12, pp. 16–21, 2013.

54. S. Khandelwal, V. Gupta, B. Raj and R. D. Gupta, "Process variability aware low leakage reliable nano scale DG-FinFET SRAM cell design technique", *Journal of Nanoelectronics and Optoelectronics*, vol. 10, no. 6, pp. 810–817, Dec. 2015.
55. V. K. Sharma, M. Pattanaik and B. Raj, "ONOFIC approach: low power high speed nanoscale VLSI circuits design", *International Journal of Electronics*, vol. 101, no. 1, pp. 61–73, 2014.
56. S. Khandelwal, Balwinder Raj and R. D. Gupta, "FinFET based 6T SRAM cell design: analysis of performance metric, process variation and temperature effect", *Journal of Computational and Theoretical Nanoscience*, vol. 12, pp. 2500–2506, 2015.
57. Sumit Singh, Shekhar Yadav, Jagdeep Rahul, Anurag Srivastava and B. Raj, "Impact of HfO_2 in graded channel dual insulator double gate MOSFET", *Journal of Computational and Theoretical Nanoscience*, vol. 12, no. 6, pp. 950–953, April 2015.
58. Vijay Kumar Sharma, Manisha Pattanaik and B. Raj, "PVT variations aware low leakage INDEP approach for nanoscale CMOS Circuits", *Microelectronics Reliability*, vol. 54, pp. 90–99, 2014.
59. B. Raj, A. K. Saxena and S. Dasgupta, "Quantum mechanical analytical modeling of nanoscale DG FinFET: evaluation of potential, threshold voltage and source/drain resistance", *Journal of Material Science in Semiconductor Processing*, vol. 16, no. 4, pp. 1131–1137, 2013.
60. Maisagalla Gopal, Siva Sankar D. Prasad and Balwinder Raj, "8T SRAM cell design for dynamic and leakage power reduction", *International Journal of Computer Applications*, vol. 71, no. 9, pp. 43–48, June 2013.
61. Manisha Pattanaik, B. Raj, Shashikant Sharma and Anjan Kumar, "Diode based trimode multi-threshold CMOS technique for ground bounce noise reduction in static CMOS adders", *Advanced Materials Research*, vol. 548, pp. 885–889, 2012.
62. Balwinder Raj, A. K. Saxena and S. Dasgupta, "Nanoscale FinFET based SRAM cell design: analysis of performance metric, process variation, underlapped FinFET and temperature effect", *IEEE Circuits and System Magazine*, vol. 11, no. 2, pp. 38–50, 2011.
63. V. K. Sharma, M. Pattanaik and Balwinder Raj, "Leakage current ONOFIC approach for deep submicron VLSI circuit design", *International Journal of Electrical, Computer, Electronics and Communication Engineering*, vol. 7, no. 4, pp. 239–244, 2013.
64. Parminder Kaur, Sandeep Singh Gill and B. Raj, "Comparative analysis of OFETs materials and devices for sensor applications", *Journal of Silicon*, vol. 14, pp. 4463–4471, 2022.
65. Sanjeev Kumar Sharma, Parveen Kumar, Balwant Raj and B. Raj· "$In_{1-x}Ga_xAs$ double metal gate-stacking cylindrical nanowire MOSFET for highly sensitive photo detector", *Journal of Silicon*, vol. 14, pp. 3535–3541, 2022.
66. B. Raj, A. K. Saxena and S. Dasgupta, "Analytical modeling of quasi planar nanoscale double gate FinFET with source/drain resistance and field dependent carrier mobility: a quantum mechanical study", *Journal of Computer*, vol. 4, no. 9, pp. 1–8, 2009.

67. S. Bhushan, S. Khandelwal and B. Raj, "Analyzing different mode FinFET based memory cell at different power supply for leakage reduction", *Seventh International Conference on Bio-Inspired Computing: Theories and Application* (BIC-TA 2012), Advances in Intelligent Systems and Computing, vol. 202, pp. 89–100, 2013.
68. Jeetendra Singh and B. Raj, "Temperature dependent analytical modeling and simulations of nanoscale memristor", *Engineering Science and Technology, an International Journal*, vol. 21, pp. 862–868, Oct. 2018.
69. Shradhya Singh, Shashi Bala, Balwant Raj and B. Raj, "Improved sensitivity of dielectric modulated junctionless transistor for nanoscale biosensor design", *Sensor Letter*, vol. 18, pp. 328–333, April 2020.
70. Vivek Kumar, Santosh Kumar Vishvakarma and B. Raj, "Design and performance analysis of ASIC for IoT applications", *Sensor Letter*, vol. 18, pp. 31–38, Jan. 2020.
71. Akanksha Jaiswal, R. K. Sarin, B. Raj and Shikha Sukhija, "A novel circular slotted microstrip-fed patch antenna with three triangle shape defected ground structure for multiband applications", *Advanced Electromagnetic*, vol. 7, no. 3, pp. 56–63, Aug. 2018.
72. Girish Wadhwa and B. Raj, "Label free detection of biomolecules using charge-plasma-based gate underlap dielectric modulated junctionless TFET", *Journal of Electronic Materials*, vol. 47, no. 8, pp. 4683–4693, Aug. 2018.
73. Gurmohan Singh, R. K. Sarin and B. Raj, "Design and performance analysis of a new efficient coplanar quantum-dot cellular automata adder", *Indian Journal of Pure & Applied Physics*, vol. 55, pp. 97–103, Feb. 2017.
74. Parminder Kaur, Vikas Pandey and B. Raj, "Comparative study of efficient design, control and monitoring of solar power using IoT", *Sensor Letter*, vol. 18, pp. 419–426, May 2020.
75. Anil Kumar Bhardwaj, Sumeet Gupta and B. Raj, "Development & analysis of compact model for double gate Schottky barrier CNTFET", *Journal of Nanoelectronics and Optoelectronics*, vol. 15, pp. 1199–1208, Aug. 2020.
76. Girish Wadhwa, Priyanka Kamboj, Jeetendra Singh and B. Raj, "Design and investigation of junctionless DGTFET for biological molecule recognition", *Transactions on Electrical and Electronic Materials*, vol. 22, pp. 282–289, 2021.
77. Sachin Kumar Verma, Shailendra Singh, Girish Wadhwa and B. Raj, "Detection of biomolecules using charge-plasma based gate underlap dielectric modulated dopingless TFET", *Transactions on Electrical and Electronic Materials*, vol. 21, pp. 528–535, June 2020.
78. Neeraj Jain and B. Raj, "Impact of underlap spacer region variation on Electrostatic and analog/RF performance of symmetrical high-k SOI FinFET at 20 nm channel length", *Journal of Semiconductors*, vol. 38, no. 12, p. 122002, Dec. 2017.
79. Shailendra Singh and B. Raj, "Analytical modeling and simulation analysis of T-shaped III–V heterojunction vertical T-FET", *Superlattices and Microstructures*, vol. 147, pp. 106717, Nov. 2020.
80. Gurmohan Singh, R. K. Sarin and B. Raj, "Design and analysis of area efficient QCA based reversible logic gates", *Journal of Microprocessors and Microsystems*, vol. 52, pp. 59–68, May 2017.

81. Shailendra Singh and B. Raj, "Parametric variation analysis on hetero-junction vertical t-shape TFET for supressing ambipolar conduction", *Indian Journal of Pure and Applied Physics*, vol. 58, pp. 478–485, June 2020.
82. Soniya, Shailendra Singh, Girish Wadhwa and Balwinder Raj, "Design and analysis of dual source vertical tunnel field effect transistor for high performance", *Transactions on Electrical and Electronics Materials*, vol. 21, pp. 74–82, Oct. 2019.
83. Manjit Kaur, Neena Gupta, Sanjeev Kumar, B. Raj and Arun Kumar Singh, "RF performance analysis of intercalated graphene nanoribbon based global level interconnects", *Journal of Computational Electronics*, vol. 19, pp. 1002–1013, June 2020.
84. Girish Wadhwa and B. Raj, "Design and performance analysis of junctionless TFET biosensor for high sensitivity", *IEEE Nanotechnology*, vol. 18, pp. 567–574, 2019.
85. Jeetendra Singh and B. Raj, "Enhanced nonlinear memristor model encapsulating stochastic dopant drift", *Journal of Nanoelectronics and Optoelectronics*, 14, 958–963, 2019.
86. Anjana Bhardwaj, Pradeep Kumar, B. Raj and Sunny Anand, "Design and performance optimization of doping-less vertical nanowire TFET using gate stack technique", *Journal of Electronic Materials*, vol. 41, no. 7, pp. 4005–4013, 2022.
87. Jeetendra Singh and B. Raj, "Tunnel current model of asymmetric MIM structure levying various image forces to analyze the characteristics of filamentary memristor", *Applied Physics A*, vol. 125, no. 3, pp. 203.1–203.11, Feb. 2019.
88. Candy Goyal, Jagpal Singh Ubhi and B. Raj, "Low leakage zero ground noise nanoscale full adder using source biasing technique", *Journal of Nanoelectronics and Optoelectronics*, vol. 14, pp. 360–370, March 2019.
89. Gurmohan Singh, R. K. Sarin and B. Raj, "A novel robust exclusive-OR function implementation in QCA nanotechnology with energy dissipation analysis", *Journal of Computational Electronics*, vol. 15, no. 2, pp. 455–465, June 2016.
90. Girish Wadhwa, Priyanka Kamboj and Balwinder Raj, "Design optimisation of junctionless TFET biosensor for high sensitivity", *Advances in Natural Sciences: Nanoscience and Nanotechnology*, vol. 10, no. 7, p. 45001, 2019.
91. Priya Bansal and B. Raj, "Memristor modeling and analysis for linear dopant drift kinetics", *Journal of Nanoengineering and Nanomanufacturing*, vol. 6, pp. 1–7, 2016.
92. D. Vaithiyanathan Ashima and B. Raj, "Performance analysis of charge plasma induced graded channel Si nanotube", *Journal of Engineering Research*, EMSME Special no. pp. 146–154, Aug. 2021.
93. Abhishek Singh Tomar, Vijay Kumar Magraiya and B. Raj, "Scaling of access and data transistor for high performance DRAM cell design", *Quantum Matter*, vol. 2, pp. 412–416, Oct. 2013.
94. Neeraj Jain and B. Raj, "Parasitic capacitance and resistance model development and optimization of raised source/drain SOI FinFET structure for analog circuit applications", *Journal of Nanoelectronics and Optoelectronics*, vol. 13, pp. 531–539, April 2018.

95. Shradhya Singh, S. K. Vishvakarma and B. Raj, "Analytical modeling of split–gate junction-less transistor for a biosensor application", *Sensing and Bio-Sensing*, vol. 18, pp. 31–36, April 2018.
96. Maisagalla Gopal and Balwinder Raj, "Low power 8T SRAM cell design for high stability video applications", *ITSI Transaction on Electrical and Electronics Engineering*, vol. 1, no. 5, pp. 91–97, 2013.
97. Balwinder Raj, Jatin Mitra, Deepak Kumar Bihani, V. Rangharajan, A. K. Saxena and S. Dasgupta, "Analysis of noise margin, power and process variation for 32 nm FinFET based 6T SRAM cell", *Journal of Computer*, vol. 5, no. 6, pp. 1–8, 2010.
98. Divya Sharma, Rajesh Mehra and B. Raj, "Comparative analysis of photovoltaic technologies for high efficiency solar cell design", *Superlattices and Microstructures*, vol. 153, pp. 106861, May 2021.
99. Pawandeep Kaur, Avtar Singh Buttar and B. Raj, "A comprehensive analysis of nanoscale transistor based biosensor: a review", *Indian Journal of Pure and Applied Physics*, vol. 59, pp. 304–318, April 2021.
100. Divya Yadav, Balwant Raj and B. Raj, "Design and simulation of low power microcontroller for IoT applications", *Journal of Sensor Letters*, vol. 18, pp. 401–409, May 2020.
101. Shailendra Singh and B. Raj, "A 2-D analytical surface potential and drain current modeling of double-gate vertical t-shaped Tunnel FET", *Journal of Computational Electronics*, vol. 19, pp. 1154–1163, April 2020.
102. Jeetendra Singh and B. Raj, "An accurate and generic window function for non-linear memristor model", *Journal of Computational Electronics*, vol. 18 no. 2, pp. 640–647, June 2019.
103. Manjit Kaur, Neena Gupta, Sanjeev Kumar, B. Raj and Arun Kumar Singh, "Comparative RF and crosstalk analysis of carbon based nano interconnects", *IET Circuits, Devices & Systems*, vol. 15, no. 6, pp. 493–503, Feb. 2021.
104. Nehru Kandasamy, Firdous Ahmad, D. Ajitha, B. Raj and Nagarjuna Telagam, "Quantum dot cellular automata based scan flip flop and boundary scan register", *IETE Journal of Research*, vol. 66, pp. 535–548, 2020.
105. Jeetendra Singh and B. Raj, "Design and investigation of 7T2M NVSARM with enhanced stability and temperature impact on store/restore energy", *IEEE Transactions on Very Large Scale Integration Systems*, vol. 27, no. 6, pp. 1322–1328, June 2019.
106. Anil Kumar Bhardwaj, Sumeet Gupta, B. Raj and Amandeep Singh, "Impact of double gate geometry on the performance of carbon nanotube field effect transistor structures for low power digital design", *Computational and Theoretical Nanoscience*, vol. 16, pp. 1813–1820, 2019.
107. Neeraj Jain and B. Raj, "Thermal stability analysis and performance exploration of asymmetrical dual-k underlap spacer (ADKUS) SOI FinFET for security and privacy applications", *Indian Journal of Pure & Applied Physics*, vol. 57, pp. 352–360, May 2019.

Chapter 13

Design of MUX-Based CMOS 8-Bit Encoder device for high-speed 5G Flash ADC binary conversion in UWB applications

Banoth Krishna, Sandeep Singh Gill, and Amod Kumar

13.1 INTRODUCTION

In the era of burgeoning 5G networks and the ever-expanding landscape of ultra-wideband (UWB) applications, the demand for high-performance analog-to-digital converters (ADCs) has become paramount. These converters play a critical role in efficiently capturing analog signals and converting them into digital formats for subsequent processing and transmission. As the need for faster and more reliable data conversion grows, ADC designs must continually evolve to meet these demanding requirements. One of the fundamental challenges in ADC design is achieving a balance between speed, accuracy, and power efficiency. The process of encoding the analog input signal into a digital representation is a pivotal stage that directly influences the overall performance of the ADC [1–4]. The choice of encoding technique can impact factors such as conversion speed, power consumption, and silicon area utilization. Traditional approaches to encoding, such as binary-ROM-based encoders, offer simplicity in design but often at the cost of speed and power efficiency. Conversely, more complex methods like the Fat Tree Encoder, which employs OR-gates in a tree structure, can provide better speed but are burdened with challenges such as increased area overhead and susceptibility to bubble errors. As the demand for higher speeds and greater efficiency continues to rise, novel solutions are required to address these challenges and optimize ADC performance.

In this context, this research aims to present an innovative solution for achieving high-speed binary conversion in the context of 5G Flash ADCs tailored for UWB applications. We propose the design and implementation of a MUX-Based CMOS 8-Bit Encoder [2–4] as shown in Figure 13.1, specifically engineered to balance the trade-offs between speed, accuracy, and power consumption [5]. By employing a carefully crafted combination of multiplexers, full-adders, and binary search algorithms, this encoder aims to provide a novel approach to efficient binary encoding [6–10]. To assess the viability of our proposed design, we utilize the Cadence Virtuoso simulator in conjunction with a 180 nm CMOS technology. This combination allows us to evaluate the encoder's performance under realistic conditions,

DOI: 10.1201/9781003612797-13

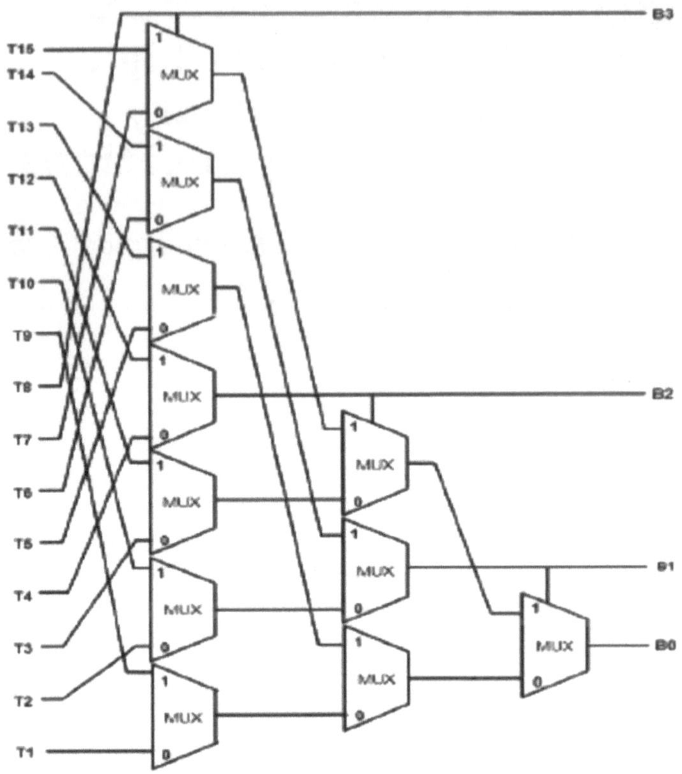

Figure 13.1 MUX-based encoder architecture.

considering the characteristics of modern semiconductor fabrication processes. Through extensive simulations and analysis, we explore how our MUX-Based CMOS 8-Bit Encoder compares to existing methods in terms of speed, power consumption, and area utilization in Table 13.1.

The subsequent sections of this chapter will delve into the design methodology, the challenges posed by traditional encoding techniques, our innovative solution using the Wallace tree encoder, the practical implementation details, simulation results, and finally, the implications of our findings. Ultimately, this research aims to contribute to the advancement of ADC

Table 13.1 Comparison of encoders architectures in terms of speed, power, and area

Encoder type	*Speed*	*Power*	*Area*
ROM Encoder	Low	Low	Low
Fat Tree Encoder	Medium	Medium	Medium
Wallace Tree Encoder	High	High	High
Proposed Multiplexer-based Encoder	Very high	Very low	Very low

design paradigms, fostering enhanced data conversion capabilities for the rapidly evolving landscape of 5G communication systems and UWB applications [11–16].

13.2 TRANSMISSION GATE

Transmission gates are bidirectional switches that can pass analog or digital signals between input and output terminals as shown in Figure 13.2. They are controlled by an enable signal, which is active or inactive. These gates have low ON resistance, minimizing signal attenuation and distortion. They are commonly used in applications with signal integrity, such as analog-to-digital converters and digital-to-analog converters [17–23].

MOS transistors operations

$$V_{gs} > V_{th}$$

$V_{ds} < V_{gs} - V_{th}$ ---------→Transistor operates at non-saturation region

$V_{ds} > V_{gs} - V_{th}$ ----------→Transistor operates at saturation region

$V_{gs} < V_{th}$ ---------→ Transistor operates at cut-off region

$$I_{ds} = \frac{\mu_{n/p} Cox}{tox} \frac{W}{L} \left(\left(V_{gs} - V_{th} \right) - \frac{V_{ds}}{2} \right) V_{ds}$$ ------→MOS transistor non-saturation current

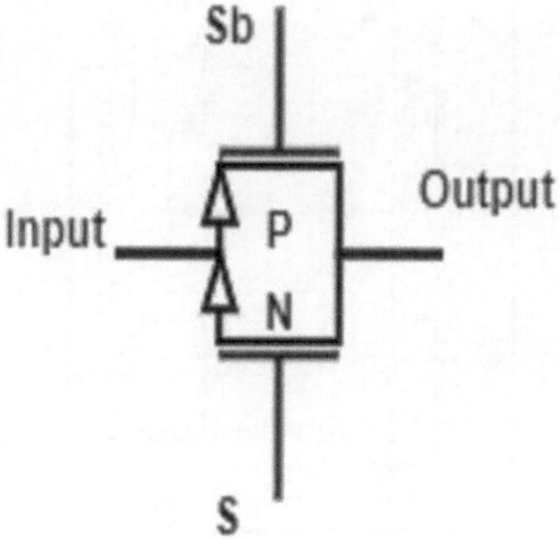

Figure 13.2 CMOS transmission gate schematic.

$$I_{ds} = \frac{\mu_{n/p} Cox}{tox} \frac{W}{L} \frac{\left(V_{gs} - V_{th}\right)^2}{2}$$ ------------→MOS transistor saturation current

Transmission gates are often implemented using complementary metal–oxide–semiconductor (CMOS) technology, which combines PMOS and NMOS transistors as shown in Figure 13.2. However, they can consume more power than simple CMOS gates, especially when in the ON state. Transmission gates can also exhibit parasitic effects, such as capacitance and leakage current, which can impact performance in high-frequency or low-power applications. Proper layout and sizing techniques are crucial for optimizing their behavior.

13.3 2:1 MUX USING TRANSMISSION GATES

A 2:1 MUX is a digital circuit that selects one input and directs it to the output based on a control signal S_0 and S_0b as shown in Figure 13.3. Transmission gate-based design utilizes N-type metal–oxide–semiconductor (NMOS) and P-type metal–oxide–semiconductor (PMOS) transistors as electronic switches, allowing low-to-high signal transmission and high-to-low transmission. The design involves signal paths, NMOS control, PMOS control, output stage, and control signal [24–29]. Advantages include low voltage drop, bidirectional capability, and simple control. Challenges include signal degradation and propagation delay. Overall, transmission

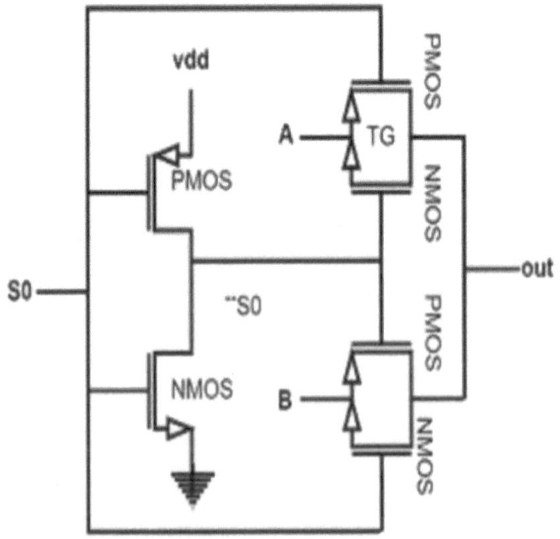

Figure 13.3 2:1 MUX schematic architecture using TG.

gate-based design provides an efficient and compact solution for signal routing and selection in digital circuits.

13.4 DESIGN OF THERMOMETER CODE-TO-BINARY ENCODER USING 2:1 MUX

The thermometer-to-binary encoder for an 8-bit Flash ADC consists of 256 comparators comparing an analog input voltage against a set of reference voltages using one-hot encoding. The design uses a transmission gate-based 2:1 MUX to implement priority encoding, generating a binary representation for each bit of the thermometer code. Control signals are generated from the thermometer-coded input, with "1" representing active voltage and "0" indicating higher bit selection. The outputs combine to form the binary representation of the ADC's digital output, indicating the highest active bit in the code [30–36]. This design offers advantages such as reduced complexity, efficiency, and signal integrity, but also challenges in signal integrity [11–15].

13.5 DESIGN METHODOLOGY

The design methodology of the proposed MUX-Based CMOS 8-Bit Encoder for high-speed 5G Flash ADC binary conversion in UWB applications involves a systematic approach that balances efficiency, accuracy, and power considerations. The primary goal is to overcome the limitations of traditional encoding techniques while capitalizing on the benefits of innovative designs like the Wallace Tree Encoder [37–43].

1. **Requirement analysis:** The design process begins with a thorough analysis of the requirements imposed by the target application. The specific needs of high-speed 5G communication systems and UWB applications are carefully considered. Conversion speed, power efficiency, and area utilization are identified as critical parameters to optimize.
2. **Encoder architecture selection:** Traditional binary-ROM-based encoders are evaluated for their simplicity but acknowledged limitations in speed and power efficiency. Similarly, the Fat Tree Encoder's potential for high-speed conversion is weighed against its complexity and susceptibility to bubble errors. The Wallace Tree Encoder is identified as a promising solution to mitigate these shortcomings [44–49].
3. **Wallace Tree Encoder design:** The Wallace Tree Encoder is designed as a cascading structure of full-adders to efficiently count ones in the thermometer-coded input. This design leverages the binary search

algorithm to reduce critical path delay and mitigate bubble errors. Careful consideration is given to the arrangement and sizing of full-adders to optimize performance.
4. **Multiplexer integration:** To manage the input selection and enable the binary search algorithm, multiplexers are strategically placed within the encoder architecture. The middle bit (bit 32) of the input signal is utilized as a control signal to guide the selection process.
5. **Control logic implementation:** Control logic is developed to determine whether the lower or upper outputs should be forwarded to the next stage of the encoder. This logic is based on the state of I16, which is derived from the input signal. The control logic ensures seamless switching between processing paths based on the condition of I16.
6. **Simulation and analysis:** The entire design is then subjected to rigorous simulation using the Cadence Virtuoso simulator, utilizing a 180 nm CMOS technology. This simulation phase assesses the performance of the encoder under realistic conditions, including process variations and transistor characteristics. Simulation results provide insights into speed, power consumption, and other key performance metrics [50–57].
7. **Optimization and trade-offs:** Simulation results are analyzed to identify areas for optimization and potential trade-offs. Parameters such as transistor sizing, power supply voltage, and signal propagation delay are adjusted to strike a balance between performance metrics. Iterative refinements are made to enhance the design's efficiency.
8. **Verification and validation:** The design is rigorously verified against expected outcomes and validated against benchmark performance metrics. Simulation results are compared with theoretical expectations to ensure the encoder's accuracy and effectiveness [58–64].

13.6 RESULT ANALYSIS

The simulation results validate the effectiveness of the proposed MUX-Based CMOS 8-Bit Encoder for high-speed 5G Flash ADC binary conversion in UWB applications. Key performance metrics, including conversion speed, power consumption, and area utilization, were evaluated to assess the encoder's efficiency.

The proposed Wallace Tree Encoder offers significant improvements in conversion speed, power consumption, area utilization, robustness against bubble errors, and comparison to traditional techniques [65–72]. Its binary search algorithm reduces critical path delay, enabling faster conversion rates and making it crucial for high-speed 5G communication systems and UWB applications. The encoder's compact architecture minimizes silicon area usage, ensuring cost-effectiveness and scalability in large-scale integrated circuits. The encoder also mitigates a bubble error problems, ensuring accurate and reliable binary conversion even in dynamic signal conditions (Figures 13.4–13.10).

13.7 MODIFIED PROPOSED ENCODER

The "Modified Proposed Multiplexer-based Encoder" refers to (Figures 13.11–13.16) a revised version or adaptation of an encoder utilizing a multiplexer in the context of an 8-bit Flash analog-to-digital converter (ADC) design. This modification has demonstrated significantly improved performance in the ADC design process [73–78].

The utilization of a multiplexer in the encoder is crucial for efficiently handling and processing the analog input signals in an 8-bit -Flash ADC. By modifying the design of this key component, the ADC can achieve enhanced accuracy, resolution, speed, or other desired performance metrics, making it highly effective for converting analog signals to digital representations.

MSB-1bit G1 = C1 .... (13.1)

MSB-2bit G2 = C2 .... (13.2)

MSB-3bit G3 = C32 .... (13.3)

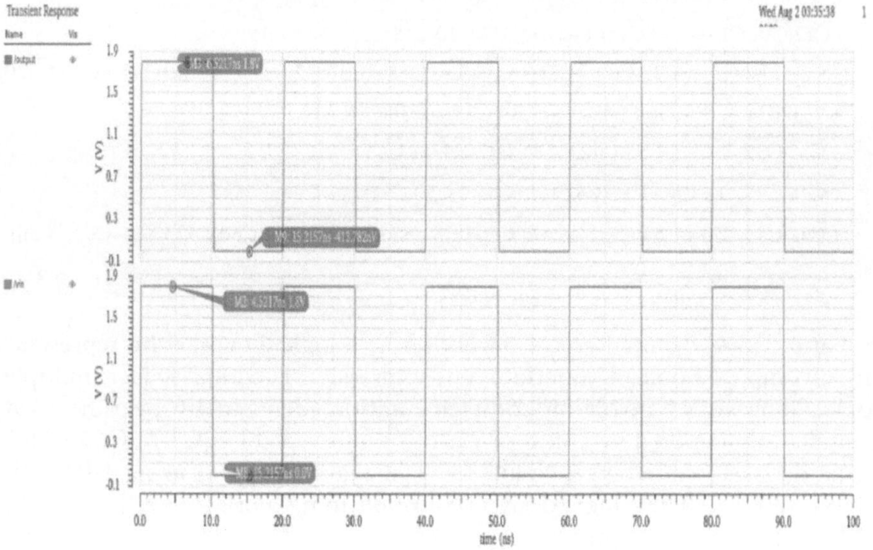

Figure 13.4 Output response of transmission gate.

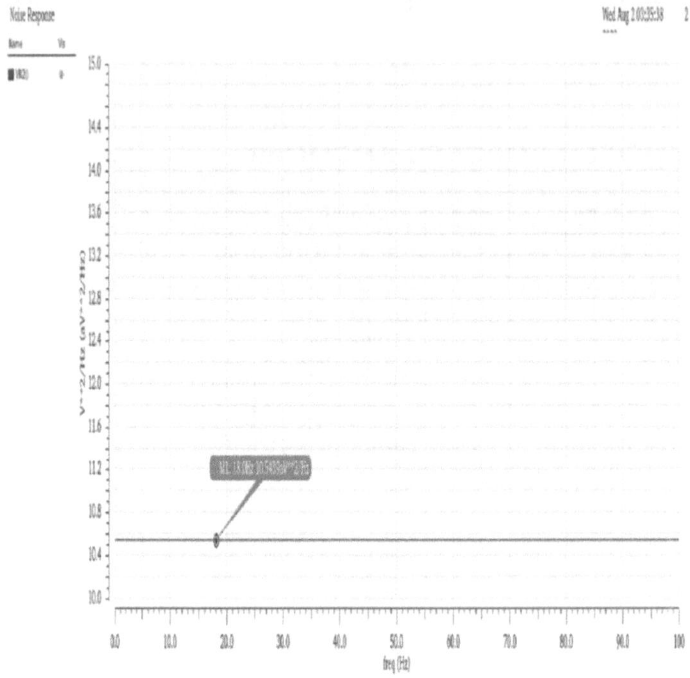

Figure 13.5 Output noise response of transmission gate.

$$\begin{aligned}G4 = {} & \overline{C32}\{\overline{C16}\{\overline{C8}[\overline{C4}(\overline{C2}.C1 + C2.\overline{C3}) + C4(\overline{C6}.C5 + C6.\overline{C7})] + C8[\overline{C12}(\overline{C10}.C9 + C10.\overline{C11}) \\ & + C12(\overline{C14}.C13 + C14.\overline{C15})]\} + C16\{\overline{C24}[\overline{C20}(\overline{C18}.C17 + C18.\overline{C19}) \\ & + C20(\overline{C22}.C21 + C22.\overline{C23})] + C24[\overline{C28}(\overline{C26}.C25 + C26.\overline{C27}) + C28(\overline{C30}.C29 + C30.\overline{C31})]\} \\ & + C32\{\overline{C48}\{\overline{C40}[\overline{C36}(\overline{C34}.C33 + C34.\overline{C35}) + C36(\overline{C38}.C37 \\ & + C38.\overline{C39})] + C40[\overline{C34}(\overline{C42}.C41 + C42.\overline{C43}) + C44(\overline{C46}.C45 \\ & + C46.\overline{C47})]\} + C48[\overline{C56}[\overline{C52}(\overline{C50}.C49 + C50.\overline{C51}) \\ & + C52(\overline{C54}.C53 + C54.\overline{C55})] + C56[\overline{C60}(\overline{C58}.C57 + C58.\overline{C59}) + C60(\overline{C62}.C61 + C62.\overline{C63})]\}\end{aligned} \tag{13.4}$$

Equation 13.4, denoted as G4 output bit, is a logical expression representing a complex Boolean function. The expression is comprised of multiple logical operations involving complementation (denoted by overline) and conjunction (denoted by a dot) of various input variables (C1 to C63). This equation is associated with Figure 13.11 in a larger context using 2:1 multiplexer, likely a digital logic design or a Boolean network [79–83].

The equation description:

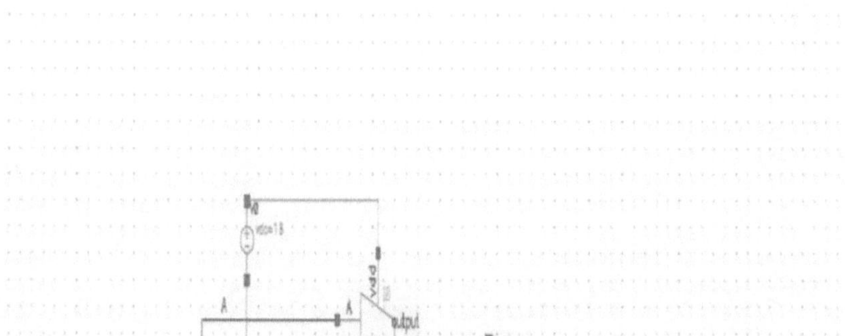

Figure 13.6 2:1 MUX schematic diagram using transmission gates.

The equation G4 is a function of C32, starting with the negation of C32. It contains a complex logical expression separated by conjunctions, with each term contributing to the final output. The equation suggests a complex logic circuit, possibly a multi-level AND–OR combination. The specific context of Figure 13.11 provides more insight into the interpretation of this equation, likely representing the logic of a specific block or component within a digital system, such as a decoder or encoder. Understanding the purpose and context of the figure is crucial for a detailed interpretation of the equation and its role in the overall system [84–89].

$$\begin{aligned} G5 = {} & \overline{C32}\{\overline{C16}[\overline{C8}(\overline{C4}.\ C2 + C4.\overline{C6}) + C8(\overline{C12}.C10 + C12.\overline{C14})] + C16[(\overline{C24}.C18 + C20.\overline{C22}) \\ & + C24(\overline{C28}.C26 + C28.\overline{C30})]\} + C32\{\overline{C48}[\overline{C40}(\overline{C36}.C34 + C36.\overline{C38}) \\ & + C40(\overline{C44}.C42 + C44.\overline{C46})] + C48[\overline{C56}(\overline{C52}.C50 + C52.\overline{C54}) + C56(\overline{C60}.C58 + C60.\overline{C62})]\} \end{aligned} \tag{13.5}$$

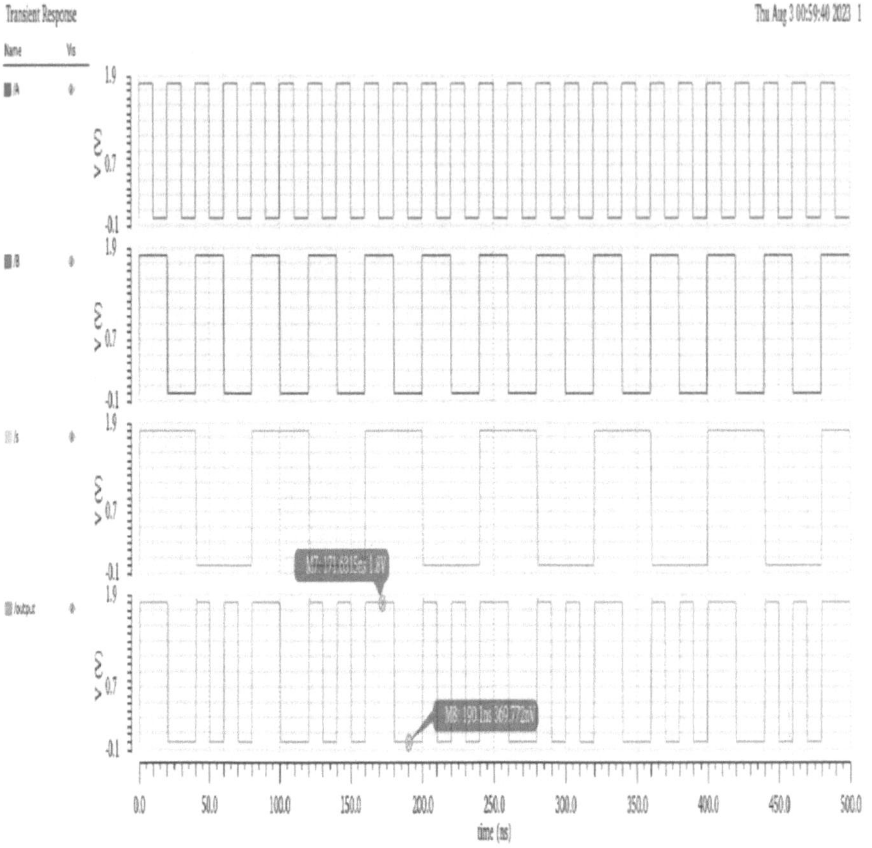

Figure 13.7 Output response of 2:1 MUX using transmission gates.

Equation 13.5, denoted as G5 output bit, represents a logical expression that defines the behavior of a specific logic circuit or component associated with Figure 13.12. This equation involves a complex combination of logical operations using multiple input variables (C2 to C62) and is designed using 2:1 multiplexer.

The equation description:

The G5 function is a complex logical expression involving various terms separated by conjunctions. The first term, C16, is complemented and nested with terms involving C8, C4, and C4. This pattern repeats for higher-order terms, including C8, C16, and C32. The second term, C32, is nested similarly, with terms involving C48, C40, and C36. Equation 13.5 suggests a logic circuit using a 2:1 multiplexer, with the selection inputs determined by

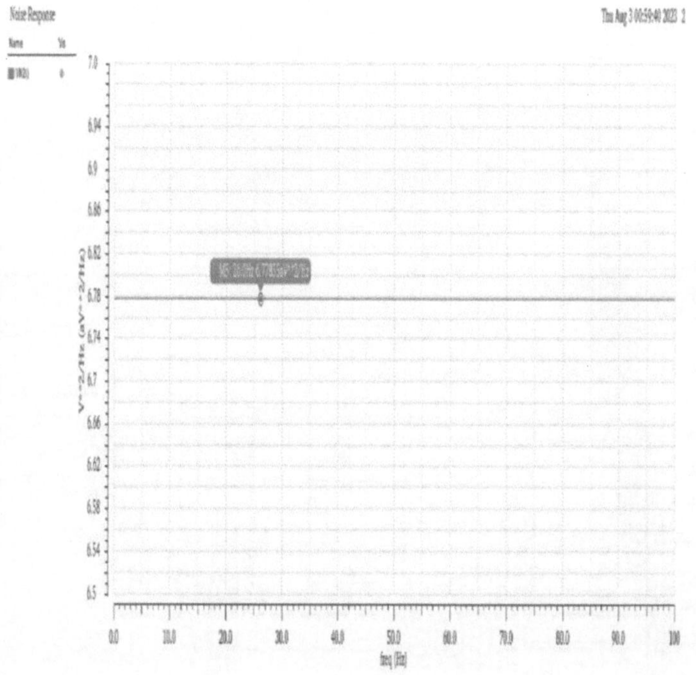

Figure 13.8 **Output noise response of 2:1 using transmission gates.**

binary values of C2 to C62. This implies a data path that can take different routes based on these control inputs [90–95].

$$\begin{aligned} G6 = {} & \overline{C32}\{\overline{C16}(\overline{C8}.\ C4 + C8.\overline{C12})\ + C16(\overline{C24}.C20 + C24.\overline{C28})\} \\ & + C32\{\overline{C48}(\overline{C40}.C36 + C40.\overline{C44}) + C48(\overline{C56}.C52 + C56.\overline{C60})\} \end{aligned} \tag{13.6}$$

Equation 13.6, denoted as G6 output bit, represents a logical expression that defines the behavior of a specific logic circuit or component associated with Figure 13.13. Similar to the previous equations, this one involves a combination of logical operations using multiple input variables (C4 to C60) and is designed using 2:1 multiplexer.

The equation description:
The G6 function is a function of C32, with the expression starting with its complement. The outermost set of parentheses contains a complex logical expression, with terms separated by conjunctions. The first term, C16, is complemented and nested with terms involving C8 and C4 [96–101]. The second term, C32, is nested similarly, with terms involving C48 and C40. Equation 13.6 suggests a logic circuit using a 2:1 multiplexer, with multiplexer inputs determined by binary values of C4 to C60.

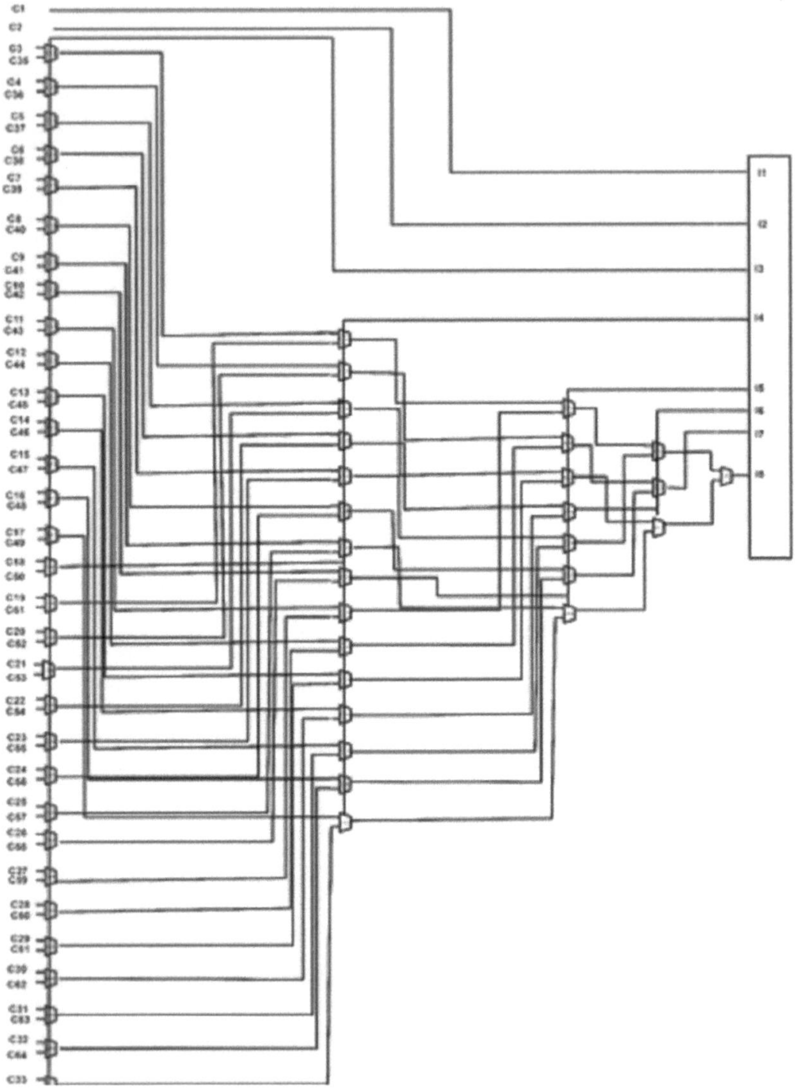

Figure 13.9 Proposed 8-Bit Multiplexer-based Encoder using 2:1 multiplexer.

$$G7 = \overline{C32}(\overline{C16}.C8 + C16.\overline{C24}) + C32\ (\overline{C48}.C40 + C48.\overline{C56}) \tag{13.7}$$

Equation 13.7, denoted as G7 output bit, represents a logical expression defining the behavior of a specific logic circuit or component associated with Figure 13.14. This equation involves logical operations using multiple input variables (C8, C16, C24, C32, C40, C48, C56) and is designed using 2:1 multiplexer.

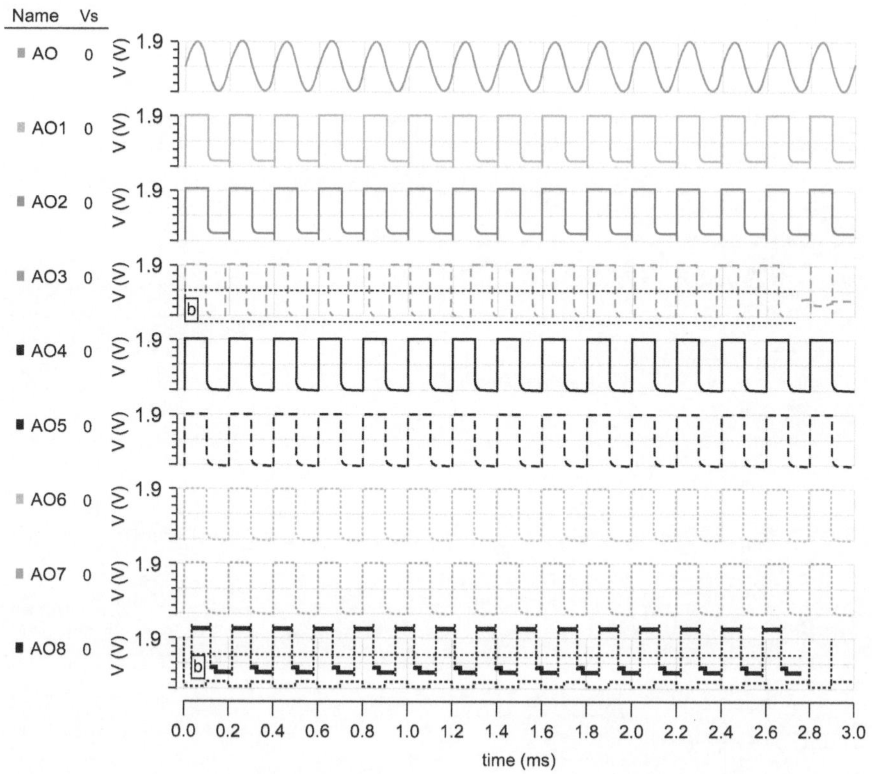

Figure 13.10 **Proposed 8-Bit Multiplexer-based Encoder using 2:1 multiplexer.**

The equation description:

The G7 function is a function of C32, with the expression starting with its complement. The outermost set of parentheses contains a complex logical expression with various terms separated by conjunctions. The first term is C16, followed by C8, C8, and C24, and repeats for the higher-order term, C32. The second term is C32, nested similarly with C48 and C40. Equation 13.7 defines a logic circuit using a 2:1 multiplexer, with multiplexer inputs determined by binary values of C8, C16, C24, C32, C40, C48, and C56.

$$G8 = \overline{C32}.C16 + C32.\overline{C32} \tag{13.8}$$

Equation 13.8, denoted as G8 output bit, represents a logical expression defining the behavior of a specific logic circuit or component associated with Figure 13.15. This equation involves logical operations using multiple input variables (C16, C32, C48) and is designed using 2:1 multiplexer [102–105].

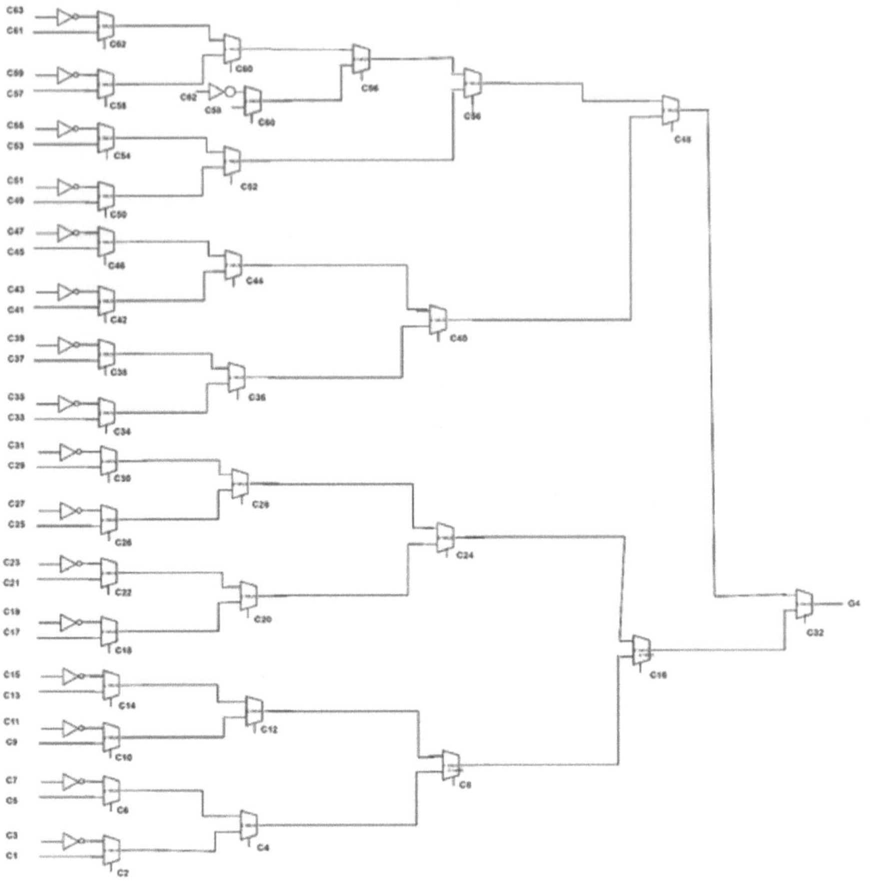

Figure 13.11 Modified Proposed 8-Bit Multiplexer-based Encoder for G4.

The equation description:

The G8 function is a function of C32, with the entire expression involving its complement. Two terms are connected by a logical OR operation, with the first term involving C16 and the second term involving C32. Equation 13.8 defines a logic circuit using a 2:1 multiplexer, with the selection inputs determined by binary values of C16, C32, and C48. This implies a data path that can take different routes based on these control inputs (Table 13.2).

The power analysis shows an increase in power dissipation from the proposed MUX-based Encoder to the existing MUX-based Encoder shown in Figure 13.17, followed by the Wallace Tree Encoder and finally the Fat Tree Encoder. Delay analysis shows an increase from the proposed MUX-based Encoder to the existing MUX-based Encoder shown in Figure 13.18, with

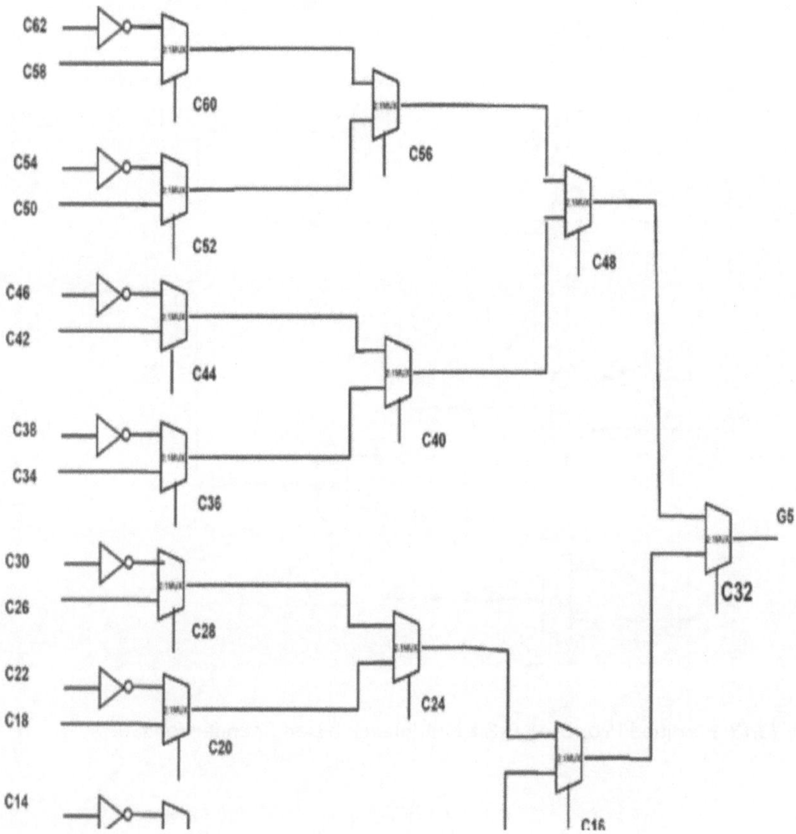

Figure 13.12 Modified Proposed 8-Bit Multiplexer-based Encoder for G5.

the Wallace Tree Encoder having the lowest delay. The figure of merit also increases shown in Figure 13.19.

13.8 CONCLUSION

This study has successfully developed and presented a MUX-Based CMOS 8-Bit Encoder, strategically designed to elevate the speed and efficiency of analog-to-digital converters (ADCs) tailored for 5G communication systems and ultra-wideband (UWB) applications. The overarching objective was to surmount the limitations inherent in traditional encoding approaches, and the research endeavored to introduce a novel solution that seamlessly integrates speed, accuracy, and power efficiency. Recognizing the shortcomings of conventional binary-ROM encoders, which compromise speed and cost-effectiveness, and the complexities associated with Fat Tree Encoders,

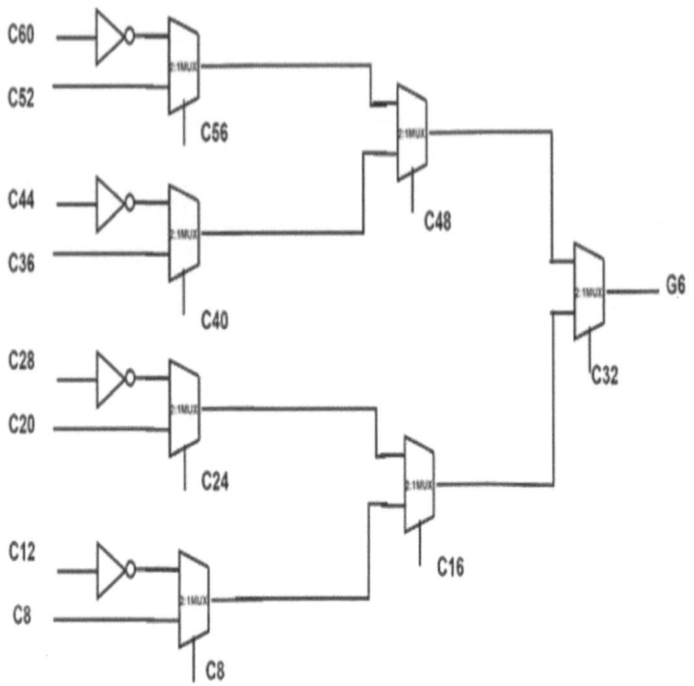

Figure 13.13 Modified Proposed 8-Bit Multiplexer-based Encoder for G6.

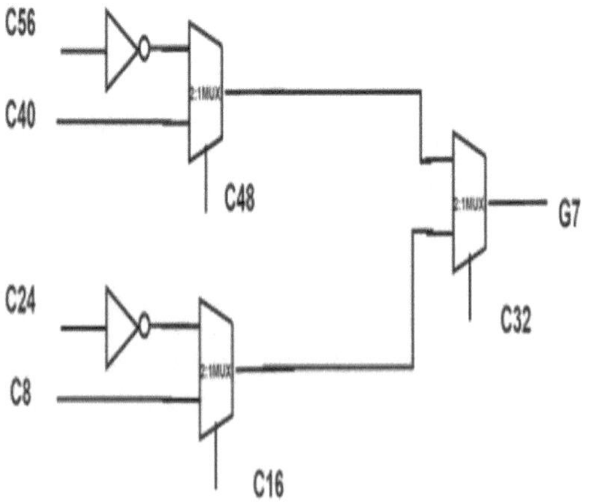

Figure 13.14 Modified Proposed 8-Bit Multiplexer-based Encoder for G8.

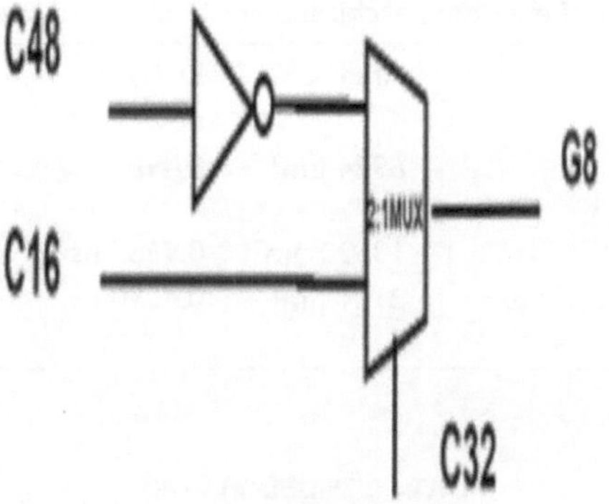

Figure 13.15 Modified Proposed 8-Bit Multiplexer-based Encoder for G8.

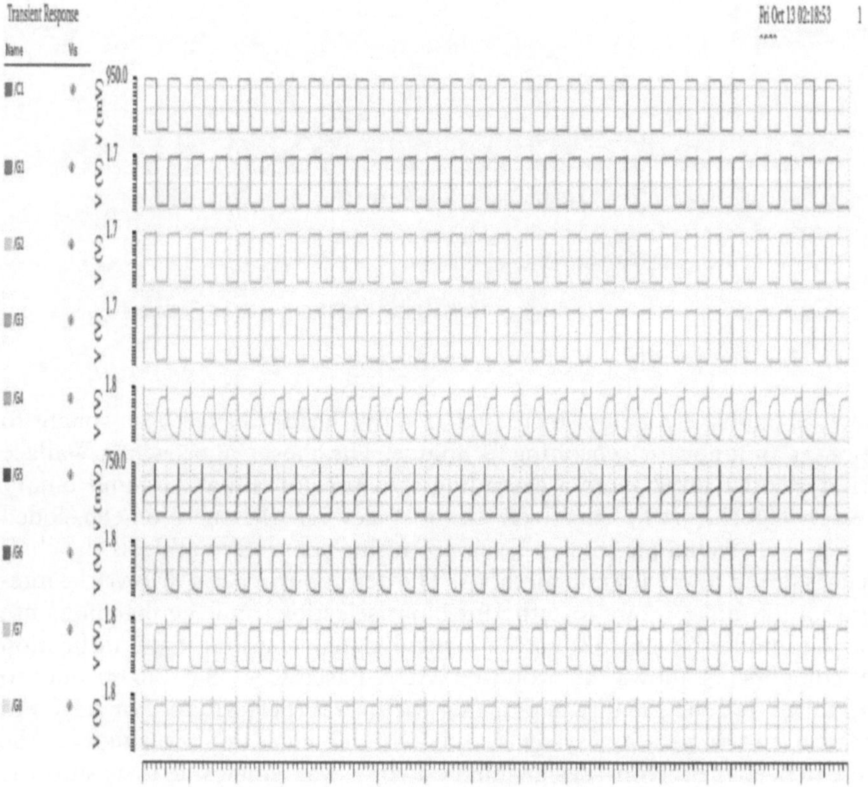

Figure 13.16 Proposed 8-Bit Multiplexer-based Encoder using 2:1 multiplexer.

Table 13.2 Comparison of encoders architectures in terms of speed, power, and area

Encoder type	Area	Delay	Power	FOM
ROM Encoder		Low	Low	Low
Fat Tree Encoder	6984 μm²	8.498	0.217 mw	1744.07 fJ
Wallace Tree Encoder	13520 μm²	0.9663 ns	62.94 μw	60.81 fJ
Modified Proposed Multiplexer-based Encoder	3161 μm²	0.4090 ns	23.32 μw	13.28 fJ

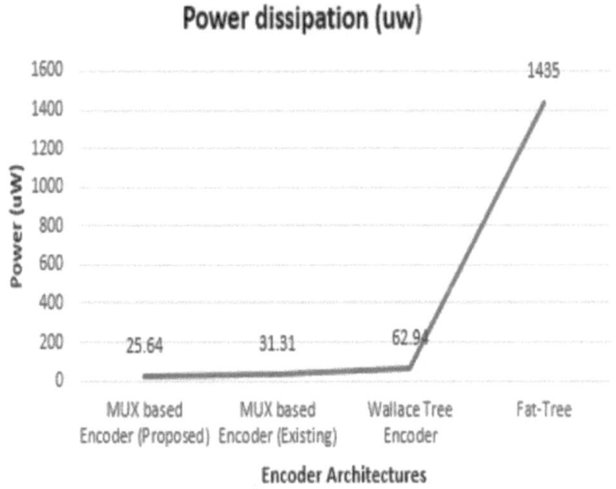

Figure 13.17 Power analysis of different encoder architectures.

such as bubble errors and increased area overhead, this research sought to pioneer an innovative encoding technique. The chosen approach, a Wallace Tree encoder, introduces a cascading structure of full-adders and binary search algorithms to effectively count "ones" in the thermometer-coded input. This technique emerged as a promising solution to address the challenges posed by traditional methods. The design intricacies involve the integration of multiplexers for efficient input selection, and the ingenious use of the middle bit (bit 32) as the control signal. This strategic utilization of components allows the proposed MUX-Based CMOS 8-Bit Encoder to optimize its functionality, enhancing the overall performance of ADCs in critical communication systems and UWB applications. Throughout the research, a meticulous design methodology was employed to ensure the reliability and effectiveness of the proposed encoder. Extensive simulations conducted in a 180 nm CMOS technology yielded impressive results, underscoring the achievements of the design. Notably, the compact footprint with an area of 3161 μm^2, a minimal delay of 0.4090 ns, a low power

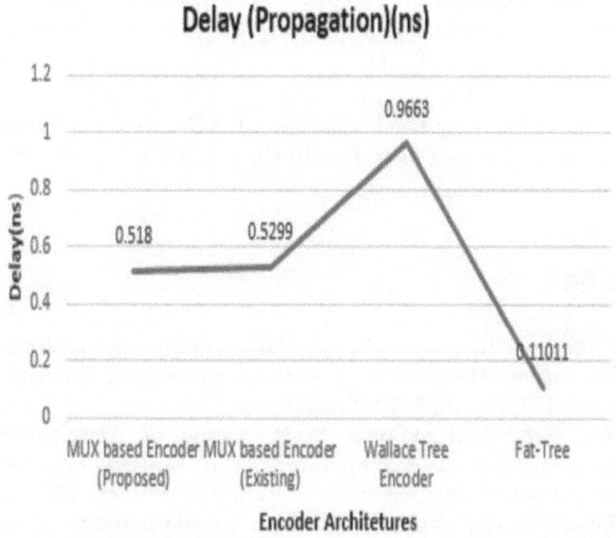

Figure 13.18 Delay analysis of different encoder architectures.

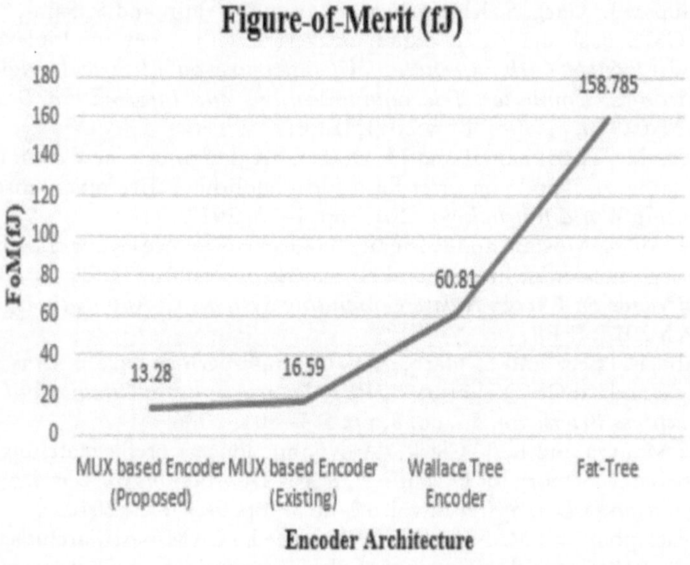

Figure 13.19 Figure-of-merit analysis of different encoder architectures.

consumption of 23.32 μW, and a commendable figure of merit at 13.28 fJ collectively demonstrate the enhanced conversion speed and reduced power consumption of the proposed design. The presented MUX-Based CMOS 8-Bit Encoder stands as a pioneering solution to the challenges faced by conventional encoding methods. By successfully addressing issues related to speed, accuracy, and power consumption, this research contributes significantly to the optimization of ADC performance in the context of 5G communication systems and UWB applications. The proposed encoder holds promise for revolutionizing the landscape of ADC design, setting the stage for more efficient and reliable systems in the rapidly advancing field of high-speed communication technologies.

REFERENCES

1. K. A. Chandra and N. S. Murty, "Low power high performance priority encoder using 2D-array to 3D-array conversion", *Procedia Computer Science*, vol. 171, pp. 1037–1045, 2020.
2. T. J. Kutre, S. N. Patil and S. K. Kore, "Design of MUX based flash ADC for reduction in number of comparators", *IOP Conference Series: Materials Science and Engineering*, vol. 925, no. 1, p. 012063, Sep. 2020.
3. S. M. Mayur, "Design of novel multiplexer-based thermometer to binary code encoder for 4-bit flash ADC", *2017 2nd IEEE International Conference on Recent Trends in Electronics, Information & Communication Technology (RTEICT)*, pp. 1006–1009, May 2017, IEEE.
4. Y. Gupta, L. Garg, S. Khandelwal, S. Gupta, S. Jain and S. Saini, "A 4-bit, 3.2 GSPS flash analog to digital converter with a new multiplexer-based encoder", *2014 11th International Conference on Electrical Engineering/Electronics, Computer, Telecommunications and Information Technology (ECTI-CON)*, pp. 1–6, May 2014, IEEE.
5. M. Chakir, H. Akhamal and H. Qjidaa, "A design of a new column-parallel analog-to-digital converter flash for monolithic active pixel sensor", *The Scientific World Journal*, vol. 2017, pp. 1–12, 2017.
6. K. M. Ali, H. Mostafa and T. Ismail, "High performance layout-friendly 64-bit priority encoder utilizing parallel priority look-ahead", *5th International Conference on Energy Aware Computing Systems & Applications*, pp. 1–4, March 2015, IEEE.
7. S. Abdel-Hafeez and S. Harb, "A VLSI high-performance priority encoder using standard CMOS library", *IEEE Transactions on Circuits and Systems II: Express Briefs*, vol. 53, no. 8, pp. 597–601, 2006.
8. S. K. Maurya and L. T. Clark, "A dynamic longest prefix matching content addressable memory for IP routing", *IEEE Transactions on Very Large-Scale Integration (VLSI) Systems*, vol. 19, no. 6, pp. 963–972, 2010.
9. M. Faezipour and M. Nourani, "Wire-speed TCAM-based architectures for multimatch packet classification", *IEEE Transactions on Computers*, vol. 58, no. 1, pp. 5–17, 2008.
10. S. Tambatkar, S. N. Menon, V. Sudarshan, M. Vinodhini and N. S. Murty, "Error detection and correction in semiconductor memories using 3D parity check code with hamming code", *2017 International Conference on Communication and Signal Processing (ICCSP)*, pp. 974–978, April 2017. IEEE.

11. X. T. Nguyen, H. T. Nguyen and C. K. Pham, (2017). "A scalable high-performance priority encoder using 1D-array to 2D-array conversion", *IEEE Transactions on Circuits and Systems II: Express Briefs*, vol. 64, no. 9, pp. 1102–1106, 2017.
12. S. Abdel-Hafeez and S. Harb, "A VLSI high-performance priority encoder using standard CMOS library", *IEEE Transactions on Circuits and Systems II: Express Briefs*, vol. 53, no. 8, pp. 597–601, 2006.
13. M. Faezipour and M. Nourani, "Wire-speed TCAM-based architectures for multimatch packet classification", *IEEE Transactions on Computers*, vol. 58, no. 1, pp. 5–17, 2008.
14. S. Tambatkar, S. N. Menon, V. Sudarshan, M. Vinodhini and N. S. Murty, "Error detection and correction in semiconductor memories using 3D parity check code with hamming code", *2017 International Conference on Communication and Signal Processing (ICCSP)*, pp. 0974–0978, April 2017, IEEE.
15. X. T. Nguyen, H. T. Nguyen and C. K. Pham, "A scalable high-performance priority encoder using 1D-array to 2D-array conversion", *IEEE Transactions on Circuits and Systems II: Express Briefs*, vol. 64, no. 9, pp. 1102–1106, 2017.
16. Girish Wadhwa and B. Raj, "Parametric variation analysis of charge-plasma-based dielectric modulated JLTFET for biosensor application", *IEEE Sensor Journal*, vol. 18, no. 15, pp. 6070–6077, 1 Aug. 2018.
17. Jeetendra Singh and B. Raj, "Comparative analysis of memristor models for memories design", *Journal of Semiconductors*, vol. 39, no. 7, pp. 074006-1–12, July 2018.
18. Divya Yadav, Shailesh Singh Chouhan, Santosh Kumar Vishvakarma and B. Raj, "Application specific microcontroller design for IoT based WSN", *Sensor Letter*, vol. 16, pp. 374–385, May 2018.
19. Gurmohan Singh, R. K. Sarin and B. Raj, "Fault-tolerant design and analysis of quantum-dot cellular automata based circuits", *IEEE/IET Circuits, Devices & Systems*, vol. 12, pp. 638–64, 2018.
20. Jeetendra Singh and B. Raj, "Modeling of mean barrier height levying various image forces of metal insulator metal structure to enhance the performance of conductive filament based memristor model", *IEEE Nanotechnology*, vol. 17, no. 2, pp. 268–267, March 2018.
21. Aakash Jain, Sanjeev Sharma and B. Raj, "Analysis of triple metal surrounding gate (TM-SG) III–V nanowire MOSFET for photosensing application", *Opto-Electronics Journal*, vol. 26, no. 2, pp. 141–148, May 2018.
22. Aakash Jain, Sanjeev Sharma and B. Raj, "Design and analysis of high sensitivity photosensor using cylindrical surrounding gate MOSFET for low power sensor applications", *Engineering Science and Technology, an International Journal*, vol. 19, no. 4, pp. 1864–1870, Dec. 2016.
23. G. Saiphani Kumar, Amandeep Singh and B. Raj, "Design and analysis of gate all around CNTFET based SRAM cell design", *Journal of Computational Electronics*, vol. 17, no. 1, pp. 138–145, March 2018.
24. Gurinder pal Singh, B. S. Sohi and Balwinder Raj, "Material properties analysis of graphene base transistor (GBT) for VLSI analog circuits", *Indian Journal of Pure & Applied Physics*, vol. 55, pp. 896–902, Dec. 2017.

25. Sunil Kumar and B. Raj, "Estimation of stability and performance metric for inward access transistor based 6T SRAM cell design using n-type/p-type DMDG-GDOV TFET", *IEEE VLSI Circuits and Systems Letter*, vol. 3, no. 2, pp. 25–39, June 2017.
26. Shashikant Sharma, Anjan Kumar, Manisha Pattanaik and B. Raj, "Forward body biased multimode multi-threshold CMOS technique for ground bounce noise reduction in static CMOS adders", *International Journal of Information and Electronics Engineering*, vol. 3, no. 3, pp. 567–572, 2013.
27. Hamendra Singh, Pankaj Kumar and Balwinder Raj, "Performance analysis of majority gate SET based 1-bit full adder", *International Journal of Computer and Communication Engineering*, vol. 2, no. 4, pp. 1–8, 2013.
28. Anil Kumar Bhardwaj, Sumeet Gupta and B. Raj, "Investigation of parameters for Schottky Barrier (SB) height for Schottky Barrier based carbon nanotube field effect transistor device", *Journal of Nanoelectronics and Optoelectronics*, vol. 15, pp. 783–791, July 2020.
29. Priya Bansal and B. Raj, "Memristor: a versatile nonlinear model for dopant drift and boundary issues," *Journal of Computational and Theoretical Nanoscience*, vol. 14, no. 5, pp. 2319–2325, May 2017.
30. Neeraj Jain and B. Raj, "An Analog and digital design perspective comprehensive approach on Fin-FET (Fin-Field Effect transistor) technology – a review", *Reviews in Advanced Sciences and Engineering*, vol. 5, pp. 1–14, 2016.
31. Balwinder Raj, A. K. Saxena and S. Dasgupta, "Analytical modeling for the estimation of leakage current and subthreshold swing factor of nanoscale double gate FinFET device", *Microelectronics International*, vol. 26, pp. 53–63, 2009.
32. Shailendra Singh Soniya, Girish Wadhwa and Balwinder Raj, "An analytical modeling for dual source vertical tunnel field effect transistor", *International Journal of Recent Technology and Engineering*, vol. 8 no. 2, pp. 603–608, July 2019.
33. Shailendra Singh and B. Raj, "Design and analysis of heterojunction vertical t-shaped tunnel field effect transistor", *Journal of Electronics Material*, vol. 48, no. 10, pp. 6253–6260, Oct. 2019.
34. Candy Goyal, Jagpal Singh Ubhi and B. Raj, "A low leakage CNTFET based inexact full adder for low power image processing applications", *International Journal of Circuit Theory and Applications*, vol. 47, no. 9, pp. 1446–1458, Sep. 2019.
35. B. Raj, A. K. Saxena and S. Dasgupta, "A compact drain current and threshold voltage quantum mechanical analytical modeling for FinFETs", *Journal of Nanoelectronics and Optoelectronics*, vol. 3, no. 2, pp. 163–170, 2008.
36. Girish Wadhwa and B. Raj, "An analytical modeling of charge plasma based tunnel field effect transistor with impacts of gate underlap region", *Superlattices and Microstructures*, vol. 142, p. 106512, June 2020.
37. Shailendra Singh and B. Raj, "Modeling and simulation analysis of SiGe hetrojunction double gatevertical t-shaped tunnel FET", *Superlattices and Microstructures*, vol. 142, p. 106496, June 2020.
38. Neeraj Jain and Balwinder Raj, "Device and circuit co-design perspective comprehensive approach on FinFET technology – a review", *Journal of Electron Devices*, vol. 23, no. 1, pp. 1890–1901, 2016.

39. Sunil Kumar and B. Raj, "Analysis of I_{ON} and ambipolar current for dual-material gate-drain overlapped DG-TFET", *Journal of Nanoelectronics and Optoelectronics*, vol. 11, pp. 323–333, June 2016.
40. Naveed Anjum, Tarun Bali and B. Raj, "Design and simulation of handwritten multiscript character recognition", *International Journal of Advanced Research in Computer and Communication Engineering*, vol. 2, no. 7, pp. 2544–2549, July 2013.
41. Karmjit Singh and B. Raj, "Performance and analysis of temperature dependent multi-walled carbon nanotubes as global interconnects at different technology nodes", *Journal of Computational Electronics*, vol. 14 no. 2, pp. 469–476, June 2015.
42. Sunil Kumar and B. Raj, "Compact channel potential analytical modeling of DG-TFET based on evanescent–mode approach", *Journal of Computational Electronics*, vol. 14, no. 2, pp. 820–827, July 2015.
43. Karmjit Singh and B. Raj, "Temperature dependent modeling and performance evaluation of multi-walled CNT and single-walled CNT as global interconnects", *Journal of Electronic Materials*, vol. 44, no. 12, pp. 4825–4835, Dec. 2015.
44. V.K. Sharma, M. Pattanaik and B. Raj, "INDEP approach for leakage reduction in nanoscale CMOS circuits", *International Journal of Electronics*, vol. 102, no. 2, pp. 200–215, 2014.
45. Karmjit Singh and B. Raj, "Influence of temperature on MWCNT bundle, SWCNT bundle and copper interconnects for nanoscaled technology nodes", *Journal of Materials Science: Materials in Electronics*, vol. 26, no. 8, pp. 6134–6142, 2015.
46. Naveed Anjum, Tarun Bali and B. Raj, "Design and simulation of handwritten Gurumukhi and Devanagri numerical recognition", *International Journal of Computer Applications*, vol. 73, no. 12, pp. 16–21, 2013.
47. S. Khandelwal, V. Gupta, B. Raj and R. D. Gupta, "Process variability aware low leakage reliable nano scale DG-FinFET SRAM cell design technique", *Journal of Nanoelectronics and Optoelectronics*, vol. 10, no. 6, pp. 810–817, Dec. 2015.
48. V. K. Sharma, M. Pattanaik and B. Raj, "ONOFIC approach: low power high speed nanoscale VLSI circuits design", *International Journal of Electronics*, vol. 101, no. 1, pp. 61–73, 2014.
49. S. Khandelwal, Balwinder Raj and R. D. Gupta, "FinFET based 6T SRAM cell design: analysis of performance metric, process variation and temperature effect", *Journal of Computational and Theoretical Nanoscience*, vol. 12, pp. 2500–2506, 2015.
50. Sumit Singh, Shekhar Yadav, Jagdeep Rahul, Anurag Srivastava and B. Raj, "Impact of HfO_2 in graded channel dual insulator double gate MOSFET", *Journal of Computational and Theoretical Nanoscience*, vol. 12, no. 6, pp. 950–953, April 2015.
51. Vijay Kumar Sharma, Manisha Pattanaik and B. Raj, "PVT variations aware low leakage INDEP approach for nanoscale CMOS circuits", *Microelectronics Reliability*, vol. 54, pp. 90–99, 2014.

52. B. Raj, A. K. Saxena and S. Dasgupta, "Quantum mechanical analytical modeling of nanoscale DG FinFET: evaluation of potential, threshold voltage and source/drain resistance", *Journal of Material Science in Semiconductor Processing*, vol. 16, no. 4, pp. 1131–1137, 2013.
53. Maisagalla Gopal, Siva Sankar D Prasad and Balwinder Raj, "8T SRAM cell design for dynamic and leakage power reduction", *International Journal of Computer Applications*, vol. 71, no. 9, pp. 43–48, June 2013.
54. Manisha Pattanaik, B. Raj, Shashikant Sharma and Anjan Kumar, "Diode based trimode multi-threshold CMOS technique for ground bounce noise reduction in static CMOS adders", *Advanced Materials Research*, vol. 548, pp. 885–889, 2012.
55. Balwinder Raj, A. K. Saxena and S. Dasgupta, "Nanoscale FinFET based SRAM cell design: analysis of performance metric, process variation, underlapped FinFET and temperature effect", *IEEE Circuits and System Magazine*, vol. 11, no. 2, pp. 38–50, 2011.
56. V. K. Sharma, M. Pattanaik and Balwinder Raj, "Leakage current ONOFIC approach for deep submicron VLSI circuit design", *International Journal of Electrical, Computer, Electronics and Communication Engineering,* vol. 7, no. 4, pp. 239–244, 2013.
57. Parminder Kaur, Sandeep Singh Gill and B. Raj, "Comparative analysis of OFETs materials and devices for sensor applications", *Journal of Silicon*, vol. 14, pp. 4463–4471, 2022.
58. Sanjeev Kumar Sharma, Parveen Kumar, Balwant Raj and B. Raj· "$In_{1-x}Ga_xAs$ double metal gate-stacking cylindrical nanowire MOSFET for highly sensitive photo detector", *Journal of Silicon*, vol.14, pp. 3535–3541, 2022.
59. B. Raj, A. K. Saxena and S. Dasgupta, "Analytical modeling of quasi planar nanoscale double gate FinFET with source/drain resistance and field dependent carrier mobility: a quantum mechanical study", *Journal of Computer*, vol. 4, no. 9, pp. 1–8, 2009.
60. S. Bhushan, S. Khandelwal and B. Raj, "Analyzing different mode FinFET based memory cell at different power supply for leakage reduction", *Seventh International Conference on Bio-Inspired Computing: Theories and Application* (BIC-TA 2012), Advances in Intelligent Systems and Computing, vol. 202, pp. 89–100, 2013.
61. Jeetendra Singh and B. Raj, "Temperature dependent analytical modeling and simulations of nanoscale memristor", *Engineering Science and Technology, an International Journal,* vol. 21, pp. 862–868, Oct. 2018.
62. Shradhya Singh, Shashi Bala, Balwant Raj and B. Raj, "Improved sensitivity of dielectric modulated junctionless transistor for nanoscale biosensor design", *Sensor Letter*, vol. 18, pp. 328–333, April 2020.
63. Vivek Kumar, Santosh Kumar Vishvakarma and B. Raj, "Design and performance analysis of ASIC for IoT applications", *Sensor Letter*, vol. 18, pp. 31–38, Jan. 2020.
64. Akanksha Jaiswal, R. K. Sarin, B. Raj and Shikha Sukhija, "A novel circular slotted microstrip-fed patch antenna with three triangle shape defected ground structure for multiband applications", *Advanced Electromagnetic*, vol. 7, no. 3, pp. 56–63, Aug. 2018.

65. Girish Wadhwa and B. Raj, "Label free detection of biomolecules using charge-plasma-based gate underlap dielectric modulated junctionless TFET", *Journal of Electronic Materials*, vol. 47, no. 8, pp. 4683–4693, Aug. 2018.
66. Gurmohan Singh, R. K. Sarin and B. Raj, "Design and performance analysis of a new efficient coplanar quantum-dot cellular automata adder", *Indian Journal of Pure & Applied Physics*, vol. 55, pp. 97–103, Feb. 2017.
67. Parminder Kaur, Vikas Pandey and B. Raj, "Comparative study of efficient design, control and monitoring of solar power using IoT", *Sensor Letter*, vol. 18, pp. 419–426, May 2020.
68. Anil Kumar Bhardwaj, Sumeet Gupta and B. Raj, "Development & analysis of compact model for double gate Schottky Barrier CNTFET", *Journal of Nanoelectronics and Optoelectronics*, vol. 15, pp. 1199–1208, Aug. 2020.
69. Girish Wadhwa, Priyanka Kamboj, Jeetendra Singh and B. Raj, "Design and investigation of junctionless DGTFET for biological molecule recognition", *Transactions on Electrical and Electronic Materials*, vol. 22, pp. 282–289, 2021.
70. Sachin Kumar Verma, Shailendra Singh, Girish Wadhwa and B. Raj, "Detection of biomolecules using charge-plasma based gate underlap dielectric modulated dopingless TFET", *Transactions on Electrical and Electronic Materials*, vol. 21, pp. 528–535, June 2020.
71. Neeraj Jain and B. Raj, "Impact of underlap spacer region variation on electrostatic and analog/RF performance of symmetrical high-k SOI FinFET at 20 nm channel length", *Journal of Semiconductors*, vol. 38, no. 12, p. 122002, Dec. 2017.
72. Shailendra Singh and B. Raj, "Analytical modeling and simulation analysis of T-shaped III–V heterojunction vertical T-FET", *Superlattices and Microstructures*, vol. 147, p. 106717, Nov. 2020.
73. Gurmohan Singh, R. K. Sarin and B. Raj, "Design and analysis of area efficient QCA based reversible logic gates", *Journal of Microprocessors and Microsystems*, vol. 52, pp.59–68, May 2017.
74. Shailendra Singh and B. Raj, "Parametric variation analysis on hetero-junction vertical t-shape TFET for supressing ambipolar conduction", *Indian Journal of Pure and Applied Physics*, vol. 58, pp. 478–485, June 2020.
75. Shailendra Singh Soniya, Girish Wadhwa and Balwinder Raj, "Design and analysis of dual source vertical tunnel field effect transistor for high performance", *Transactions on Electrical and Electronics Materials*, vol. 21, pp. 74–82, Oct. 2019.
76. Manjit Kaur, Neena Gupta, Sanjeev Kumar, B. Raj and Arun Kumar Singh, "RF performance analysis of intercalated graphene nanoribbon based global level interconnects", *Journal of Computational Electronics*, vol. 19, pp.1002–1013, June 2020.
77. Girish Wadhwa and B. Raj, "Design and performance analysis of junctionless TFET biosensor for high sensitivity", *IEEE Nanotechnology*, vol.18, pp. 567–574, 2019.
78. Jeetendra Singh and B. Raj, "Enhanced nonlinear memristor model encapsulating stochastic dopant drift", *Journal of Nanoelectronics and Optoelectronics*, vol. 14, pp. 958–963, 2019.

79. Anjana Bhardwaj, Pradeep Kumar, B. Raj and Sunny Anand, "Design and performance optimization of doping-less vertical nanowire TFET using gate stack technique", *Journal of Electronic Materials*, vol. 41, no. 7, pp. 4005–4013, 2022.
80. Jeetendra Singh and B. Raj, "Tunnel current model of asymmetric MIM structure levying various image forces to analyze the characteristics of filamentary memristor", *Applied Physics A*, vol. 125, no.3, pp. 203.1–203.11, Feb. 2019.
81. Candy Goyal, Jagpal Singh Ubhi and B. Raj, "Low leakage zero ground noise nanoscale full adder using source biasing technique", *Journal of Nanoelectronics and Optoelectronics*, vol. 14, pp. 360–370, March 2019.
82. Gurmohan Singh, R. K. Sarin and B. Raj, "A novel robust exclusive-or function implementation in QCA nanotechnology with energy dissipation analysis", *Journal of Computational Electronics*, vol. 15, no. 2, pp. 455–465, June 2016.
83. Girish Wadhwa, Priyanka Kamboj and Balwinder Raj, "Design optimisation of junctionless TFET biosensor for high sensitivity", *Advances in Natural Sciences: Nanoscience and Nanotechnology*, vol. 10, no. 7, p. 45001, 2019.
84. Priya Bansal and B. Raj, "Memristor modeling and analysis for linear dopant drift kinetics", *Journal of Nanoengineering and Nanomanufacturing*, vol. 6, pp. 1–7, 2016.
85. D. Vaithiyanathan Ashima and B. Raj, "Performance analysis of charge plasma induced graded channel Si nanotube", *Journal of Engineering Research*, EMSME Special no. pp. 146–154, Aug. 2021.
86. Abhishek Singh Tomar, Vijay Kumar Magraiya and B. Raj, "Scaling of access and data transistor for high performance DRAM cell design", *Quantum Matter*, vol. 2, pp. 412–416, Oct. 2013.
87. Neeraj Jain and B. Raj, "Parasitic capacitance and resistance model development and optimization of raised source/drain SOI FinFET structure for analog circuit applications", *Journal of Nanoelectronics and Optoelectronics*, vol. 13, pp. 531–539, April 2018.
88. Shradhya Singh, S. K. Vishvakarma and B. Raj, "Analytical modeling of split-gate junction-less transistor for a biosensor application", *Sensing and Bio-Sensing*, vol. 18, pp. 31–36, April 2018.
89. Maisagalla Gopal and Balwinder Raj, "Low power 8T SRAM cell design for high stability video applications", *ITSI Transaction on Electrical and Electronics Engineering*, vol. 1, no. 5, pp. 91–97, 2013.
90. Balwinder Raj, Jatin Mitra, Deepak Kumar Bihani, V. Rangharajan, A. K. Saxena and S. Dasgupta, "Analysis of noise margin, power and process variation for 32 nm FinFET based 6T SRAM cell", *Journal of Computer*, vol. 5, no. 6, pp. 1–8, 2010.
91. Divya Sharma, Rajesh Mehra and B. Raj, "Comparative analysis of photovoltaic technologies for high efficiency solar cell design", *Superlattices and Microstructures*, vol. 153, p. 106861, May 2021.
92. Pawandeep Kaur, Avtar Singh Buttar and B. Raj, "A comprehensive analysis of nanoscale transistor based biosensor: a review", *Indian Journal of Pure and Applied Physics*, vol. 59, pp. 304–318, April 2021.
93. Divya Yadav, Balwant Raj and B. Raj, "Design and simulation of low power microcontroller for IoT applications", *Journal of Sensor Letters*, vol. 18, pp. 401–409, May 2020.

94. Shailendra Singh and B. Raj, "A 2-D analytical surface potential and drain current modeling of double-gate vertical t-shaped tunnel FET", *Journal of Computational Electronics*, vol. 19, pp. 1154–1163, April 2020.
95. Jeetendra Singh and B. Raj, "An accurate and generic window function for non-linear memristor model", *Journal of Computational Electronics*, vol. 18 no. 2, pp. 640–647, June 2019.
96. Manjit Kaur, Neena Gupta, Sanjeev Kumar, B. Raj and Arun Kumar Singh, "Comparative RF and crosstalk analysis of carbon based nano interconnects", *IET Circuits, Devices & Systems*, vol. 15, no. 6, pp. 493–503, Feb. 2021.
97. Nehru Kandasamy, Firdous Ahmad, D. Ajitha, B. Raj and Nagarjuna Telagam, "Quantum dot cellular automata based scan flip flop and boundary scan register", *IETE Journal of Research*, vol. 66, pp. 535–548, 2020.
98. Jeetendra Singh and B. Raj, "Design and investigation of 7T2M NVSARM with enhanced stability and temperature impact on store/restore energy", *IEEE Transactions on Very Large Scale Integration Systems*, vol. 27, no. 6, pp. 1322–1328, June 2019.
99. Anil Kumar Bhardwaj, Sumeet Gupta, B. Raj and Amandeep Singh, "Impact of double gate geometry on the performance of carbon nanotube field effect transistor structures for low power digital design", *Computational and Theoretical Nanoscience*, vol. 16, pp. 1813–1820, 2019.
100. Neeraj Jain and B. Raj, "Thermal stability analysis and performance exploration of asymmetrical dual-k underlap spacer (ADKUS) SOI FinFET for security and privacy applications", *Indian Journal of Pure & Applied Physics*, vol. 57, pp. 352–360, May 2019.
101. Neeraj Jain and Balwinder Raj, "Dual-k spacer region variation at the drain side of asymmetric SOI FinFET structure: performance analysis towards the analog/RF design applications", *Journal of Nanoelectronics and Optoelectronics*, vol. 14, pp. 349–359, March 2019.
102. Jeetendra Singh, Sanjeev Sharma, B. Raj and Mamta Khosla, "Analysis of barrier layer thickness on performance of In1-xGaxAs based gate stack cylindrical gate nanowire MOSFET", *Journal of Nanoelectronics and Optoelectronics*, vol. 13, pp. 1473–1477, Oct. 2018.
103. Neeraj Jain and B. Raj, "Analysis and performance exploration of high-k SOI FinFETs over the conventional low-k SOI FinFET toward analog/RF design", *Journal of Semiconductors*, vol. 39, no. 12, pp. 124002-1–7, Dec. 2018.
104. Candy Goyal, Jagpal Singh Ubhi and B. Raj, "A reliable leakage reduction technique for approximate full adder with reduced ground bounce noise", *Journal of Mathematical Problems in Engineering*, vol. 2018, p. 16, 15 Oct. 2018.
105. Anuradha, Jeetendra Singh, B. Raj and Mamta Khosla, "Design and performance analysis of nano-scale memristor-based nonvolatile SRAM", *Journal of Sensor Letter*, vol. 16, pp. 798–805, Oct. 2018.

Index

D

E

F

G

H

I

J

K

L

M